Unravelling Complexities in Genetics and Genomics

Impact on Diagnosis Counseling and Management

Unravelling Complexities in Genetics and Genomics

Impact on Diagnosis
Counseling and Management

Moyra Smith

UC Irvine, USA

World Scientific

NEW JERSEY · LONDON · SINGAPORE · BEIJING · SHANGHAI · HONG KONG · TAIPEI · CHENNAI · TOKYO

Published by

World Scientific Publishing Co. Pte. Ltd.

5 Toh Tuck Link, Singapore 596224

USA office: 27 Warren Street, Suite 401-402, Hackensack, NJ 07601

UK office: 57 Shelton Street, Covent Garden, London WC2H 9HE

Library of Congress Cataloging-in-Publication Data
Names: Smith, Moyra, author.
Title: Unravelling complexities in genetics and genomics :
    impact on diagnosis counseling and management / Moyra Smith.
Description: New Jersey : World Scientific, 2016. |
    Includes bibliographical references and index.
Identifiers: LCCN 2016027776 | ISBN 9789814759465 (hardcover : alk. paper)
Subjects: | MESH: Genomics | Genetic Counseling | Disease Management
Classification: LCC QH438.7 | NLM QU 460 | DDC 572.8/6--dc23
LC record available at https://lccn.loc.gov/2016027776

British Library Cataloguing-in-Publication Data
A catalogue record for this book is available from the British Library.

Typeset by Stallion Press
Email: enquiries@stallionpress.com

Preface

During the 19th and 20th centuries, clinicians documented signs and symptoms of specific disorders and clinicians and geneticists analyzed family histories. Together they defined syndromes with specific patterns of inheritance.

Much progress was made in the 20th and in the first decade of the 21st century in the elucidation of genes and mutations that contribute to specific Mendelian disorders. In studies in human diseases and in studies on model organisms, the effects of loss of function and of gene knockouts on phenotype were particularly emphasized. In a few disorders gain of function mutations were identified. A reductionist approach was adopted and a specific syndrome was ascribed to effects of impaired function a specific single gene or to the additive effects of dysfunction of a few genes.

For complex common diseases, including diabetes, heart disease and psychiatric disorders, few examples of loss of function mutations and genes of major effect were obtained. Techniques such as genome wide association studies were developed to evaluate genetic and genomic variation in large populations. Difficulties that emerged in genome wide analysis of genetic variants associated with diseases included the realities of linkage disequilibrium within specific segments in the genome. In addition, evidence emerged that disease-associated variants frequently had small effect sizes.

The increased sophistication of instruments to analyze phenotypes and genotypes in depth and the availability of the internet that has provided databases to store information and facilitate communication and information sharing, have led to more refined classifications of syndromes and disorders and to expanded lists of causative gene defects in specific syndromes. Expanded genome analyses and bioinformatics have also led to greater insights into pathogenesis of cancers.

Through application of enhanced DNA sequence analysis and bioinformatics resources to document the occurrence of variations in gene sequences and genomic architecture, it has become clear that rare variants, intermediate frequency variants, common variants and segmental genomic variants, and somatic mutations all contribute to disease. It has also become clear that epigenetic factors, including DNA and chromatin modifications, play important roles in gene regulation and may in specific instances contribute to disease pathogenesis.

In this book I review efforts to establish genotype phenotype correlations and to elucidate molecular pathogenesis of specific Mendelian and complex common disorders. I also review progress in analysis of DNA sequence, DNA and chromatin modifications and progress in analysis of gene regulation that continue to provide new insights into physiological processes and into disease pathogenesis. Progress then in all of these arenas contribute to unravelling complexities in genetics and genomics and will impact diagnosis, counselling and management.

Acknowledgements §

I wish to thank World Scientific Publishing and especially editor Catherine Yeo for her guidance and work on this book. I am grateful to the Department of Pediatrics in the School of Medicine at the University of California Irvine for the privilege of working and learning in that department for 34 years. I thank our patients and the faculty and students in the Division of Genetic and Genomic Medicine for presenting problems and complexities to solve. I am particularly grateful for the University of California, Library and Internet resources. My special thanks are also due to Doctor Simon Prinsloo for his encouragement.

Contents

1. Expanding Insights into Genes and Expression

1.1 Human Genome Structure: Protein Coding and Non-protein Coding Segments

During the late 1970s, it became clear that eukaryotic protein coding genes contain protein coding segments, exons, interspersed with non-protein coding segments, introns (Leder, 1978). Transcription of messenger RNA is initiated from the 5′ end of the gene from the transcription initiation site that is located downstream of the promoter elements of the gene. Gene promoter sequence is not transcribed; however, core DNA sequence elements in promoters contain binding sites for transcription factors and for RNA polymerase II (POLII) that is required to generate mRNA transcripts.

The primary mRNA transcripts generated from an eukaryotic protein coding gene undergo splicing, a cleavage process that removes primary transcript regions that correspond to introns and rejoins transcript regions that correspond to exons.

Comprehensive sequencing has revealed that the human genome is much more widely transcribed than was previously thought. Through activities of the ENCODE project, Djebali *et al.* (2012) produced a catalog of the RNAs produced by 15 different human cell lines. They analyzed nuclear and cytosolic RNAs. Their studies revealed that between 62.1% and 74.7% of the human genome gives rise to primary or processed RNA transcripts. Their studies therefore reduced the proportion of the human genome that is intergenic. They documented evidence for significant overlapping of neighboring genes and discussed the concept of an interleaved genome.

Derrien and members of the ENCODE consortium (2012) reported results of the annotation of 9,277 long non-coding RNAs (lncRNAs). They reported that lincRNAs are generated by RNA POLII transcription. They are distinct from other messenger RNAs in that they lack open reading frames and their nucleotide sequences cannot generate peptides. Lnc RNAs have similarities with coding RNAs in that they have promoters, splice sequences and undergo histone modification. Lnc RNAs tend to have lower levels expression than coding RNAs and they manifest more tissue specific expression. Derrien *et al.* noted that a large fraction of lnc RNAs is expressed in the brain. Lnc RNAs occur primarily in the nucleus, and to a lesser extent, in the cytoplasm.

Mercer and Mattick (2013) reported that lnc RNAs sometimes also function as antisense RNAs and can also regulate mRNA stability. It is important to note that lnc RNAs can act in cis and in trans. There is also evidence that lnc RNAs containing complexes can generate small RNAs. They noted further that lnc RNAs can bind proteins and provide platforms for chaperones and transport proteins.

Pelechano and Steinmetz (2013) emphasized the role of lnc RNA in antisense transcription and gene regulation. They noted that antisense transcripts arise from promoters and in some cases from bidirectional promoters. Antisense transcripts can also regulate transcription by affecting methylation of the gene. One example of this is the aberrant antisense transcript LUC7L that is antisense to the hemoglobin subunit alpha 1 (HBA1) hemoglobin locus and leads to methylation of the CpG island in the HBA1 promoter. This methylation impairs HBA1 expression and leads to alpha1 thalassemia.

X inactive specific transcript (XIST) is the best known antisense transcript that modifies chromatin and methylation and leads to X chromosome inactivation. Pelechano and Steinmetz reported that the antisense transcript ANRIL represses expression of the cyclin dependent kinase (CDK) inhibitors cyclin-dependent kinase inhibitor 2B (CDKN2B) and cyclin-dependent kinase inhibitor 2A (CDKN2A) that are also tumor suppressors. This repression is due to histone lysine 27 methylation. They noted that specific antisense transcripts can also impact isoform generation from specific loci. They also considered the possibility that genomic arrangement with antisense RNAs with transcription oriented opposite to that of sense RNAs might constitute a self-regulatory circuit that facilitates genes regulating their own transcription.

1.1.1 *Lnc RNAs and diseases*

As knowledge about lnc RNAs increases, evidence grows about the role they play in diseases. One interesting example is the lnc RNA DBET described by Cabianca *et al.* (2012) who discovered that expression of DBET is increased in cases of fascioscapulohumeral muscular dystrophy (FSHD). This disorder maps to chromosome 4q35 and is associated with deletions of repeats in the D4Z4 locus. In healthy individuals with the normal number of repeats in D4Z4, the Polycomb repressive group complex (PcG) binds to chromatin within the D4Z4 flanking regions and represses expression of genes in those regions. In patients with deletion of D4Z4 repeats there is expression of the lnc RNA DBET. This lnc RNA then causes trithorax complex related proteins to replace polycomb complex. The trithorax complex promotes activation of gene expression in the D4Z4 regions. These genes include ANT1 (SLC25A4 adenine translocator), FRG1, FRG2 (FSH region genes) and DUX4 (homeobox gene).

1.1.2 *Repertoire of RNAs produced by human cells*

Djebali *et al.* (2012) through activities in the ENCODE project documented presence or absence of RNA at the 7-methyl-guanosine cap at the 5′ end and the number of polyadenylation signals at the 3′ termini of mRNAs. Their studies revealed that genes express many different isoforms (10–12 on average) and that different isoforms were expressed simultaneously. However they noted that not all isoforms were expressed at the same level.

One third of the RNAs identified in their study were novel antisense RNAs and RNAs derived from intergenic regions. Novel RNAs identified include long-non-coding RNAs and also previously unannotated short RNAs. Many of the unannotated short RNAs were derived from promoters and terminator regions of genes.

Djebali *et al.* documented production of RNAs from distal enhancers and labeled these eRNAs. The transcribed distal enhancers bind RNA POLII. Transcribed enhancers showed different patterns of chromatin modifications than non-transcribed enhancers.

1.2 Promoters and Transcription

1.2.1 *Promoters, alternate promoters, core elements and RNA POLII*

Sequence elements within promoter regions include transcription factor binding sites and RNA POLII binding sites. TATA boxes are sequence elements present in about 24% of promoters. TATA boxes and TATA box binding proteins help position RNA POLII.

Many genes have several possible promoters (Pal *et al.*, 2011). At different stages of development and in different cells and tissues different promoters are used. The brain derived neurotrophic factor gene BDNF has nine different promoters, and different transcripts arise that result from the use of different promoters (Autry and Monteggia, 2012).

1.2.2 *Transcription factors*

Transcription factors bind to core sequences in promoters of genes and they also bind to enhancer sequences. Promoters are often located 100–1,000 nucleotides upstream from the transcription start sites. Promoters have binding sites for transcription factors and for RNA POLII.

Transcription factors bind to specific sequence elements in DNA. These elements are typically 6–12 nucleotides in length and the DNA sequence specificity for a specific transcription factor is often low, and a number of different transcription factors may bind to a specific DNA sequence element (Spitz and Furlong, 2012). Furthermore priming of specific elements may occur, i.e. binding of one transcription factor to a specific site to prepare that site for subsequent binding of other factors.

There is also evidence that binding of a transcription factor to a specific cis regulatory element in DNA, e.g. an enhancer, subsequently impacts chromatin structure and nucleosome occupancy so that gene transcription is facilitated.

In a review of transcription factors, Vaquerizas *et al.* (2009) summarized data on 1,391 manually curated transcription factors. In the human genome 20% of transcription factors map in high density clusters; the short and long arms of chromosome 19 harbor particularly striking clusters of transcription factors. They reported that approximately one third of transcription factors were expressed primarily in one tissue.

Wingender *et al.* (2015) reported on a specific database TFClass that provided classification of 1,558 human transcription factors. Somatic mutations in transcription factors play important roles in certain cancers. There is growing evidence that germline mutations in genes that encode specific transcription factors are key factors in a number of developmental defects. Germline mutations in a number of different genes that encode Forkhead transcription factors occur in specific developmental defects.

1.2.3 *Forkhead (FOX) transcription factors*

The different forkhead transcription factors share homologous 100 amino acid regions. They are encoded by FOX genes. Benayoun *et al.* (2011) reported that in humans 26 of the 50 FOX genes are arranged in 9 clusters, indicating that they likely arose through duplication. Some of the forkhead transcription factors show widespread expression in different tissues. However the majority exhibit tissue specific expression.

Forkhead transcription factors interact with other proteins, including other transcription factors, transcription co-activators and co-repressors. The forkhead transcription factor proteins undergo extensive post-translational modification and these modifications impact their interactions.

Benayoun *et al.* emphasize that forkhead transcription factors act as terminal effectors for a number of major signaling pathways including the transforming growth factor B (TGFB) pathway, the mitogen-activated protein kinase (MAPK) pathway, the sonic hedgehog pathway, and the insulin-insulin-like growth factor pathway. The signaling cascades most often act through inducing post-translational modifications in the forkhead proteins.

Examples of human developmental defects due to mutations in forkhead transcription factors include cleft lip and palate due to forkhead box E1 (FOXE1) defect. Forkhead box G1 (FOXG1) defects occur in Rett-like syndrome and in epilepsy. Forkhead box P1 (FOXP1) and forkhead box P2 (FOXP2) defects occur in autism and in certain patients with language impairments. Forkhead box P3 (FOXP3) defects occur in a specific immune dysregulation disorder with polyendocrinopathy and enteropathy (IPEX).

1.2.4 *Properties of enhancers*

Shlyueva *et al.* (2014) reviewed properties of transcriptional enhancers. They noted that the initial step in gene expression involves binding of RNA POLII to genomic sequence in the vicinity of the transcription start site. However active transcription requires the activity of enhancers and cis regulatory elements. They noted further that enhancer elements contain DNA motifs with sequences that bind specific transcription factors; those in turn bind co-activators and co-repressors. In addition specific chromatin modifications occur in the vicinity of active enhancers. These include histone lysine 4 monomethylation (H3K4me1) and H3K27 acetylation (H3K27ac).

They emphasized that enhancers may be located at large distances from the genes they impact and that looping brings the core promoters of genes into contact with distant enhancers.

Shlyueva *et al.* noted that enhancers differ from other transcription factor binding sites in that they frequently bind multiple transcription factors. There is also evidence that enhancers are activated by combinations of transcription factors and that binding of only one transcription factor is insufficient for their activation.

New technologies including chromatin capture and interaction analyses and paired end sequencing have facilitated identification of interactions between distant enhancers and promoters associated with the formation of chromatin loops. Heidari *et al.* (2014) reported that enhancer promoter interactions were highly cell type specific.

1.2.5 *The mediator complex (MED)*

The MED plays key roles in RNA POLII mediated transcription. Napoli *et al.* (2011) reviewed functions of the MED and its relation to human disease. The MED interacts with the large subunit of RNA POLII. In mammals there are at least 30 distinct MED subunits. Electron microscopy revealed that MED has an elliptical structure with head middle and tail sections. The head section interacts with RNA POLII. In addition to the MED subunits the mammalian complex contains CDK molecules.

There are large and small MEDs and isoforms so that all the subunits are not present in each MED. Napoli *et al.* reported that each sub-type of the MED is formed in response to signals from transcription factors that are recruited in response to changes in the cellular environment. Complexes with the CDK module differ in function from those without CDK. These different forms determine whether transcription is directly activated or whether it is stalled. The CDK module plays a regulatory role; presence of the CDK model in the complex represses transcription. They noted evidence that when the MED interacts with activators it undergoes reorganization leading to loss of the CDK module and then transcription can go on. In some instances RNA POL II and MED are stalled at approximately 50 nucleotides from the transcription start site.

Napoli *et al.* reported that MED also regulates non-protein coding RNA genes and that this involves interaction between MED and histone acetyl-transferases. In addition the MED associates with heterochromatin at telomeres and impacts life span.

1.2.6 *Interactions of specific MED subunits*

Napoli *et al.* reported that specific MEDs interact with specific transcription factors. Specific MED subunits therefore regulate expression of distinct groups of genes during development and differentiation. This explains how defects in specific MED subunits are associated with specific developmental defects or specific disease manifestations. For example, MED13 and MED13L defects impact the heart and defects in these subunits lead to heart disease and transposition of the great vessels. MED17 defects lead to infantile cerebral and cerebellar atrophy. MED15 deficiency likely plays a role in the DiGeorge velocardiofacial syndrome that occurs in cases with deletion in chromosome 22q11.21. MED25 defects lead to neuromuscular disease and sensory defects that constitute one form of Charcot Marie Tooth disease. MED1 is an essential subunit for transcription activation that involves nuclear receptor recruitment. It therefore plays roles in metabolism and adipogenesis.

1.2.7 *Defects of MED12*

MED12 defects lead to syndromic forms of X-linked intellectual disability. Four differently named clinical conditions are associated with defects in MED12 that is encoded by a gene on Xq13 (Graham and Schwartz, 2013). The Opitz–Kaveggia syndrome, also called FG syndrome, is associated with intellectual disability, microcephaly, hypotonia and imperforate anus. A recurrent MED12 mutation occurs in this syndrome c.2881C>T, p.R961W.

Lujan syndrome patients have tall stature and are lean; they have macrocephaly, a high narrow palate, dental crowding, receding chin, mild to moderate intellectual disability and behavioral abnormalities. Brain imaging studies reveal dysgenesis of the corpus callosum. A common MED12 mutation that occurs in these patients is c.3020 A>G, p.N1007S.

Ohdo syndrome (also known as OSMB syndrome) patients have blepharophimosis, ptosis, narrow eyes, intellectual disabilities and small mouth. There are several variants of this syndrome and some patients have microcephaly and epilepsy. A number of different MED12 mutations have been observed in these patients (Graham and Schwartz, 2013).

Spaeth *et al.* (2011) reported that at least 19 different CDK subunits were present in MED and that defects in CDK 19 led to microcephaly, intellectual disability and retinal defects. They noted that MED plays essential roles in coupling developmentally coded signals with precise gene expression. It is important to note that individual MED subunits can have activator or repressor function and impact the binding of the core MED to RNA POL II.

Napoli *et al.* noted that specific MED subunits have altered expression in certain cancers. CDK8 acts as an oncogene. CDK8 expression is altered in colorectal cancer cells.

1.3 Splicing, Spliceosomes and Alternate Splicing

1.3.1 *Splicing, spliceosomes, and splice regulatory factors*

Splicing involves removal of introns from mRNA transcripts and it occurs co-transcriptionally. Key nucleotides at splice sites include the donor splice site GU (guanine uridine) in mRNA (GT, guanine thymine in DNA sequence) at the start of the intron adjacent to the 5′ exon, and the splice acceptor site AG (adenine guanine) at the end of the intron and close to the 3′ exon. Upstream of the AG acceptor site is a polypyrimidine tract and upstream of that there is the splice branch site.

Consecutive transesterification reactions are essential for splicing. The first involves a trans-esterification between the phosphodiester group at the splice donor site and the hydroxyl group (OH) at the branch point. This reaction results in cleavage at the 5′ splice donor site and formation of a lariat structure. A second transesterification reaction then occurs that involves the G at the downstream splice acceptor site adjacent to the 3′ exon. This reaction results in excision of the lariat and joining of the two exons (Will and Lührmann, 2011). Mutations at splice sites or at branch sites can lead to disease.

Documentation of the range of 3′ and 5′ splice sequences was enabled through the ENCODE project (Harrow *et al.*, 2012). Invariant sequence at the 5′ end of the intron in pre-mRNA is GU, other less highly conserved sequences surround this. The splice acceptor site at the 3′ end of the intron has almost invariant AG.

Mercer *et al.* (2015) identified 59,359 high confidence branch points in more than 10,000 genes in the human genome. They reported that branch point nucleotides are predominantly adenosine (A). To isolate branch point sequences they purified lariats following exoribonuclease digestion of linear mRNA. Branch point junctions were then captured with oligonucleotide probes. They reported that splicing of a specific exon might occur at more than one branch point. Nucleotides that flank branch points bind specific snRNAs and these nucleotides show conservation between species. They referred to the conserved sequence elements that overlap branch points as B boxes. Sequences within B boxes are enriched for CG and U nucleotides in mRNA and A is excluded since it represents the key branch point nucleotide.

1.3.2 *Spliceosome*

Matera and Wang (2014) reviewed the biogenesis and function of the spliceosome, the ribonucleoprotein complex that utilizes ATP hydrolysis to bring about splicing. The RNA components of the spliceosome are non-coding and non-polyadenylated and are referred to as snRNAs (small nuclear RNAs). Specific sequences within the snRNAs interact with protein co-factors. Different classes of snRNAs are defined based on their specific interacting sequences and the proteins they bind.

The snRNAs are transcribed from highly specialized RNA POLII binding promoters and specific modifications of RNA POLII are required for promoter binding. Both general and specialized transcription factors are required; in addition, snRNA activating complexes must be present. These contain synaptosome associated proteins encoded by *SNAP* genes 1 through 5. In addition processing of the 3′ ends of snRNAs also requires specific factors.

Nuclear encoded snRNPs are transported from the nucleus to the cytoplasm where they undergo additional modification before being retransported to the nucleus. Matera and Wang noted evidence that precursor snRNA transcripts pass through specific nuclear structures, the Cajal bodies, as they leave the nucleus. There is evidence that specific complexes of snRNPs form in the Cajal bodies in the nucleus.

In the cytoplasm, snRNPs interact with the survival motor neuron complex (SMN complex) that contains components that modify the snRNAs prior to their re-transportation to the nucleus. Following return to the nucleus the snRNPs interact with cofactors.

Within the nucleus spliceosome assembly takes place on primary mRNA transcripts. Matera and Wang reported that there is a specific order in which snRNPs assemble on primary mRNA transcripts. Initially U1 snRNA pairs with the 5′ splice site. The ATP dependent splicing factor SF1 binds to the branch and is then replaced by U2 snRNA; subsequently U2AF (U2 snRNA auxiliary factor) binds to the polypyrimidine tract and to the AG at the 3′ end of the intron. Later the 3′ acceptor splice site is brought into proximity with the branch site as U4, U5 and U6 snRNAs join the complex.

1.3.3 *Alternative splicing*

Chen and Manley (2009) reviewed mechanisms of alternative splicing. They reported that almost 95% of multi-exon genes undergo alternative splicing. Specific exons may be constitutively included in all mature transcripts of a specific gene. In a particular gene specific exons may only be included in a subset of the mature RNA transcripts.

Chen and Manley reported that cis-regulatory elements determine which exons are to be included in the final transcript. They identified four categories of cis regulatory elements involved in splicing. These included exonic splicing enhancers, exonic splicing silencers and intronic splicing enhancers and silencers.

RNA and protein interactions play key roles in the splicing process. Serine-arginine proteins bind to exonic splice enhancers. At least 18 different serine arginine proteins have been identified. These proteins bind to exonic enhancers and play key roles in 5′ splice site recognition. They also play roles in other aspects of RNA metabolism. Heterogeneous ribonucleoproteins (HNRNPs), bind to exonic and intronic splice silencers. Intronic splice enhancers bind neuro-oncological ventral antigen (NOVA), FOX1 (RBFOX1), FOX2 (RBFOX2) (RNA binding proteins) and specific forms of HNRNPs (HNRNPF and HNRNPH).

Tissue specific alternative splicing is achieved partly through expression of tissue specific splicing factors. For example more than 300 different RNA binding proteins occur in mouse brain and influence alternative splicing patterns. Chen and Manley reported that post-translational modifications of splicing factors (for e.g., through phosphorylation) are also known to impact the activity of these factors.

Wang *et al.* (2008) used high throughput mRNA sequencing to assess alternative mRNA isoforms in a range of different tissues. Their studies revealed pervasive tissue specific regulation of isoform generation. Mechanisms that led to these differences include alternative splicing, skipping of exons, mutually exclusive exon inclusion, retention of introns, use of alternative 5' splice sites, use of alternative 3' splice sites. In addition isoform differences can arise due to inclusion of alternative first exons, alternative promoter use, and alternative transcription start site use. Their studies also revealed a high frequency of tissue regulated differences in polyadenylation site usage and occurrence of differences in terminal exons and differences in the length of 3-prime untranslated regions. They proposed that tissue specific RNA binding factors regulate both mRNA splicing and alternate polyadenylation site use.

Their studies revealed that for 52–80% of genes alternative splicing events differed between different tissues. Wang *et al.* hypothesized that alternative splicing was a principal contributor to the evolution of phenotypic complexity in mammals.

They also found that polymorphisms altered the relative abundance of tissue specific isoforms in different individuals. However they concluded that the relative abundance of tissue specific variation in isoform formation was much more prominent than individual variation.

Alternative splicing leads to the production of distinct protein isoforms. The different protein isoforms generated are most suitable for the functional requirements of specific tissues and at specific time-points

(Zhang and Manley, 2013). Factors that play roles in spatial and temporal regulation include alterations in cellular concentration of splice regulatory factors and mutation in components of the splicing machinery. It is important to note that alternative splicing of a specific gene transcript may lead to generation of the protein isoforms that differ in their cellular location or in their specific functions. For example in the case of the FLT1 (FMS-related tyrosine kinase 1) the long isoforms are transmembrane proteins, whereas the short isoforms generated from transcripts that are missing one or more exons, are soluble proteins that function in the cytoplasm. Different isoforms of caspase 3 (CASP3) protein generated through alternative splicing of the CASP3 pre-mRNA transcript have different functions some are pro-apoptotic while others are anti-apoptotic.

Licatalosi and Darnell (2010) put forward the concept that "biological complexity has RNA at its core". They based this conclusion on the fact that great diversity can be generated at the RNA level.

1.3.4 *Next-generation RNA sequencing*

Irimia *et al.* (2014) developed a RNA sequencing pipeline and carried out deep RNA sequencing on 50 different cell and tissue types derived at different developmental stages and they specifically identified neural regulated alternative splicing events. In these studies they determined that micro-exons are frequently included in neuronal transcription products. The inclusion of micro-exons in neuronal cells and tissues was particularly dependent on expression of the neural specific serine/arginine related splicing factor nSR100/SRRM4. They analyzed mRNA from glial cell types including astrocytes, microglia and oligodendrocytes, and mRNA from neurons. The rate of micro-exon inclusion was greatest in neuronal mRNA.

Their detailed subsequent studies confirmed that micro-exons frequently encoded disordered protein regions. Micro-exons that did not encode disordered protein regions frequently encoded protein adjacent to folded protein domains. The micro-exon encoded regions were enriched in peptide and lipid binding domains and in domains involved in cellular signaling (e.g. SH3 (SRC homology domains)) and PH domains (plekstrin homology domains).

Irimia *et al.* (2014) also carried out analyses of micro-exons in mRNA derived from post-mortem samples from autistic spectrum disorder patients. They analyzed superior temporal gyrus derived mRNA. Their studies revealed frequent misregulation of micro-exon inclusion in autism patient derived samples. They also determined that expression of nDR100, the brain specific splicing factor, was reduced in those samples.

There is evidence that micro-exon inclusion occurs particularly at late stages of neuronal differentiation that are associated with synaptogenesis and axonogenesis.

Li *et al.* (2014) defined micro-exons as sequence elements less than 51 nucleotides in length. They surveyed sets of deeply sequenced genomic DNA, cDNA and RNA to identify micro-exons. They reported that micro-exons usually require activity of splicing enhancers in order to be included in transcripts. They also determined micro-exon inclusion was facilitated by activity of RNA binding proteins encoded by RBFOX1 gene and by polypyrimidine tract binding protein 1 (PTBP1). It is interesting to note that the functions of both of these proteins were reported as impaired in autism. RBFOX1 recognized sequence elements (U) GCAUG (guanine cytosine adenine uracil guanine) sequence in RNA transcripts and PTBP1 is a polypyrimidine tract binding protein.

1.3.5 *MRNA splicing and human diseases*

Singh and Cooper (2012) reviewed aberrant mRNA splicing leading to human diseases. They noted availability of a comprehensive database on alternative splicing and aberrant splicing associated human diseases (http://www.dbass.org.uk).

There is evidence that interference with 5′ or 3′ splice sites commonly results in exon skipping. Mutations at these sites sometimes lead to intron retention. Splice silencers (inhibitors) and splice enhancers influence which exons are present in specific transcripts. In addition mutations that impact the splicing machinery, core spliceosome components or splice regulators can lead to disease.

It is interesting to note that mutations in the splice regulator RBFOX1 were identified in some cases of autism spectrum disorders. This splice regulator plays critical roles during neuronal differentiation (Fogel *et al.*, 2012).

Singh and Cooper (2012) noted that modified antisense oligonucleotides were being investigated as therapeutic targets for aberrant splicing sequences associated with disease.

1.4 Polyadenylation of Transcripts

Pre-RNA transcripts generated through activity of RNA POLII undergo splicing and they also undergo cleavage at their 3′ end and polyadenylation.

Gruber *et al.* (2014) reviewed aspects of polyadenylation. Many genes have multiple polyadenylation sites and a number of different factors are involved in polyadenylation site selection. The use of proximal polyadenylation sites, i.e. sites close to the end of the protein coding sequence, lead to shorter RNA transcripts than are found when distal polyadenylation sites are used.

It is interesting to note that there is evidence that highly expressed genes in rapidly proliferating cells tend to have shorter 3′ untranslated regions (Ji *et al.*, 2011). There is evidence that shorter 3′ untranslated regions are associated with increased protein output, possibly because they are less susceptible to microRNA binding and inhibition of translation (Sandberg *et al.*, 2008).

1.4.1 *Sequence elements at polyadenylation sites*

Computational analyses have identified sequence elements in the 3′ UTR regions that promote cleavage and polyadenylation. Adenosine (A) residues predominate in regions upstream of cleavage and are often located 21nucleotides upstream of the cleavage site. The A nucleotide in that site is frequently followed by a series of uridine (U) nucleotides (T in DNA).

Gruber *et al.* (2014) reported that the sequence motif most reproducibly found at polyA sites is AAUAAA. There is evidence that point mutations at specific polyadenylation sites can alter the isoforms generated. Higgs *et al.* (1983) reported that mutations at a specific site in hemoglobin A gene led to generation of altered transcripts with altered expression.

The highly complex cleavage and polyadenylation complex has been characterized. This complex is composed of a core component of cleavage/polyadenylation specific (CPSF) units. In addition this core complex is composed of 50 cleavage stimulation factor (CSTF) units. Polymerases alpha, beta and gamma add poly A tails comprised of up to 250 adenosine nucleotides.

Gruber *et al.* reported that epigenetic factors likely influence polyadenylation site selection and processing. Chromatin histone modification differences occur in regions around polyadenylation sites.

Ji *et al.* (2011) reported that highly expressed genes have lower levels of nucleosomes but higher levels of modified histones H3K4me3 and H3K36me3 at proximal polyA sites relative to distal sites.

1.5 Post-Transcription Modifications

Specific enzymes, adenosine deaminases (ADAR1, 2 and 3), act on coding and non-coding RNA transcripts to convert adenosine to inosine. In this reaction the amine group NH3 in the 6th position on adenosine is replaced by oxygen. Slotkin and Nishikura (2013) reported that ADAR1 and ADAR2 are expressed in many tissues in humans while ADAR3 is expressed primarily in brain. In coding mRNA the alteration of the adenosine nucleotide can alter the codon for amino acid; the altered mRNA transcript may also alter splicing.

Slotkin and Nishikura reported that adenosine to inosine editing of mRNA can lead to blocking of microRNA binding and blocking of micro-RNA activity.

1.5.1 *Translation of proteins from MRNA transcripts*

Sonenberg and Hinnebusch (2009) reviewed translation initiation in eukaryotes. They noted that translational control of stored mRNA transcripts facilitates cellular homeostasis and facilitates rapid changes in the protein concentration in cells. There is evidence that specific genes have transcripts that are particularly subject to regulation of expression through control of translation initiation. The genes encode products involved in signal transduction, development and neural functions.

Regulation is primarily due to control at the first stage of translation initiation that involves recognition of the AUG translation start codon by methionyl-tRNA (Met-tRNA). Sonenberg and Hinnebusch reported that in eukaryotes a pre-initiation complex PIC that contains the small 40S ribosomal subunit binds to the mRNA transcript close to its 5′ end and scans the mRNA transcript for the AUG codon.

Important factors prime the mRNA transcript for binding of the pre-initiation complex (PIC). These factors include eukaryotic initiation factors (EIFs) that recognize the m7G cap structure at the 5′ end of the mRNA structure. The m7G cap structure is guanine linked to the mRNA via an unusual triphosphate linkage at the 7th position. This guanine is methylated through activity of a methyl transferase. In addition RNA POLII is required for 5′ cap activity. The 5′ end of the RNA binds EIF4E and its partners EIF4A and EIF4G.

The PIC complex contains EIF1A, EIF2, 3 and 5. Interaction occurs between EIF3 in the PIC and EIF4G and IEF4B at the 5′ cap site. Met-tRNA is anchored to the PIC by EIF2. Binding of the PIC to AUG results in hydrolysis of EIF2 GTP to GDP and release of other EIFs. Following this release, the large 60S ribosomal subunit joins the 20S subunit to form the 80S initiation. Successive triplets then enter the decoding ribosome and appropriate aminoacyl tRNAs then enter the 80S complex to generate peptides.

There is also evidence that interaction of polyA binding protein PABP with EIFG occurs and this facilitates formation of an mRNA loop structure.

1.5.2 *Regulation of EIF4E activity*

The 4E binding protein 4EBP binds to EIF4E and represses translation. Phosphorylation of 4EBP through activity of the mechanistic target of rapamycin (mTOR) dissociates the 4EBP from EIF4E and thereby facilitates translation. Under starvation conditions mTOR activity is blocked. Then 4EBP blocks EIF4E and its interaction with EIF4G and the interactions of EIF4E with its partners and the 5' cap site.

Hypophosphorylated 4EBP binds strongly to EIF4E and inhibits it. Phosphorylation of 4EBP weakens interactions with EIF4E and thereby releases EIF4E and partners to bind to the 5' RNA and to facilitate translation.

EIFs also bind to the polyA tail at the end of the 3' end of the mRNA transcript. Specific RNA binding proteins complex to the 3' end of the mRNA and facilitate binding of the EIFs.

MicroRNAs are loaded into the RNA induced silencing complex (RISC) and together they bind to the 3' end of mRNA and generally repress translation.

1.5.3 *Learning, memory and translation*

Sonenberg and Hinnebusch noted that long lasting forms of neuronal plasticity and memory formation depend on the synthesis of new proteins. This involves general protein synthesis and synthesis of specific proteins. They noted further that phosphoinositide 3-kinase (PI3K), mTOR and MAP-ERK signaling are also important for the control of protein synthesis and for long term neuronal potentiation and memory. Translation homeostasis is likely for adequate neuronal function.

The fragile X locus protein FMRP binds to EIF4E via the protein encoded by the CYFIP1 locus on chromosome 15q11.2. This binding displaces EIF4E from EIF4G and inhibits translation.

Feoktistova *et al.* (2013) reported that EIF4A has helicase activity and that this plays an important role in unwinding the secondary structure in the 5' untranslated region of mRNA. Furthermore the EIF4A unwinding properties are stimulated by EIF4G and EIF4B.

1.5.4 *Cancer and translation*

A number of studies have revealed increased ribosome genesis, increased phosphorylation of 4EBP and increased levels of translation. Many oncogenes and tumor suppressor gene impact translation.

1.6 Insight Into Gene Regulation through Genome Wide Association Studies (GWASs) in Common Multifactorial Traits

Knight (2014) emphasized that GWASs studies in common multi-factorial traits have most frequently revealed associations between diseases and non-protein coding regions of the genome and that this finding indicates that regulatory loci play important roles in common disorders. He noted further that the role of regulatory elements must be considered in the context of the epigenomic landscape. Furthermore there is growing evidence for the importance of regulatory loci that alter the levels of gene expression; i.e. the regulatory elements act as quantitative trait loci. Regulatory variants found associated with common diseases have most frequently been single nucleotide variants. However copy number variants in regulatory loci may also impact levels of gene expression. Knight emphasized that regulatory loci may exert their effects at a distance. This is achieved in some cases because of the phenomenon of chromatin looping and the interaction of distant loci.

Knight noted that regulatory sequence elements that impact expression of a particular gene could occur at various positions, and these key positions of regulatory elements are:

(i) At DNAse 1 hypersensitivity sites where chromatin regulatory functions are important

(ii) Regulatory elements frequently occur in enhancer sequences. Enhancer elements are often transcription factor binding sites and can occur upstream of genes or within promoter regions. Sometimes enhancer sequences are present within introns of genes

(iii) Elements at the 3′ end of genes that impact microRNA binding may act as regulatory elements

(iv) Variants that alter splicing sometimes represent regulatory elements

(v) Regulatory elements may alter gene expression levels through alteration of methylation

Knight presented examples of different functional regulatory elements. Variants at the chromosome 11q13 renal cancer susceptibility locus were shown to impact an enhancer of CCND1 (cyclin D1). Variants impacted the binding of hypoxia inducing factor to the CCND1 enhancer. Higher CCND1 expression likely increases susceptibility to renal cell carcinoma.

A specific variant within an enhancer at the BCL11A encoding locus diminishes BCL11A expression and results in elevated hemoglobin F expression. Bauer and Orkin (2015) proposed that this enhancer represents a possible therapeutic target to achieve continued expression of fetal hemoglobin in cases of beta thalassemias. An example of a functional variant in a promoter that alters gene expression occurs in sequence upstream of the alpha globin locus where a new single nucleotide variant leads to creation of a new promoter and GATA transcription factor binding site that interferes with normal function of the downstream promoter.

An example of a variant that impacts splicing occurs in the tumor necrosis factor receptor gene TNFRSF1A. The alteration in splicing induced by the variant leads to generation of a soluble form of the receptor that blocks the functioning of the normal receptor. This variant was found to be associated with multiple sclerosis (Gregory *et al.*, 2012).

A DNA variant that increased promoter methylation led to silencing of expression of the HNF1B transcription factor. Knight reported that the regulatory epigenetic landscape and chromatin accessibility can be investigated through identification of factors that bound chromatin and chromatin capture experiments.

It is important to note that short and long non-coding RNAs play roles in regulating gene expression. These RNAs impact gene expression through modification of chromatin structure. They frequently recruit modification complexes including histone and DNA methyl transferases (Holoch and Moazed, 2015).

1.6.1 *Expression quantitative trait loci (eQTL)*

EQTL are genetic variants that impact levels of gene expression. Characterization of these variants provides insights into gene regulation and is important in interpretation of results of GWASs. The majority of disease associated loci identified in GWASs map to non-protein coding (non-coding) regions of the genome. The eQTL may be located on the same chromosome as the gene it impacts and is then a cis QTL; eQTLs are also sometimes located on a different chromosome than the genes they impact and are then referred to as trans eQTLs.

Nica and Dermitzakis (2013) reported that standard eQTL analyses involve direct associations of genome variants and levels of gene expression in large numbers of individuals. They noted further that cis eQTLs frequently map within 1 megabase of the gene they impact and they may be located upstream or downstream of the transcription start site of that gene.

Correlations of genome variants and gene expression levels determined by transcriptome sequencing have provided more accurate and more comprehensive data on eQTLs. Studies to determine the relationship of eQTLs to a specific disease depends in some cases on availability of relevant tissues for transcription analyses.

Nica and Dermitzakis reported examples of GWASs where a specific disease was found to be associated with a genome region that harbored an eQTL. Crohn's disease (a form of inflammatory bowel disease) was found to be associated within a region devoid of coding genes (a gene desert) on chromosome 5. Subsequently a cis regulator 270kb distant from the progesterone receptor 4 locus (PTERG4) was found to impact quantitative expression of that gene. In addition Prager *et al.* (2014) reported that a specific nucleotide variant within PTERG4 also increases susceptibility to Crohn's disease.

A possibility to be considered is whether specific regulatory loci, including cQTLs, contribute to expression of networks of genes. Schadt (2009) proposed that the constellation of genetic and environmental factors affect the molecular state of networks of genes and that these in turn affect disease risk.

1.6.2 *EQTLs and splicing quantitative trait loci (sQTLs)*

Li *et al.* (2014) used high quality genome sequencing and RNA sequencing in a 17 individual 3 generation family to define eQTLs and sQTLs. SQTL studies measure the relative abundance of alternative transcript isoforms derived from a particular gene. DNA genome sequencing and RNA sequencing studies have demonstrated that specific nucleotide variants impact the inclusion level of specific exons in transcripts.

Studies in this large family enable analyses of patterns of Mendelian segregation of alleles and correlations with transcript and expression levels. Li *et al.* emphasized the importance of non-coding (non-protein coding) variants in impacting gene expression levels, allele specific expression levels and splicing patterns.

1.6.3 *Genome architecture, regulatory variation and eQTLs*

Some eQTLs are close to promoters, overlap enhancers or promoters of the specific genes they impact. There is growing evidence that higher order chromatin structure and chromosome looping bring distant eQTLs spatially close to their target genes (Duggal *et al.*, 2014).

Knight (2014) presented an example of a QTL with clinical relevance. A specific variant on chromosome 1p13 rs12740374 was reported to be strongly associated with levels of plasma low density lipoproteins (LDL) and LDL cholesterol and myocardial infarction. This variant altered the expression of the SORTL1 gene.

This variant occurs in non-protein coding DNA; it creates a transcription factor binding site that alters liver expression of the SORTL1 protein. SORTL1 functions modulate cholesterol trafficking in liver. Extended GWASs revealed that homozygosity for minor allele of this polymorphism was associated with 16mg/dl lower LDLC as compared with homozygosity for the major allele. The most significant difference occurred in the levels of the small dense LDLs that are considered to be most atherogenic. The minor allele increases sortilin production. Strong and Rader (2012) reported that sortilin localizes to the Golgi and that it facilitates trafficking to the lysosome.

Knight noted that quantitative trait loci might impact protein levels; they may also impact functionality of proteins through effects on post-translational modifications including phosphorylation, glycosylation and sulfation. Knight also reported that there is evidence that 15–20% of autosomal genes show heritable allele specific differences in expression.

1.7 Insights Into Gene Function through Analysis of Effects of Mutations and through Gene Editing

Genome editing is defined as the process of making modifications in the genome, in its output or in its epigenetic marks.

Defining gene function can often best be carried by disrupting that gene or altering its expression and then observing the phenotype. Genome editing includes deletions, insertion or modifications of nucleotide in specific genes or in regulatory elements.

Boettcher and McManus (2015) reported that despite availability of the human genome sequence the function of many genes remains unknown. They noted that gene editing facilitates functional analysis of genes. They reviewed methods to achieve gene editing.

1.7.1 *RNA inhibition (RNAi)*

RNAi described by Fire *et al.* (1998) became an important method to achieve alteration of gene function. Boettcher and McManus noted that RNAi usually led to reduction in gene expression rather than to complete knockdown. Short inhibitory RNAs (siRNAs) or short hairpin RNAs (shRNAs) introduced into cell are loaded into the endogenous RISC system and this then leads to their degradation in the cytoplasm. However this then reduces the availability of the RISC components in cells. In addition the presence of sequences in endogenous RNAs with limited homology to the loaded siRNAs or shRNAs may also undergo degradation, and endogenous microRNAs in the cells that also utilize the RISC system can be negatively impacted.

Additional gene editing tools were subsequently developed. These included technologies based on incorporation of programmable endonucleases. Hsu *et al.* (2014) reviewed these technologies. Kim (1996) designed zinc finger nuclease as transcription factor bound to zinc. Each zinc finger recognized three bases in DNA; thus an array of six zinc finger nuclease bound 18 bases in DNA and zinc fingers were bound at one end to FOK1 endonuclease.

Zinc finger nucleases bind to single stranded DNA. In order to achieve cleavage, zinc fingers were designed to bind staggered sequences with one zinc finger set binding the sense strand and the other set binding to the antisense strand. FOK1 then cleaved between the staggered bound zinc fingers.

Gaj *et al.* (2013) reported on the use of Talens for gene editing. Talens are based on plant transcription factors. Each Talen contains 33–35 amino acids including two amino acids specific for a particular nucleotide in DNA. Talens were designed to recognize 15–20 DNA bases and to bind to sense and antisense strands in a particular gene region. It is therefore necessary to synthesize two Talens for cleavage of each desired site. The central endonuclease between the two DNA bound Talens then cleaves DNA between the Talens.

DNA breaks induced by zinc finger nuclease or Talens could then be repaired by endogenous processes, including non-homologous end joining or homologous recombination. An exogenous DNA template could also be added to facilitate repair.

Although some successes in gene editing using zinc finger nucleases or Talens were reported, the design of these editing tools proved problematic in many cases.

It is interesting to note that zinc finger nuclease editing was used to knockout the CCR5 receptors in hematopoietic stem cells derived from a patient with HIV AIDS. Tebas *et al.* (2014) reported that following irradiation of the patient to destroy bone marrow, the CCR5 knockout cells were transplanted in the patient. The HIV AIDs infection was significantly reduced and the patient's condition improved dramatically.

1.7.2 *CRISPR Cas9 editing*

The CRISPR Cas9 was initially identified in bacteria and archaea where it plays critical roles in eliminating invading DNA (e.g. from viruses and bacteriophage). The CRISPR locus is comprised of genes that encode RNAs, crRNA and tracRNA and it is rich in interspersed repeat sequences. Cas loci that encode endonucleases are located in close proximity to CRISPR.

Cas endonucleases, including Cas9, are multifunctional proteins with two nuclease domains and they are capable of introducing double stranded breaks in DNA.

DNA from invading organism is cleaved and fragments of the invading DNA are incorporated into the CRISPR locus. Transcription of the CRISPR locus and incorporated invading DNA fragments generates an RNA that is linked to the Cas endonuclease sequence. Subsequent targeting of this transcript and its hybridization to the CRISPR locus with incorporated invading DNA brings the Cas endonuclease into proximity with the aberrant locus and leads to its cleavage from the bacterial genome.

For gene editing experiments, the crRNA and tracRNA genes of CRISPR were engineered to derive a single guide RNA (sgRNA). The sgRNA for a specific purpose is designed so that 20 nucleotides at the 5′ end will undergo Watson–Crick binding to the selected target DNA. At the 3′ end of the at the 3′ end of the targeted DNA a PAM sequence must be located. The PAM sequence is protospacer adjacent motif. PAM sequence motifs are NGG or NAG. The targeting sequence and PAM must bind Cas9. Cas9 and RNA are then targeted to matched DNA sequence N20-NGG.

In summary, the sgRNA with 5′ nucleotides targeting DNA is located upstream of the PAM protospacer and these guide Cas9 to induce double stranded breaks in DNA at the binding site. Cas9 nuclease cuts 3-nt upstream of the PAM site.

Double stranded DNA cleavage can then be repaired by non-homologous end joining or homologous recombination. Exogenous templates can be used for insertion via homologous recombination.

The Cas9 system was originally purified from streptococcus pyogenes. Ran *et al.* (2015) isolated smaller Cas orthologues from *Staphylococcus aureus*. They demonstrated that these smaller Cas orthologues together with guide RNA could be inserted into an adenoviral vector to target sequence in mouse liver. This is an important improvement since the original Cas9 guide RNA combinations are too large to be targeted in an adeno-associated viral vector.

Cong *et al.* (2013) developed a mutant form of Cas 9, Cas9D10A, that only cleaves single stranded DNA, and it generates nicks in DNA. Two staggered guide RNAs can then be used to guide adjacent nicks (single stranded breaks can be repaired by base excision repair).

Hsu *et al.* (2014) reported that use of CRISPR reagents for editing often results in the generation of mosaics and techniques need to be utilized to distinguish between different sequences generated through editing.

1.7.3 *Application of genome editing*

Gene editing techniques are currently primarily used to interrogate gene function and to analyze effects of gene variation on biological function. Hsu *et al.* reported that CRISPR based editing has been used to recapitulate in cell systems genetic mutations found in patients. Cell systems used include induced pluripotent stem cells (IPS cells). They noted further that multiplexing numbers of different sgRNAs linked to Cas9 enabled analysis effects of polygenic variants in cell systems. Effects of targeting regulatory elements, including enhancers and promoters, have also been investigated. They also noted that tethering fluorescent tags to CRISPR Cas sgRNAs have also been used to labile DNA segments and to analyze nuclear organization.

Current applications of CRISPR Cas technologies have primarily been used for research purposes. However application of CRISPR Cas technologies for therapeutic purposes remain possibilities. Hsu *et al.* discussed possibilities of using CRISPR Cas to inactivate mutant alleles, or to remove damaged or duplicated segments of DNA. Problems of delivery and questions of off target effects need to be fully addressed.

2. Genomic Integrity and Relevance to Disease

This chapter includes sections on structural chromosome variants, DNA damage and repair, telomeric chromosome regions and consequences of abnormal structure and function.

2.1 Structural Chromosome Variations

2.1.1 *Detection of structural chromosome abnormalities*

The development of microarray technologies over recent decades has facilitated detection of segmental chromosome deletions and duplications that lead to dosage changes in genomic DNA. These changes are referred to as copy number variants (CNVs). Microarray studies that detect CNVs do not, however, detect structural rearrangements such as inversions of chromosome segments or translocations between chromosomes unless these lead to genomic dosage changes. Increasingly, DNA sequencing and particularly long-range DNA sequencing methods are being applied to analysis of structural chromosome abnormalities.

There is evidence that significant structural chromosome variation occurs in the genomes of healthy people. Extensive information on structural variation in the genomes of healthy individuals in different populations was made available in 2015 through publications of studies in the 1,000 genomes project (Sudmani *et al.*, 2015).

In this project, whole genome sequencing was used to develop an integrated map of structural variants in 68,818 healthy unrelated individuals with ancestry in 26 populations. The categories of structural variants detected through whole genome long read sequencing included deletions, duplications, inversions, and insertions that did not map to the reference genome including sequence of mobile genetic elements and nuclear mitochondrial translocations.

The majority of structural variants occurred at low frequency (variant allele frequency (VAF) <0.2%). In addition rare structural variants were encountered and found to be specific to particular populations. African populations showed a higher number of rare population specific variants than other populations.

Of particular interest regarding function was the 1,000 genomes project finding that homozygous deletions of 240 genes occurred in normal individuals, indicating that loss of these genes did not impact function.

2.1.2 *Elements in DNA that predispose to structural variation*

Large blocks of repetitive DNA sequences occur in the human genome and the DNA sequence in many repetitive regions is highly homologous. Ji *et al.* (2000) identified blocks of repetitive DNA in the human genome and reported that because of the high sequence homology of the DNA in these regions, they are predisposed to structural variation through a defined mechanism known as non-homologous recombination.

2.1.3 *Structural chromosome rearrangements leading to pathology*

There is evidence that rearrangements leading to pathology result primarily from rearrangements in a subset of chromosome regions. These regions then constitute hotspots of recurrent rearrangements. Watson *et al.* (2014) documented more than 20 regions involved in recurrent microdeletions and microduplications (CNVs) that give rise to specific syndromes often associated with developmental delay, intellectual disability and congenital malformations. The CNVs that result in pathology are most frequently 400kb or larger in size. It is however important to note that for several of these regions there are individuals who carry the rearrangement but do not manifest symptoms. However the frequency of structural rearrangements is higher in individuals with pathology than in the normal population.

In addition to recurrent structural variants, there are also non-recurrent structural variants that result from different mechanisms. The initial event in creating these variants is single or double stranded DNA break. Single stranded breaks in particular may be repaired by non-homologous end joining. DNA breaks with loss of DNA sequence may also be repaired by insertion of sequence segments with microhomology and these segments may be derived from a position elsewhere on the chromosome or from a different chromosome. Breakage and repair may also involve repeat sequences in endogenous retroviral derived sequences that are scattered throughout the human genome. LINE and HERV retroviral derived repeat sequences are particularly important in this regard (Weckselblatt and Rudd, 2015).

Structural rearrangements may lead to pathology through altering gene dosage (see Figure 2.1), through disruption of sequence in a specific gene and through generation of fusion genes that function differently from the parent genes.

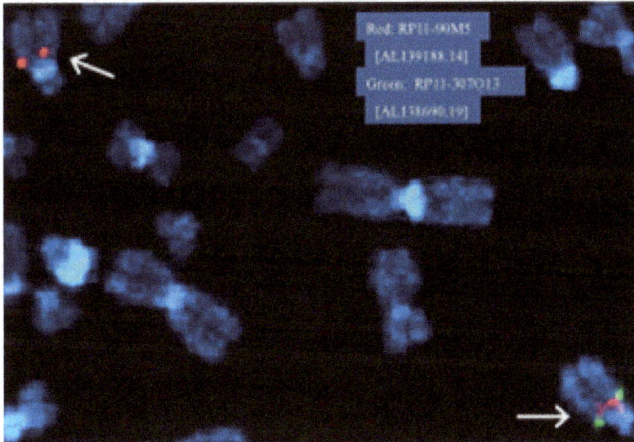

Figure 2.1. Deletion within chromosome 13q12. Two probes for chromosome 13q12 genes were used in FISH analysis one labeled with red dye and another labeled with green dye. Note that the red labeled probe hybridizes to both chromosomes. The green labeled probe hybridizes to only one chromosome.

2.2 RNA Damage and DNA Damage and Repair

2.2.1 *Oxidative damage to nucleotide triphosphates*

Reactive oxygen species can damage DNA and RNA. There is growing evidence that conditions of rapid cell proliferation lead to oxidative damage not only in nucleic acids but also in nucleotide-triphosphates in the nucleotide pool. Free nucleotides of guanosine triphosphate are oxidized to 8-hydroxydeoxyguanosine triphosphate, 8-hydroxyguanosine triphosphate, and 8-oxo-guanosine-5′ triphosphate (8 oxoGTP). Free nucleotides of ATP are oxidized to 2-hydroxy-deoxyadenosine (2OHATP).

Dominissini and He (2014) noted that NTPs (nucleotide triphosphates) are more than 13,000 times more susceptible to oxidative damage. They noted further that the incorporation of oxidized NTPs into nucleic acids can lead to hundreds of times more damaged bases in DNA than mutations. There is evidence that the MTH1 enzyme (NUDT1) hydrolyzed 8OXO GTP and 2OHATP, thereby converting them to mono-phosphates that cannot be used for DNA synthesis. Gad *et al.* (2014) emphasized that although MTH1 is non-essential for sanitizing the nucleotide pool in normal cells, it is required by rapidly proliferating cells, such as cancer cells. Specific inhibitors of MTH1 are being explored in cancer therapy. These investigators emphasize that use of MTH1 inhibitors addresses the problem of intra-tumor heterogeneity in target mutations and is also applicable to the treatment of a range of different tumors.

Zauri *et al.* (2015) reported that over-expression of the enzyme cytidine deaminase protects cells from accumulation of the damaging oxidation products of cytidine, namely 5-OH-2 deoxycytidine. Over expression of cytidine deaminase also protects cells from 5-formyl deoxycytidine that forms in response to therapy with the cytidine analog gemcitabine. It is likely that inhibitors of cytidine deaminase will enhance effects of chemotherapy with cytidine analogs.

2.2.2 *Oxidative damage to RNA*

Nunomura *et al.* (2012) emphasized the effects of oxidative damage to RNA, including protein coding RNA and non-protein coding

RNA. They noted that 8-hydroxy guanosine can be incorporated into RNA. They noted that because RNA is single stranded its bases are not protected by hydrogen bonding and RNA may be more susceptible to oxidative insults. Oxidation of the bases can occur directly in the RNA strand or bases can be oxidized in the nucleotide pool. They noted further that although 8-hydroxyguanine is most common, 5-hydroxycytosine and 5-hydroxyuridine have also been identified in RNA. Nunomura *et al.* reported that oxidized mRNA could potentially lead to defective proteins.

The brain is particularly vulnerable to oxidative damage due to its high oxygen consumption rate. Nunomura *et al.* postulated that RNA oxidation likely plays an important role in neurodegeneration.

2.2.3 *DNA damage and repair: Introduction*

In 2015, the Nobel Prize in Chemistry was awarded to Tomas Lindahl, Paul Modrich and Aziz Sancar, pioneers in studies on DNA damage and repair. The 2015 Lasker award was given to two other pioneers in studies on DNA damage and repair: Stephen J. Elledge and Evelyn M. Witkin.

DNA damage can arise due to endogenous processes and as a result of environmental exposures. Endogenous processes leading to DNA damage can arise during DNA replication. In addition DNA damage can occur due to endogenous processes such as depurination and deamination of bases and modification by alkylation. Reactive oxygen species may lead to oxidation of DNA bases and to DNA breaks.

Environmental factors leading to DNA damage include exposure to ultra-violet light that leads to the formation of pyrimidine dimers. Radiation can induce oxidation of DNA and to single stranded or double stranded breaks. Chemotherapeutic drugs can act as alkylating agents and can also induce cross-linking.

Ciccia and Elledge (2010) described the DNA damage response (DDR) as a signal transduction pathway that senses and responds to DNA damage. They noted that this pathway utilizes kinases (including ataxia telangiectasia ATM, ATR and DNAPK) and it utilizes polyADP polymerases. DNA damage response factors are located to the site of DNA damage by specific damage sensing proteins. Single and double stranded DNA breaks are sensed by PARP1 and PARP2 (polyADP ribose polymerases). PARP1 and PARP2 activation leads to the synthesis of polyADP ribose chains, structures that act as scaffolds to recruit factors that promote DNA repair. Matsuoka *et al.* (2007) investigated proteins in the DNA damage response network (DDR). They defined this as the network that senses different of types of DNA damage and co-ordinates responses that include activation of transcription, cell cycle control, DNA repair processes, apoptosis and senescence. Matsuoka *et al.* concluded that their findings indicated a broad landscape of DNA damage response and evidence for multiple interacting modules involved in DNA repair.

Insights into the DNA damage response have been obtained in part through studies on the pathogenesis of DNA repair defects in a number of rare genetic syndromes characterized by abnormal DNA repair processes and increased sensitivity to DNA damaging agents. These syndromes include Fanconi syndrome, ataxia telangiectasia, Bloom syndrome, Werner syndrome and Rothmund Thompson syndrome. In each of these syndromes the DNA repair defect occurs at a different position in the repair pathway. The types of DNA damage are listed in Table 2.1., and the schematic of DNA damage is shown in Figure 2.2.

2.2.4 *Types of DNA damage and associated repair mechanisms*

2.2.4.1 *DNA repair mechanisms*

There are differences in the types of DNA damage and different repair mechanisms are required for the different types of damage. Damaged bases are removed by base excision repair. Nucleotide excision repair corrects pyrimidine dimers (e.g. introduced by ultraviolet light) and intra-strand cross links. Interstrand cross links are repaired by proteins in the Fanconi pathway. Specific processes are utilized to repair breaks, these include single strand break repair enzymes and double strand break repair factors. Double stranded breaks require repair through processes such as non-homologous end joining or homologous recombination.

Table 2.1. DNA damage types and repair mechanisms.

Type of DNA damage	Repair mechanism
Base mismatches small deletions and insertions	Mismatch repair MMR
Damaged bases, oxidized, alkylated, de-aminated	Base excision repair BER
UV damaged DNA, bulky DNA adducts	Nucleotide excision repair NER Global NER, Transcription coupled NER
Single stranded DNA breaks	Single stranded break repair SSBR
Double stranded DNA breaks	Ataxia telangiectasia pathway (ATM) Non-homologous end-joining Homologous recombination
Collapsed replication forks	Ataxia telangiectasia and Rad3 (ATR) RecQ helicases
Interstrand cross-linking	Fanconi pathway Non-homologous end- joining Homologous recombination
Multiple forms of DNA damage	RecQ helicases

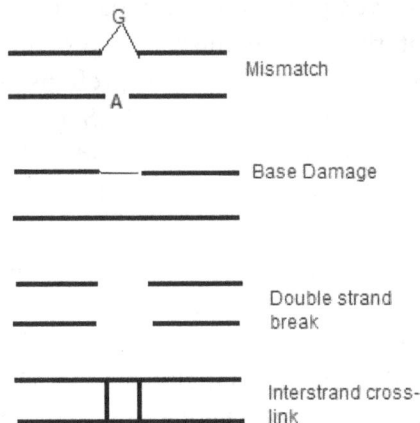

Figure 2.2. Schematic of types of DNA damage.

Specific DNA repair processes will be presented below followed by discussion of pathologies that arise as a result of defects in specific repair processes. In the case of DNA damage syndromes different syndromes arise due to mutations in the different proteins involved in specific processes in the detection and repair of DNA damage.

2.2.4.2 *Mismatch repair (MMR)*

Base-base mismatches or small deletions or insertions may be introduced into DNA during replication by polymerases. The mismatch repair system is designed to detect and excise such lesions and then to repair DNA. Key proteins important in recognition of mismatches include the Mut S homologs MSH2, MSH3 and MSH6. The MutL homologs in human MLH1, MLH3 and PMS1 and PMS2 are important in repair of mismatches.

MMR corrects single stranded DNA defects that arise during polymerase based DNA replication when the wrong nucleotide is incorporated into the newly synthesized strand. MMR can also correct short insertion or deletion defects that occur during DNA replication. Defects in MMR result in increased mutation rates. Germline MMR defects result in a higher cancer risk. There is also evidence that somatic MMR defects occur more commonly in sporadic cancers.

MMR systems show a high degree of conservation across life forms and similar MMR proteins exist in bacteria, yeast and eukaryotes. A number of the human mismatch proteins were identified based on their homologies with *E. coli* proteins (Li, 2008). These include human homologs of *E. coli* MutS (Fishel *et al.*, 1993). Human homologs of *E. coli* MutS2 include MSH2, MSH3 and MSH6. Human homologs of *E. coli* MutL include PMS1, PMS2. Human EXO1 is the homolog of the *E. coli* exonuclease Exo1. In addition human enzymes involved in mismatch excision repair include RPA1 replication factor, HMGPI upstream binding transcription factor (UBTFL) and polymerase delta that are also homologs of *E. coli* enzymes. Finally human DNA ligase 1 is a homolog of *E. coli* DNA ligase. There is now evidence that chromatin state and chromatin remodeling influence mismatch repair.

Li (2008) reported that MMR proteins also play roles in repair of other types of DNA lesions including interstrand cross-links.

2.2.4.3 *Base excision repair*

Base excision repair is important for removal of bases in DNA that are oxidized, alkylated or deaminated (Kim and Wilson, 2012). Steps in the base excision repair involve removal of the damaged bases by DNA glycosylase to create an intermediate; the sugar free intermediate is then excised with phosphodiesterase or lyase. Gap filling then takes place with DNA polymerase and strands are ligated by DNA ligase. Key enzymes include DNA glycosylases 8-oxoguanine DNA glycosylase (OGG1), nei-like glycosylase (NEIL1), endonucleases APEX1, APEX2, endo-processing polynucleotide kinase-3-phosphatase (PNKP) and polymerase B. Ligation involves activity of DNA ligase III (LIG3) and its cofactor XRRC1 or DNA ligase 1 (LIG1).

2.2.4.4 *Nucleotide excision repair*

This process is used to repair DNA damaged by UV light and chemical reactions that create bulky DNA adducts. Nucleotide excision repair (NER) follows recognition of bulky helix distorting lesions in DNA including pyrimidine dimers induced by UV exposure. Steps in NER involve damage recognition, opening of the DNA helix, incisions on both sides of the DNA lesion, removal of oligonucleotides with the lesion, gap filling synthesis and ligation. More than 30 different proteins are required for NER. There are two separate pathways of nucleotide excision repair: global genome NER and transcription coupled NER.

Transcription coupled NER removes bulky distorting lesions from the DNA of transcribed genes; in the presence of the lesion RNA polymerase is stalled. Global Genome NER removes bulky helix distorting lesions from the rest of the genome.

Nucleotide excision repair defects occur in several different human syndromes, including xeroderma pigmentosum (XP), and will be discussed further below.

2.2.4.5 *Single stranded DNA breaks*

Single stranded DNA breaks usually include loss of a single nucleotide and damaged 5' and 3' termini at the site of the break (Caldecott, 2008). These breaks result from the effects of reactive oxygen species or may result from ineffective base excision repair. Ineffective topoisomerase activity may also lead to DNA nicks and single stranded breaks.

Caldecott reported that processing of single stranded DNA breaks involves activity of APEX1 multifunctional endonuclease activity, PNKP polynucleotide kinase-3-phosphatase and APTX aprataxin has nucleotide binding activity and hydrolase activity. It is important that effective hydrolysis occurs and this requires prior structural changes in DNA folding induced by topoisomerase. Polymerase B then inserts appropriate missing nucleotide and ligase 3 (Lig 3) together with its cofactor XRCC, and then completes the ligation process.

Genetic diseases associated with single stranded breaks include ataxia oculomotor apraxia syndrome (AOA1). This syndrome is associated with uncoordinated gait and movements, choreoathetosis, limited eye movements and intellectual disability. The most frequent mutation involved the aprataxin encoding gene (APTX). A founder mutation leading to this disorder occurs in the Portuguese population. Caldecott reported that another condition associated with single stranded break repair defects is spinocerebellar ataxia with axonal neuropathy (SCAN1). This condition is apparently due to defects in required topoisomerase (TOP1) activity or defective activity of TDP1. TDP1 tyrosyl-DNA phosphodiesterase 1 repairs stalled topoisomerase activity to ensure hydrolysis of the phosphodiester group. Rulten and Caldecott (2013) reported that PARP1 (poly-ADP ribose polymerase 1) is the sensor of chromosomal single stranded breaks.

2.2.4.6 *Replication fork stalling*

Replication stress and replication fork stalling can be caused by nucleotide insufficiency and RNA lesions. ATR (ataxia telangiectasia and RAD3 related) co-ordinates response to replication stress and is activated by replication stalling. CHK1 (checkpoint kinase) is also activated by replication stalling and provides a signal for cell cycle arrest. ATR activity leads to phosphorylation of many substrates that act to promote the repair of stalling. These factors include helicases, translocases, nucleases and polymerases. Repair processes include lesion by-pass, template switching, and double stranded break induction. An important substrate that is recruited to stalled replication forks is SMARCAL, a chromatin regulator. This is recruited to stalled forks through replication protein A (RPA). There is evidence that SMARCAL function at replication forks is dependent upon its phosphorylation by ATR (Couch *et al.*, 2013).

2.2.4.7 *Double stranded DNA breaks*

Double stranded breaks involve breakage of both strands of DNA. The only physiological double stranded breaks occur in lymphocytes during immunoglobulin gene maturation (VDJ joining). Other double stranded breaks in the genome are pathological. They can be caused endogenously by reactive oxygen species and exogenously through exposure to mutagens including ionizing radiation. Following double stranded breaks, checkpoint mechanisms are initiated to arrest cell cycle replication (Dasika *et al.*, 1999).

Double stranded breaks are processed by exonucleases. They are then repaired either by non-homologous end-joining or by homologous recombination. In non-homologous end joining (NHEJ), the free ends join directly and this may lead to generation of small deletions of DNA sequence. Proteins and enzymes required for NHEJ include XRCC6, XRCC5, XRCC4 (X-ray repair cross complementing proteins), NHEJ1 (non-homologous end joining factor 1), DCLREC1 (DNA cross linking repair protein), DNA dependent protein kinases, NBS1 (Nijmegen breakage syndrome 1), ligases1 and IV, and polymerases POLM and POLG, may also be used. NHEJ may occur throughout the cell cycle.

Repair of double stranded breaks by homologous recombination involves activation of ATM serine threonine kinase and there is recent evidence that the COP9 signalosome is also an important player (Meir *et al.*, 2015). ATM phosphorylates many of the downstream effectors of repair by homologous recombination including the MRN complex (MRE11-RAD50-NBS1). Activation of CHK2 (checkpoint kinase 2) inhibits cell cycle activity during the repair. Other important effectors of repair by homologous recombination include RAD51, 52 and 54 (repair proteins), XRRC2, XRRC, BRCA1, BRCA2, and resolvase. During the repair process homologous DNA is identified and subsequently strand invasion exchange occurs.

2.2.4.8 Interstrand cross links

Interstrand cross links can be caused by chemicals including aldehydes and platinum. These cross links prevent DNA strand separation and can be blocks to DNA replication and transcription. Protein components in the Fanconi pathway play key roles in recognition and repair of interstrand cross links.

2.2.5 Specific pathologies and syndromes associated with DNA damage and defective repair

2.2.5.1 Mismatch repair defects in colon cancer and other cancers

In 1993, Fishel *et al.* reported association of MSH2 mutations with hereditary non-polyposis colon cancer. In the same year, the Vogelstein group (Leach *et al.*, 1993) reported MUTS mutations in individuals with hereditary non-polyposis colorectal cancer. In 1994, Papadopoulos *et al.* reported association of mutations in the human homolog of MutL in hereditary colon cancer.

There is now evidence that mismatch repair defects occur in a number of other cancers including cancers of endometrium, ovary, breast, stomach, bladder and prostate. Within tumors mismatch repair defects lead to instability of microsatellite repeat polymorphisms in DNA (Boland *et al.*, 1998).

Guidelines for genetic studies to predict colon cancer risk in individuals where there is a family history of colon cancer and guidelines for treatment and management of patients with hereditary colorectal cancer syndrome were published by the American Society of Clinical Oncology (Stoffel *et al.*, 2015). These guidelines followed closely the 2013 the European Society of Medical Oncology guidelines.

Stoffel *et al.* (2015) reported that 5–6% of patients with colorectal cancers have germline mutations that predispose them to colon cancer. Lynch syndrome is the most common hereditary non-polyposis cancer syndrome and is due to mutations in mismatch repair genes. Patients with Lynch syndrome have germline mutations in one of the following genes: MLH1, MSH2, MSH6, PMS2 and EPCAM (epithelial adhesion cell molecule). These mutations lead to tumors with microsatellite instability. In addition, studies in the tumors in these patients frequently reveal absence of the protein encoded by the specific mismatch repair gene that has mutations. Stoffel *et al.* reported that the lifetime risk for colorectal cancer in individuals with germline MMR gene mutations increases by 30–70%. The most common cancers in Lynch syndrome patients include colorectal cancers and cancers of the endometrium. However Lynch syndrome patients also have elevated risks for cancers of the urinary tract, small intestine and ovary.

A number of other hereditary conditions are associated with increased risk of colon cancer due to mutations in tumor suppressor genes. These include familial adenomatous polyposis due to mutations in the APC gene and Peutz–Jeghers syndrome (STK11, serine threonine kinase 11).

2.2.5.2 *Nucleotide excision repair impairments*

DiGiovanna and Kraemer (2013) reviewed three well-known syndromes that result from impaired NER and that lead to ineffective removal of lesions in DNA that result from ultra-violet (UV) exposure. These syndromes include xeroderma pigmentosum (XP), Cockayne syndrome (CS) and trichothiodystrophy (TTD). More recently two syndromes with overlapping features were added, these included UV sensitive syndrome and cerebro-oculo-facial-skeletal syndrome (COFS). DiGiovanna and Kraemer (2012) reported that in XP, CS and TTD, photophobia and skin sun sensitivity occur. However each of the three syndromes has unique features. Hyperpigmented freckles (lentiginous hyperpigmentation), occurs only in XP. Pigmentary retinal degeneration occurs only in CS. Short brittle hair occurs in TTD.

COFS is associated with microcephaly, facial dysmorphology and joint contractures (arthrogryposis). In UV sensitive syndrome there is mild photosensitivity without pigmentary abnormalities.

DiGiovanna and Kraemer emphasized however that there can be overlap of manifestations in the different syndromes. CS patients may manifest some features of XP and may have features of TTD.

The genes defective in CS CSA and CSB play key roles in transcription coupled nucleotide excision repair. When specific DNA lesions lead to stalling of RNA polymerase II (RNAPll) stable interactions occur between RNAPII and CSA, additional proteins UVSSA (UV stimulated scaffold protein A) and CSB bind to RNAPll. UVSSA then recruits deubiquitinating protein USP7 to the complex. Deubiquitination leads to degradation of the complex and opening up at the lesion site and chromatin remodeling so that repair can begin. Dijk *et al.* (2014) reported that mutations in the UVSSA protein have been identified in UV sensitive syndrome.

2.2.5.2.1 XP

Global genome nucleotide excision repair defects occur in XP. The different XP defined proteins were identified by means of *in vitro* complementation studies. The XP condition can result from defects in any one of seven genes. Recognition of UV induced lesions including pyrimidine dimers is dependent on proteins XPE (DDB2 damage specific DNA binding protein) and XPC. There is now evidence that DDB2 recruited to the lesions may be in the form of a complex of Cullin4A, DDB2 and RBX1.

The proteins XPB (ERCC3) and XPD (ERCC2) then unwind DNA surrounding and adjacent to the lesion. CHD1L chromatin helicase along with PARP1 facilitate chromatin unwinding. XPF and XPG (ERCC5) act as endonucleases that incise the lesions and the surrounding DNA to yield a fragment of approximately 30 nucleotides. The gap is then filled in with *de novo* DNA synthesis. Dijk *et al.* report that VCP (valosin containing protein) plays a role in the removal of XPC from the DNA strand to promote downstream repair.

2.2.5.2.2 Gene defects in XP

Deleterious mutations in any one of the following genes can lead to XP: XPE (DDB2), XPA, XPB (ERCC3), XPC, XPD (ERCC2), XPF (ERCC4), and POLH (polymerase eta).

XP is an autosomal recessive disorder characterized by extreme sun sensitivity, freckle like pigmentary changes, skin atrophy telangiectasias, increased frequency of skin cancers and melanoma (DiGiovanna and Kraemer, 2012). Progressive neurological degeneration due to neuronal loss occurs, likely from oxidative damage, in 25% of patients.

2.2.5.3 *The ATM pathway and molecular pathogenesis of ataxia telangiectasia*

McKinnon (2012) reviewed the molecular pathogenesis of ataxia telangiectasia. Ataxia telangiectasia is an autosomal recessive condition due to loss of activity or to diminished activity of the ATM serine kinase. The key manifestation of this disorder is ataxia, the lack of coordinated muscle movements particularly during walking. Ataxia first manifests during childhood and becomes progressively worse over the years. Other manifestations include telangiectasias, dilated small blood vessels are usually visible in the eye. Other findings in patients include immune dysfunction, increased sensitivity to radiation and predisposition to cancer. The most common cancers in these patients are leukemias and lymphomas and these can originate in T cells or B cells.

McKinnon reported that levels of immunoglobulins IGA, IGE and IGG2 were decreased in patients but that frequent infections were not common. He also noted that levels of alpha-feto-protein were elevated in patients.

McKinnon emphasized the sequence similarities between ATM and other specific protein kinases including ATR, DNA dependent protein kinases (DNAPKcs encoded by PRKDC). ATM, ATR and DNA dependent kinases each contain FAT, (FRAP/ATM/TRAPP), and FATC kinase domains toward their carboxyterminal ends. The N terminals of each of the three bind activating proteins. These three proteins also have structural similarities with the MTOR protein. McKinnon emphasized that despite their structural similarities the biological roles of ATM, ATR and DNA dependent protein kinases differ and defects in each of the proteins lead to different diseases. ATR functional defects lead to growth retardation and microcephaly. Defects in DNA dependent protein kinases due to mutations in the PRKDC gene lead primarily to immunodeficiency. ATM protein kinase differs from the homologous proteins with respect to phosphorylation sites. Phosphorylation of serine 1981 in ATM is particularly important for its function.

2.2.5.3.1 DNA damage and activation of the ATM signaling pathway

Double stranded DNA breaks activate the ATM signaling pathway. The first step in this pathway is binding of the MRN complex (MREll, RAD50, and NBS) to DNA flanking the breaks and activation of the monomeric form of ATM. McKinnon noted that ATM activation involves post-translational modification through phosphorylation and ubiquitination of ATM. Subsequently other proteins are activated including MDC (mediator of DNA damage and checkpoint control), p53 binding protein, KAP1 (TRIM28 transcriptional control protein), PIAS1-4 proteins (inhibitors of STAT) and RNF8 (ring finger protein ubiquitin ligase).

Activated ATM can phosphorylate a large number of substrates and these include proteins involved in cell cycle arrest and DNA repair. To achieve cell cycle arrest ATM can phosphorylate specific proteins in each stage of the cell cycle. These include p53, and p21 (CDKNA1) cell cycle inhibitors. In the S phase of the cell cycle ATM phosphorylation impacts function of Fanconi protein FANCD2,

SMC1 and NBS1 (nibrin). Proteins in the G2 phase of the cell cycle that are functionally impacted by ATM include checkpoint kinase CHK1, DC25, CDC25C and BRCA1.

DNA repair is facilitated through activity of TDP1 (Tyrosyl DNA phosphodiesterase 1 that impacts topoisomerase helicase function and KAP1 (TRIM28 chromatin associating protein).

ATM phosphorylated proteins can also trigger apoptosis, particularly in situations where DNA repair fails. Apoptosis of cells is facilitated through activation of CHK2 and p53. Cell cycle regulators are key ATM substrates.

McKinnon noted evidence that ATM functions as a sensor of oxidative stress and that defects in ATM function impact mitochondrial function.

2.2.5.3.2 Neurodegeneration in ataxia telangiectasia and related conditions

McKinnon emphasized that neurodegeneration is the key feature of ataxia telangiectasia. Brain MRI studies reveal cerebellar atrophy and histological studies reveal loss of Purkinje cells and loss of granule cells. Furthermore Purkinje cells have less complex arborizations than normal. It is important to note that neurodegeneration is apparently ongoing throughout life.

2.2.5.3.3 Neurological clinical manifestations in ataxia like conditions

McKinnon reported clinical features in patients with ataxia like conditions. These included ataxia with oculo-motor apraxia (apraxia inability to perform a specific task). Oculo-motor apraxia in these patients results from defects in apraxatin, a nucleotide hydroxylase that pays a role in single stranded DNA repair. Another form of ataxia with oculomotor apraxia AOA2 is due to defective function of sentaxin that is encoded by the gene SETX. Sentaxin functions as a DNA-RNA helicase. Defects in tyrosyl-DNA phosphodiesterase impact the function of the helicase topoisomerase, leads to spinocerebellar ataxia with axonal neuropathy.

2.2.5.4 *Fanconi syndrome and failure to repair interstrand cross links*

2.2.5.4.1 Fanconi syndrome

Fanconi syndrome, sometimes referred to Fanconi anemia, represents an extreme example of a disease that results from defects in any one of the subunits in a multi-subunit complex. This syndrome most frequently follows an autosomal recessive inheritance pattern; however one of the genes responsible for this condition, FANCB maps to the X chromosome.

Patients often first present with aplastic anemia or myelodysplastic syndrome due to bone marrow failure. In some cases children with this syndrome have abnormalities of the thumb and absence of the radius. In addition patients may have abnormal skin pigmentation, endocrine defects, hearing defects, and gastro-intestinal abnormalities (Fanconi, 1927; see Moustacchi, 2003).   Patients with Fanconi syndrome are increased risk for acute myelogenous leukemia and for solid tumors, frequently of the head and neck.

The molecular basis of this syndrome lies in defects in the repair of DNA damage and particularly in the failure to repair abnormal interstrand cross-linkage. Laboratory tests for Fanconi anemia are based on the quantitative assessment of hypersensitivity of Fanconi syndrome patient cells to cross-linking agents, for e.g. diepoxybutane (Kupfer, 2013). Exposure to these reagents results in increased occurrence of chromosome abnormalities including breaks, translocations and radial chromosomes.

Kim and D'Andrea (2012) reported that the Fanconi anemia pathway involves three DNA repair processes, nucleolytic incision, translesion synthesis and homologous recombination. Together the components of this pathway repair interstrand cross links that arise from exposure to chemicals, including aldehydes and platinum. They noted further that function pf FANCD2 and FANCI and monoubiquitination are key regulatory steps. Monoubiquitin complexes move to the DNA lesion and co-ordinate activities of downstream repair factors. Ubiquitin activity impacts nuclease and nucleolytic incision steps. Incision is followed by creation of a double stranded DNA break. Translesion DNA synthesis is facilitated through activity of a heterodimer of polymerase zeta and REV1 that acts as a scaffold to recruit polymerases. Homologous recombination is then utilized to resolve the double stranded break. BRCA1 plays an important role in homologous recombination.

2.2.5.4.2 Complementation groups in Fanconi anemia

Functional complementation assays were utilized to distinguish between forms of Fanconi anemia due to different underlying molecular defects due to mutations in different sub-units of the multi-subunit complex (Strathdee *et al.*, 1992). Fanconi complementation groups and genes are listed in Table 2.2.

In 2013, 16 different complementation groups were defined from A to Q. It is important to note that 5 of the genes responsible for specific Fanconi complementation groups turned out to be genes implicated in the etiology of breast cancer, BRCA2, PALB2, RAD51C, SLX4, and BRIP1. Kupfer (2013) listed the Fanconi complementation groups and genes and their chromosome locations.

Walden and Deans (2014) reviewed the structures and functions of the components of the Fanconi DNA repair pathway. They described three key features of this pathway as follows:

(i) An anchor complex that recognizes interstrand cross-links
(ii) A multi-subunit ubiquitin ligase complex that is active against substrates
(iii) Downstream repair proteins

Walden and Deans reported that unrepaired DNA damage in Fanconi patients induced raised levels of p53 and p21 and they proposed that these led to decreases in the hematopoietic stem cell pool. The failure to repair interstrand cross links leads to impaired DNA replication and transcription.

2.2.5.4.3 Anchor complex

The anchor complex is comprised of FANCM and additional proteins FAAP24 (Fanconi anemia core complex associated protein 24),

Table 2.2 Fanconi complementation groups, genes and their chromosomal locations.

Complementation group	Symbol	Chromosome location
A	FANCA	16q24.3
B	FANCB	Xp22.31
C	FANCC	9p22.3
D1	FANCD1/BRCA	13q12.13
D2	FANCD2	3p25.3
E	FANCE	6p21.22
F	FANCF	11p15
G	FANCG/XRCC9	9p13
I	FANCI/KIAA1794	15q25-q26
J	FANCJ/BRIP1	17q22-q24
L	FANCL/BHF9	2p16.1
M	FANCM	14q21.3
N	FANCN/PALB2	16p13.3
O	FANCO/RAD51C	17q23
P	FANCP/SLX4	16p13.3
Q	FANCQ	16p13.2

MHF1 and MHF2 (histone fold proteins). The anchor complex detects strand cross links and stalled replication forks. The MHF1 and MHF2 proteins bind to histone in chromatin that surrounds the sites of damage.

2.2.5.4.4 Core complex

The core complex is activated by FANCM, FAAP, MHF1 and MHF2 binding to damaged sites. However Walden and Deans (2014) noted that other proteins may also activate the core complex. The function of the core complex is to add ubiquitin to FANCD2 and FANC1 during the S phase of the cell cycle if DNA damage is detected. Fanconi anemia mutations occur predominantly in the nine proteins of the core complex. The nine proteins in the complex are

FANCA, FANCB, FANCC, FANCE, FANCF, FANCG and FANCL. Assembly of the complex requires associated proteins FAAP20 and FAAP100.

The different proteins in the core complex differ in their length and structure. Walden and Deans emphasize that the roles of the different components of the core complex are not yet clear; however more is known about the structure of the FANCD2 and the FANCI subunits that undergo mono-ubiquitination. The FANCD2 protein shows a high degree of homology across mammalian species. The protein has helices and exhibits a high degree of folding into solenoids. A number of the FANCD2 mutations impact protein folding while others impact DNA binding. FANCD2 and FANCI proteins heterodimerize to form the ID2 complex. Mono-ubiquitination of the ID2 complex is dependent on FANCL (PH9) and UBE2T. Walden and Deans note that ubiquitination of FANCD2 is most important for ubiquitin conjugation however the presence of FANCI apparently accelerates ubiquitination of FANCD2. The ID2 complex and FANCD2 ubiquitination are essential for unhooking the intra-strand cross links. Additional nucleases likely play roles in this process and FANCP (SLXQ), is apparently a dual purpose nuclease and it also repairs damage in XP.

2.2.5.4.5 Repair factors downstream in the Fanconi pathway

These include FANCD1 (BRCA2), FANCJ (BRIP1), FANCN (PALB2) and FANCO (RAD51C). These factors are activated by mono-ubiquitination. Walden and Deans reported that mutations in these proteins are rare in Fanconi anemia patients, however when present mutations in these downstream factors are associated with severer phenotypes and cancer.

They reported that the downstream components likely co-ordinate the homologous recombination processes that follow unhooking of the intra-strand links. FANCJ is an ATPase helicase; FANCO (RAD51) functions as a recombinase. FANCD1 (BRCA2) recruits FANCO (RAD51) to double stranded breaks. Gaps in our knowledge still exist regarding the final steps in the Fanconi anemia pathway.

2.2.5.5 *RecQ helicases and heritable DNA repair defects*

RecQ helicases play key roles in recombination, DNA replication, transcription and genome maintenance. There are five well studied RecQ helicases in mammals and humans. Croteau *et al.* (2014) reviewed the structure and function of these helicases. They emphasized interaction between the helicases and also noted that each helicase has a number of interacting protein partners that influence specific functions. Croteau *et al.* reported that the key function of DNA helicases is to utilize energy derived from ATP hydrolysis to unwind double stranded DNA. There are five RECQ helicases in humans these include BLM, WRN, RECQL1, RECQL4 and RECQL5.

The BLM helicase is defective in Bloom syndrome. The WRN helicase is defective in Werner syndrome. RECQL4 defects occur in rare syndromes, such as the Rothmund–Thomson, Rapadilino and Baller–Gerold syndromes. RECQL4, BLM, and WRN all bind to each other, and RECQL5 binds to WRN. RECQL1 does not bind to other helicase proteins. All helicases bind to RPA1 replication protein.

Croteau *et al.* reported that each of the helicases contains a helicase domain and close to that domain there is a RQC domain. The RQC domain is important for binding to DNA structures and most likely it binds to the phosphate backbone of DNA. In RECQL5 the RQC domain is reduced in size. WRN and BLM helicases also contain an HRDC domain that promotes localization of the protein to double stranded DNA breaks.

Croteau noted that DNA helicases play roles in base excision repair and in double stranded DNA break repair. They have less impact on nucleotide excision repair or on mismatch repair. Four helicases RECQ1, BLM, WRN and RECQL4 are all involved in double stranded DNA repair. BLM, WRN, RECQL4 are involved in telomere maintenance. BLM, WRN, RECQL5 are involved in transcription. RECQL4 is unique in that it localizes to mitochondria. Croteau noted that there is some evidence that RECQ helicase may also be involved in the repair of DNA interstrand cross-links.

WRN has an additional function; Croteau reported that it stimulates activity of the glycosylase NEIL1. Glycosylase such as NEIL1 cleave damaged nitrogenous bases from the DNA sugar-phosphate backbone. Following this, endonucleases then act to produce a single stranded break in DNA that is subsequently repaired through activity of polymerase and ligase.

RECQ helicases are involved in DNA replication and in restart replication following the collapse of replication forks. They are also involved in a process described as decatenation. Coteau *et al.* reported that decatenation occurs during the S and G2 phases of the cell cycle when parental and new DNA strands are linked together and sister chromatid exchanges and recombination occur. Helicases and topoisomerases play roles in untangling of DNA strands.

BLM, WRN and RECQL4 play important roles in telomere maintenance through interaction with the shelterin complex proteins and with the telomere repeats. The shelterin complex protects telomeres and regulates telomerase activity. Proteins in this complex include TRF1, TRF2, POT1, BLM and WRN, and they are particularly important in preventing telomere loss.

2.2.5.5.1 Manifestations of Bloom syndrome

Bloom syndrome is a rare autosomal recessive disorder. Most patients are of Ashkenazi Jewish descent. Manifestations include growth retardation in height and weight, sun sensitivity with redness of skin and development of telangiectasias in sun exposed areas, often with a butterfly distribution over the nose and cheeks. Patients have unusual facial features including a small lower jaw, they have unusual speech and learning difficulties may be present. Other manifestations include immune deficiency and chronic obstructive pulmonary disease. Adult patients with Bloom syndrome may develop diabetes and there is an increased frequency of cancer particularly lymphomas and leukemia. Males are often infertile.

2.2.5.5.2 Werner syndrome manifestations

This syndrome occurs with a frequency of 1 in 200,000 in most populations except in Japan where the frequency is 1 in 20,000. Manifestations often first appear during adolescence when the normal growth spurt fails to occur. Patients develop early aging signs. They have unusual distribution of body fat with thin arms and legs and thick trunks; they develop osteoporosis, graying of hair, cataracts, arteriosclerosis, hyperlipidemia and cataracts.

2.2.5.5.3 Rothmund–Thomson syndrome manifestations

Patients manifest redness of skin particularly on the face. Later they develop telangiectasias, and thinning and atrophy of the skin with patches of altered skin color. They have sparse hair, defects of skin and nails, cataracts, osteopenia and osteoporosis. They also are at increased risk for cancer, particularly basal cell carcinoma of the skin and osteosarcoma.

2.2.6 *Neurodevelopmental defects related to DNA breaks and non-homologous end-joining*

Rulten and Caldecott (2013) noted that during the stages of neuronal differentiation and migration there was greater dependency on non-homologous end joining (NHEJ) of DNA breaks since in these stages cells were in the G0 or G1 stage of the cell cycle. Defects in NHEJ during neural differentiation and migration stages resulted in loss of cortical neurons and microcephaly. Examples of syndromes associated with defects in NHEJ associated with microcephaly and growth retardation include the following:

- Nijmegen breakage syndrome (NBS); short stature, facial dysmorphology, immunodeficiency
- NBSLD Nijmegen breakage syndrome like disorder
- MCSZ microcephaly early onset seizure disorder
- NEHJ1 immune deficiency, microcephaly growth retardation
- LIG4 syndrome microcephaly, facial dysmorphology, impaired immune response

2.2.7 *Progeroid premature aging syndromes*

Progeroid syndromes are characterized by premature onset of aging. One of these syndromes, Werner syndrome, is due to defects in DNA repair and specifically to defects in nucleotide excision repair. Hutchinson–Guilford progeria syndrome (HGPS) is a premature aging syndrome due to mutation in the A type lamin proteins. Cells from HGPS patients have multiple nuclear defects including increased DNA damage and altered histone modifications. Prokocimer *et al.* (2013) reported that loss of telomeres and telomere dysfunctions also occur.

Clinical features of Progeria syndrome include severe growth retardation and premature aging symptoms including alopecia, lipo-dystrophy, decreased joint mobility, atherosclerosis with thickening and calcification of aortic and mitral valves. Ishida *et al.* (2014) questioned whether the cardio-vascular disease could be associated with DNA damage. They proposed the DNA damage in vascular wall cells and incomplete repair of this damage caused cells to undergo senescence and apoptosis. They noted further that senescent cells secrete pro-inflammatory cytokines and chemokines. These factors together accelerate atherosclerosis and increase plaque instability.

DNA damage and telomere erosion and accumulation of somatic mutations are thought to be important factors that contribute to aging and mitochondrial dysfunction contributes to senescence.

2.2.8 Cohesinopathies

In the cohesinopathies different but overlapping phenotypes result from mutations in each of the proteins that contribute to the formation of a specific structure and the loading of this structure onto chromatin.

The cohesin complex has a number of important functions, including sister chromatid cohesion, maintenance of genome stability, participation in repair of DNA defects and there is recent evidence that it participates in gene regulation.

The multimeric cohesin complex forms a ring structure, with the ring composed of SMC1A and SMC3. The N and C terminals of SMC1A and SMC3 meet to form globular head and stranded hinge domains (Ball *et al.*, 2014). The head domains have ATPase function and interact with non-SMC proteins RAD21 and STAG. There is evidence that the ring structures encircle chromatids. The protein NOPBL plays a role in loading cohesion onto chromatids. HDAC8 is involved in removal of cohesion from chromatin and its recycling.

Mutations in any one of five genes SMC1A, SMC3 (structural maintenance of chromosomes), RAD21, NIPBL (nipped homolog) and HDAC8 (histone deacetylase 8) lead to Cornelia de Lange syndrome (CDL) or to CDL or CDL-like phenotypes; all mutations have not yet been identified (Gil- Rodríguez *et al.*, 2015). Features of CDL include distinct facial dysmorphology, often with heavy-set eyebrows that meet in the middle (synophrys) or heavy set arched eyebrows; intellectual disability is a common feature. Other features include prenatal and post-natal growth retardation, limb malformation, organ defects including congenital heart defects. Patients are usually heterozygous with mutations on only one member of the gene pair; the mutations are *de novo* in origin in most cases.

2.2.9 *Telomeres*

Telomeres cap human chromosomes and are comprised of repetitive arrays of the nucleotide sequence 5′TTAGGG3′ and associated proteins. These repeats extend over several kilobases (kb). The average length varies between 5 and 15 kb. In cord blood cells the average telomere is on average 10kb (Armanios, 2013).

The telomeric region is double stranded except for the terminal 20–200 nucleotide regions where there is a single stranded DNA overhang. This single stranded overhang frequently folds back on itself in a T loop structure that anneals to complementary nucleotides in the duplex telomeric DNA. The T loops vary in size. There is definite evidence that progressive shortening of telomeres occurs during repetitive cell divisions. The degree of telomeric shortening during cell replication is impacted by environmental factors and also by specific cellular factors.

Different repetitive elements are abundant in subtelomeric regions. Sub-telomeric regions contain repeats of TTTGGG, TGAGGG, and TCAGGG.

Telomeres are protected by a group of proteins referred to as the shelterin complex. This complex is composed of six proteins. These include telomere repeat binding proteins TRF1 and TRF2, repressor/activator proteins RAP1, TIN2 (TRF2 interacting protein), TPP1 (TIN2 interacting protein) and POT1 (protector of telomeres). These six proteins protect telomere ends from degradation and they protect against telomere fusions.

The double stranded region of the telomere binds shelterin protein TRF1 and TRF2, and RAP1; the TIN2 protein links the double stranded and single stranded telomeric region. TIN2 binds TPP1 and POT1 binds to TPP. POT1 also recruits telomerase to the telomeres (Schmidt and Cech, 2015).

Chen *et al.* (2012) described an additional complex, the CST complex that associates with telomeres and reduces telomerase activity. Three proteins comprise the CST; they include CTC1, STN1 suppressor of CDC13 and TEN1.

Schmidt and Cech (2015) reported that telomere length is determined by a number of factors including absolute telomerase levels, frequency of telomerase-telomere interactions and number of telomeres added during recruitment of telomerase. The POT1 and TPP1 levels correlate with telomerase recruitment. The CTC1 complex displaces telomerase from the telomeres.

Uncapped telomeres are recognized by ATM or ATM DNA damage response proteins, and phosphorylated histone H2AX may localize to damaged telomere. Significant shortening of telomeres leads cells to be arrested in the G1 phase of the cell cycle (Reddel, 2014).

2.2.9.1 *Telomere elongation*

Genome stability requires maintenance of telomeres at chromosome ends. Telomerase functions to extend the 3' overhang at chromosome ends.

One telomere elongation process involves the telomerase complex and is particularly active in stem cells and in rapidly proliferating cells. Telomerase activity is high in stem cells but is present at low levels in somatic cells. A number of human cancers exhibit high levels of telomerase. The second elongation process is referred to as the alternative process of telomere replication.

Telomerase participates in telomere maintenance and elongation. Telomerase has an RNA component TERC that serves as a template and TERT that functions as a reverse transcriptase and adds TTAGGG. Ribonucleoproteins bind to telomerase, and these include NOP10, NHP2, GAR1, and DKC1 (dyskerin).

REL1 encodes a DNA helicase that prevents formation of deleterious secondary structures during telomere lengthening (Vannier *et al.*, 2014).

TERT and TERC assemble in the Cajal body in the nucleus. Cajal body proteins facilitate assembly and subsequent trafficking. Two molecules of TERT bind one molecule of TERC. In addition to the RNA component TERC and TERT telomere reverse transcriptase, many additional proteins, ribonucleoproteins, and nucleolar proteins are required to assemble the telomerase holoenzyme complex (Vogan and Collins, 2015). These include the small nucleolar riboproteins NHP2, NOP10, dyskerin, and GAR1. The protein TCAB1, also known as WRAP53, determines localization of telomerase in the Cajal body. TCAB1 impacts telomerase assembly and function.

TCAB1 (WRAP53) is a dual function protein. In addition to its function related to telomerase it is an antisense transcript to TP53 5' untranslated region and regulates TP53 expression levels.

The telomerase holoenzyme cannot interact with the telomere until the inhibitor TRF1 is released (Reddel, 2014). Telomeric regions, telomerase, and telomeric sequences are illustrated in Figure 2.3.

2.2.9.1.1 Alternative telomere lengthening

In this process lengthening of telomeres is achieved through non-homologous recombination. This may involve another region of the telomere or the telomere that is present on another chromosome. In addition, there is evidence that extra-chromosomal telomeric repeats derived from circular DNA elements may be used. Reddel (2014) reported that levels of circular telomeric DNA correlate with the level of alternate telomere lengthening. He noted further

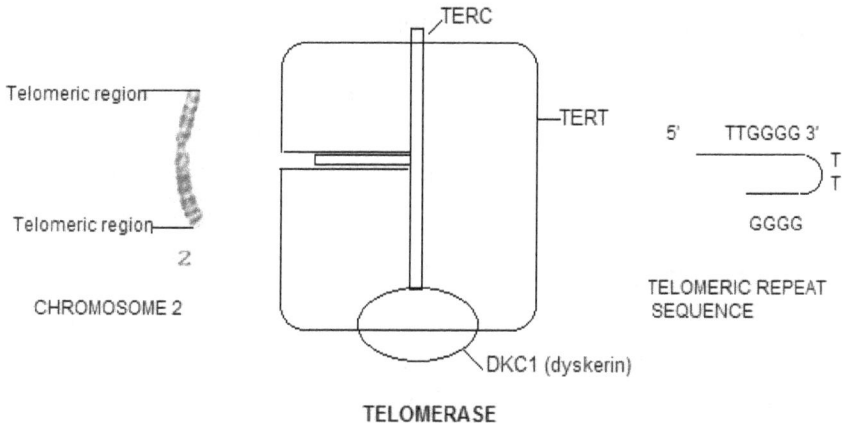

Figure 2.3. Telomeric regions on chromosome, telomerase structure and sequence at telomeres.

that extra-chromosomal telomere DNA is often associated with the PML bodies in the nucleus.

The alternative process of telomere lengthening has particular therapeutic relevance since inhibition of telomerase activity is unlikely to be successful in treatment of tumors with high levels of alternate telomeric lengthening processes.

Telomere trimming involves excision of telomeric DNA and generation of circularized telomeric DNA. This process is mediated by XRCC3 (DNA homologous repair enzyme).

2.2.9.1.2 Telomeres and chromatin

There is evidence that chromatin is present at telomeres. Furthermore there is evidence that alternative lengthening of telomeres is influenced by chromatin and by a decrease in histone 3 lysine 9 trimethylation (H3K9me3) (Episkopou *et al.*, 2014).

2.2.9.1.3 TERRA (TER)

TERRA (also known as TER) is a long non-protein coding RNA that is transcribed from most telomere ends and its expression is regulated by surveillance factor that detect shortened telomeres. Cusanelli *et al.* (2013) reported that when expression of TERRA is induced by telomere shortening, TERRA molecules accumulate in nuclear foci. Telomerase molecules then bind to the TERRA clusters and are subsequently recruited to telomeres and promote telomere lengthening.

2.2.9.2 *Upregulation of telomerase in cancer cells*

Reddel (2014) reported that the precise mechanisms involved in upregulation of telomerase in cancer cells are not known. In some tumors there is evidence for translocation of the TERT gene to an active oncogene, for e.g. MYC. In other cancers mutations are present in the promoter of the gene that encodes TR (TERC).

2.2.9.3 *Low levels of telomerase associated with stress*

High levels of perceived stress in women led to shorter telomere length and low telomerase activity. Shorter telomere lengths have also been reported in psychiatric disease including major depression and anxiety disorders and in association with childhood adversity. Mitchell *et al.* (2014) determined that disadvantaged social environments led to decreased telomere length in 9 year olds. They demonstrated that African American boys who grew up in a disadvantaged environment had shorter telomeres than those who grew up in an advantaged environment. They also determined an interaction between telomere length, social environment and genetic variation in components of the serotonin and dopamine pathways. They studied genetic variation in 5HTTLPR (serotonin transporter linked polymorphic region), in TPH2 (tryptophan hydroxylase 2) and in the dopamine pathway they studied sensitizing alleles in DRD4, DRD2 and COMT.

Saliva DNA was used to measure telomere length. They demonstrated significant association with family structure and telomere length. Harsh environment and harsh parenting were particularly associated with telomere shortening. Sensitizing alleles in the serotonin and dopamine pathways led to more marked degrees of telomere shortening.

The Converge study (Cai *et al.*, 2015) reported that individuals with current major depression or with a history of childhood stress have shorter telomere length and increased mitochondrial number. Hovatta (2015) proposed that the increase in mitochondrial number and decrease in telomere length were mediated at least in part by activity of the hypothalamic pituitary axis (HPA). There is also evidence for reversibility of telomere length changes after remission of anxiety.

Reduced telomere length is also a feature of aging and there is evidence that longevity is associated with less pronounced telomere shortening. Obesity, diabetes and lack of exercise apparently lead to shorter telomeres and to age related illness.

Telomere shortening occurs in diseases associated with impaired repair of DNA damage. These diseases include Fanconi syndrome, Werner syndrome, Bloom syndrome, Nijmegen breakage syndrome, ataxia telangiectasia and ataxia telangiectasia like syndrome. Telomere shortening also occurs in specific vascular metabolic and inflammatory disorders. These include cardio-vascular disease, ischemic heart disease, chronic liver disease and hepatitis (Kong *et al.*, 2013) and diabetes (Nilsson *et al.*, 2013). Premature aging syndromes may also be associated with telomere shortening and cellular senescence (Blasco, 2005).

2.2.9.4 *Disorders associated with primary telomere disruption or dysfunction*

It is interesting to note that many of the genes involved in telomere generation have been found to be mutated in patients with diseases in which bone marrow dysfunction occur.

Dyskeratosis congenital (DKC1) was the first disorder that proved to be associated with abnormal telomere function. The first defect defined was mutation in the dyskerin gene. However there is now evidence that this disorder can arise due to defects in any one of a number of genes that encode the telomerase complex or associated proteins. The manifestations characteristic of DKC1 are oral leukoplakia, skin hyper-pigmentation, nail dystrophy and bone marrow failure. Later other organs manifest functional defects.

DKC1 may be in inherited as an X-linked disorder or as an autosomal disorder and some forms have autosomal dominant inheritance. Mutations occur in any one of the following genes: TERC, TIN2 (TINF2), NOP10, NHP2, and TCAB1 (WRAP53).

Other forms of telomeric diseases may share features with DKC1. These include Revesz syndrome and Hoyeraal–Hreidarsson syndrome, and Coats syndrome (Holohan *et al.*, 2014).

Hoyeraal–Hreidarsson syndrome can present with intra-uterine growth retardation, microcephaly and cerebellar hyperplasia. In Revesz syndrome, an additional manifestation is exudative retinopathy. In Coats syndrome cerebral calcifications are additional manifestations.

Hoyeraal–Hreidarsson syndrome is characterized by accelerated telomere shortening, bone marrow failure, and immune deficiency. In patients with this condition mutations have been identified in TRF1, in TIN2 (TINF2). Deng *et al.* (2013) identified mutations in RTEL1 DNA helicase in patients with this syndrome.

2.2.9.4.1 Idiopathic pulmonary fibrosis

Mutations in a number of different genes may give rise to this disorder. These include genes that encode surfactants A or C (SFTPA, SFTPC), mutations in ABCA3 (ATP binding cassette A3) or mutations that encode telomerase reverse transcriptase TERT or TERC, the RNA component of telomerase.

Armanios (2012) reported that idiopathic pulmonary fibrosis is one of the most common manifestations of telomere functional disorders and that TERC or TERT mutations occur in up to one-sixth of families with this disorder. Armanios also noted that in individuals with idiopathic pulmonary fibrosis due to telomere dysfunction, bone marrow failure also occurs.

In familial idiopathic pulmonary fibrosis the mutations are often transmitted as autosomal dominant mutants with variable penetrance. Specific variants in the gene that encodes mucin 5B (MUC5B) also increase the risk for idiopathic pulmonary fibrosis. These MUC5B variants are associated with increased expression (Steele and Schwartz, 2013).

Environmental factors also impact the risk for this disorder. The occurrence of idiopathic pulmonary fibrosis is higher in individuals engaged in certain occupations, including stone-cutters and polishers, in agricultural workers, hairdressers and people exposed to wood dust. Cigarette smoking is also associated with a higher incidence of idiopathic pulmonary fibrosis.

Idiopathic pulmonary fibrosis is frequently initially diagnosed as pneumonia that is unresponsive to treatment. It is characterized by dyspnea on exertion and cough. Patients may exhibit clubbing of the fingers and hypoxemia. Lung sounds may be abnormal with rales. Distinct findings on CT scan of the lungs include prominence of interstitial lung tissue. Lung histology reveals dense collagen and extra-cellular matrix deposition and alveolar collapse. There is frequently a family history of similar problems (Kropski *et al.*, 2013).

3. Epigenetics and Epigenomics

Understanding the regulation of gene expression requires analysis of the nuclear architecture and linear structure of DNA in the genome. It also requires analysis of regulatory element composition, and epigenetic features including chromatin and DNA modifications. Progress in the Epigenome Roadmap project was documented in a collection of papers (Romanoski *et al.*, 2015).

3.1 Nucleosomes

Nucleosomes are comprised of 147 base pairs of DNA coiled around a complex of 8 histones. The histone octomer is composed of two copies each of histones H2A, H2B, H3 and H4. Histones have core structures and N terminal protruding tails. The individual nucleosomes are separated by linker strands of linear DNA and H1 histone. Tightly packed nucleosomes limit accessibility of DNA to transcription factors and regulatory proteins and this then limits gene expression. To expedite gene expression, chromatin remodeling must occur. Chromatin remodelers regulate nucleosome positioning and DNA accessibility. Chromatin remodeling complexes recognize covalent modifications in histone tails.

Epigenetic factors that impact gene expression are classified into four categories (Borrelli *et al.*, 2008). Writers modify DNA and histones, erasers remove chromatin modifications, and readers interact with specific modifications on histone. Chromatin remodelers interact with chromatin and nucleosome to modify chromatin structure.

3.2 Histones, Modifications and Readers

Bannister and Kourzarides (2011) reviewed histone modifications and the roles they play in regulating chromatin and gene activity. They noted that modifications occur primarily in histone tails. Histone tails between nucleosomes make can contact with each other and impact the density of chromatin; also the modifications of histones can impact the binding of chromatin complexes to histones.

The impact of a specific type of modification is dependent on the type of histone modified, the amino acid residue modified, its position within the histone sequence and the specific type of modification. Key histone modifications that impact gene transcription include acetylation, methylation and phosphorylation. It is important to note that modifications occur at specific amino-acid residues. However specific modifications can also be removed.

Other forms of modification occur on histone residues that apparently have less impact on transcription though they may have other functions. These include ubiquitination, sumoylation, ADP ribosylation and addition of beta acetylglucosamine.

Bannister and Kourzarides (2011) noted that clipping of histone tails also occurs and this serves to remove modification. This clipping likely takes place through activity of cathepsin enzymes. Modifications of histone residues frequently lead to alteration in chromatin charge and to subsequent structural changes; in addition, the modifications lead to altered protein interaction capacities.

3.2.1 *Histone acetylation*

This takes place primarily on lysine residues through the activity of histone acetyltransferase enzymes (HATs) that utilize acetyl coenzyme A as cofactor. Acetylation primarily involves amino acids in histone tails though histone core residues can also be acetylated. Acetylation of lysine residues in histone tails promotes transcriptional activity. Key histones and lysine residues that undergo acetylation include histone 3 on lysine residues K14 and K56; in histone 4 lysine residues K5, K8 and K16.

Deacetylation of histones occurs through activity of histone deacetylases (HDAC). Bannister and Kourzarides (2011) found that HDAC enzymes can remove acetyl groups; however, HDAC enzymes are also present in multi-enzyme complexes, for e.g. in NURD, SIN3A and COREST complexes.

3.2.2 *Histone methylation*

Methylation in histone side chains primarily impacts lysines and to a lesser degree arginine. Different degrees of methylation occur, such as mono, di, or tri-methylation. Transcriptional activity is impacted by the type of histone methylated, the specific position of the amino acid (usually lysine) within histone, and by the degree of methylation. Histone lysine methylating enzymes are generally referred to as HKMTs. Important specific methyltransferases include KMT1 (also known as SUV39H1) that specifically methylates lysine residue 9 in histone 3 (H3K9). Specific lysine residues tend to be methylated primarily by a specific methyl transferase.

Key histones and lysine residues that undergo methylation include in histone 3 lysines K4, K9, K23, K27, and K36; in histone 4, lysines K20 and in histone 1 K26.

Methylation of arginine residues in histone requires methylation by PRMT methyl transferases that transfer methyl groups from S-adenosyl methionine to arginine.

There is evidence that demethylation of histones can occur. Important lysine demethylase involved in this process include KDM1, KDM5B, KDM4C and KDM5C and the Jumonji domain enzyme JMJD6. Figure 3.1 illustrates lysine methyl transferase and lysine demethylases that act on histone 3 lysine 4.

3.2.3 *Histone phosphorylation*

Phosphorylation of amino groups in histone residues occur primarily on hydroxyl groups in serines, threonines and tyrosine. Kinases transfer phosphate from ATP. There are also phosphatases that remove

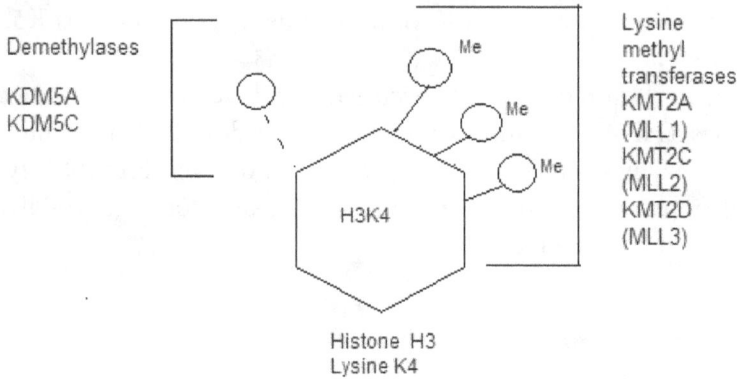

Figure 3.1. Histone3 Lysine 4 (H3K4) methylation, methyl transferases and demethylases.

phosphate modifications; Bannister and Kourzarides (2011) noted that phosphorylation primarily induces structural changes. They note that kinases and phosphatases are abundant in nuclei; however the precise functional impact of histone phosphorylation is not clear.

The enzymes that bring about the modifications referred to above are sometimes referred to as histone writers when modifying residues are added; enzymes that remove modifying residues are referred to as erasers.

3.2.4 *Readers of histone modifications*

Yun *et al.* (2011) reported that reader proteins generally have accessible cavities or grooves that can accommodate the histone modifications, and thereby protein-protein interactions are facilitated. In addition the sequences that flank the modified histone residue likely impact the interaction with the reader.

Yun *et al.* reported that acetylated lysine residues are recognized by reader proteins that contain bromodomains, specific 50 amino acid domains. They are also recognized by reader proteins that have tandem PHD domains, plant homeodomains, and domains 50–80 amino acids in length that have a globular fold.

Methylated lysines are read by a number of proteins with different domains including proteins with PHD domains, chromo domains, Tudor domains, ankyrin domains, PWWP domains or MBT domains.

Chromodomains are specific 40–50 amino acid domains that bind chromatin. Tudor domains are specific 50 amino acid domains that form anti-parallel beta sheets. Ankyrin domains have repeats of 33 specific amino acids. PWWP domains are specific 135 amino acid domains with a central core composed of sequence proline-tryptophan-tryptophan-proline. MBT domains are specific amino acid domains that form two arms around a central globular domain.

3.2.5 *Histone modifications of enhancer elements*

Djebali *et al.* (2012) noted that chromatin modification of transcribed enhancer regions manifested H3K4me1 (histone3 lysine 4 monomethylation), H3K79me2 (histone 2 lysine 79 dimethylation) and H3K27ac (histone 3 lysine 27 acetylation). Figure 3.2 illustrates nucleosome positioning and histone H3 modifications in open chromatin and closed chromatin.

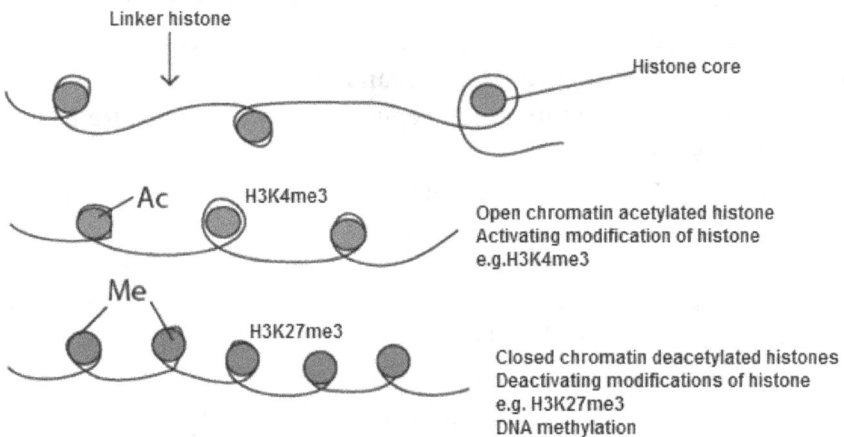

Figure 3.2. Nucleosomes, open and closed chromatin and modifications.

3.3 Long Non-coding RNAs (LncRNAs) and Intergenic LncRNAs

LncRNAs have important regulatory functions in the genome. Mercer *et al.* (2009) proposed that lncRNA function primarily as epigenetic modulators. LncRNAs recruit chromatin modifying enzymes and proteins that impact the expression of neighboring genes. There is also evidence that lincRNAs can seed heterochromatin formation. Heterochromatin is often highly methylated.

A comprehensive catalog of lncRNAs was developed as part of the ENCODE project (Derrien *et al.*, 2012). LncRNAs play important roles in inactivation of specific imprinted genes (Kanduri, 2015). The roles of long noncoding RNAs in X chromosome inactivation will be discussed later in this chapter.

Amin *et al.* (2015) analyzed the expression of lncRNAs. They determined that there was tissue specific epigenetic regulation of expression of lncRNA. They reported that H3K4me1 marked a larger fraction of lncRNA transcription start sites.

3.4 DNA Methylation

Smith and Meissner (2013) reviewed DNA methylation. They reported that methylation of cytosine occurs at the 5th position of the molecule and it occurs primarily in the CpG context. They noted that less than 10% of CpG dinucleotides occur in CpG dense regions referred to as CpG islands. CpG islands occur frequently at transcription start sites and in housekeeping genes where they are resistant to methylation.

In many regions of the genome DNA methylation is dynamic and is essential for developmental transitions.

DNA methyltransferases DNMT1, DNMT3a and DNMT3b play key roles in DNA methylation. Smith and Meissner reported that DNMT1 is most abundantly expressed when cells enter the S phase of the cell cycle. Following DNA replication DNMT1 and regulatory factors are recruited to hemimethylated DNA to correctly methylate the new DNA strand. Following this methylation E3 ubiquitin ligases target DNMT1 and trigger its polyubiquitination and degradation.

FOLIC ACID
⇩
DIHYDROFOLATE
⇩
TETRAHYDROFOLATE
⇩
5-10-METHYLENE TETRAHYDROFOLATE

⇩

5-METHYLTETRAHYDROFOLATE
⇩
METHIONINE

METHIONINE
⇩ Vitamin B12
S-ADENOSYLMETHIONINE
⇩ Vitamin B12
S-ADENOSYLHOMOCYSTEINE

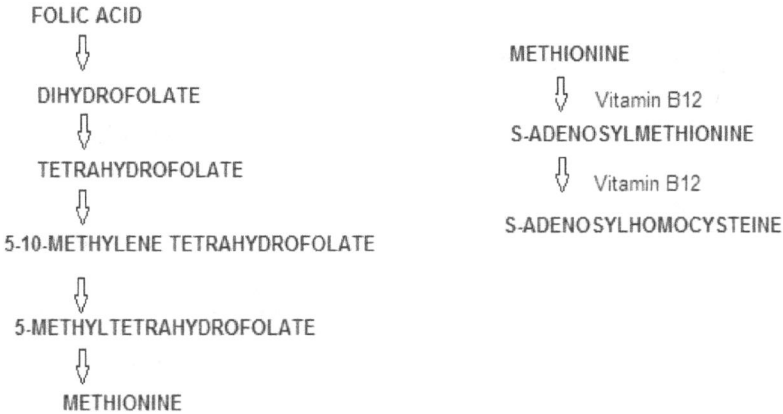

Figure 3.3. Synthesis of methyl donors for DNA methylation.

3.4.1 *Exclusion of DNA methyltransferases from promoter CpG islands*

Smith and Meissner emphasized the importance of exclusion of DNA methyltransferases from CpG islands at active promoters and noted that transcription factor binding plays important roles in this exclusion. Furthermore specific histone modifications at CpG islands are important in limiting CpG methylation at promoter sites. Histone H2AZ occurs at active promoters. They reported that some promoters are silenced following specific developmental phases and that H3K9 methylation and DNA methylation are required for stable promoter silencing.

Figure 3.3 illustrates the generation of methyl groups through metabolism. Methyl groups are essential for DNA methylation. Folic acid and vitamin B12 are important in methyl group generation.

3.4.2 *Non-CG methylation with special reference to brain*

Methylation of cytosine in DNA occurs primarily in the context of cytosine-guanine dinucleotides when cytosine is 5' to guanine, abbreviated 5mCG, mCG or CpG. In non-CG methylation cytosine methylation occurs when cytosine is 5' to nucleotides adenine, cytosine or thymine and this is often abbreviated as mCH.

There is evidence that mCH occurs at low levels in various tissues and cells in various parts of the body. It is however particularly prevalent in the brain and in pluripotent stem cells. Lister *et al.* (2013) reported methylation of cytosine in CH dinucleotides occurs during post-natal development and that mCH levels progressively increase during synaptogenesis in the first two years of life.

He and Ecker (2015) reported that in neurons the predominant sequence context for mCH is 5'CAC and that in embryonic stem cells the predominant mCH sequence is mCAG. They also reported that mCH is undetectable in fetal cortex and that it increases in postnatal life during establishment of neural circuits. The methyltransferase DNMT3A predominantly methylates CG nucleotides. Both DNMT3A and DNMT3B can methylate non-CG nucleotides. Both DNMT3A and DNMT3B bind DNMT3L as coregulators. He and Ecker report that histone modification in the genome can impact DNMT3 activity and that when lysine 4 in histone 3 (H3K4) is methylated, DNMT3A-DNMT3L cannot bind to CG in that region. Readers of epigenetic modification can bind to methylated CG (mCG) and to methylated CH (mCH). The specific impact of MECP2 binding to mCG or to mCH is currently not fully elucidated. Gabel *et al.* (2015) reported that MECP2 binds to mCH and particularly to mCA.

He and Ecker reported that TET enzymes do not apparently bind to and oxidize mCH. They emphasized that fact might explain the abundance of mCH in neurons.

There is evidence that mCH distribution is not uniform throughout the genome; it is greatly enriched in gene bodies, and is enriched at some splice sites and may play roles in alternative splice site usage. There is also evidence that mCH is enriched in inactive enhancers, absent from active enhancers, promoters and from transcription factor binding sites. In addition mCH occurs in regions of low CG density.

He and Ecker reported that depletion of methylated forms of CH and CG from promoter and proximal regions and exclusion of DNMT3A and DNMT3B from these regions coincides with transcription factors binding there and recruitment of H3K4 methyltransferase to those regions.

He and Ecker report that within neurons CG and CH residues in gene bodies are hypermethylated in genes that are normally expressed in glial cells. Similarly in glial cells CG and CH residues in the gene bodies in glial expressed genes are hypomethylated. These findings imply a close relationship between CG and CH methylation state and gene expression. They noted that in transposable elements non-CG residues are abundant and that they are methylated.

Aberrant hypermethylation of CH in promoter regions plays roles in specific diseases. Hypermethylation of CH residues in the promoter region of the PGC1 alpha gene was reported in cases of type 2 diabetes. This hypermethylation led to lower transcription of PGC1alpha and to reduced mitochondrial biogenesis.

3.4.3 *Hydroxymethylcytosine*

There is evidence that hydroxymethylcytosine (hmCG) is substantially enriched in neurons in the nervous system in the cerebral cortex and cerebellum. Kinde *et al.* (2015) reported that levels of hmCG increase post-natally in parallel with increases in mCH. Most hydroxymethylcytosine in brain is in the CG context.

Hydroxymethylcytosine results from the activity of TET enzymes. Kinde *et al.* noted that regulation of hydroxymethylcytosine activity has been implicated in learning and memory and in brain plasticity. They noted further that generation of hmCG in brain likely increases binding sites for proteins.

3.5 TET Enzymes

These enzymes modify 5-methyl cytosine through oxygenation. One TET enzyme had initially been identified through its location close to chromosome translocation rearrangement breakpoints and fusions between chromosomes 10 and 11 in hematologic malignancy. Subsequently three TET enzymes were identified and shown to be paralogous to oxygenases JBP1 and JBP2 in *Trypanosoma brucei*.

In mammals the TET enzymes recognize 5-methylcytosine and they bring about interactive oxidation of cytosine leading to generation of

5-formylcytosine and 5-carboxycytosine DNA methylation. TET1 and TET2 are expressed in embryonic stem cells and there is evidence that loss of TET enzymes leads to dysregulated DNA demethylation. TET enzymes can demethylate 5-methylcytosine and 5-hydroxymethyl cytosine.

Li *et al.* (2015) report that the three TET enzymes require iron Fe II as cofactor and that loss of iron binding sequences in TET enzymes inhibits their activity. Since TET enzymes demethylate DNA they play important roles in gene regulation. Deficiencies of TET2 or TET3 lead to impaired neuronal differentiation.

Other pathways for demethylation of cytosine include passive demethylation and base excision repair. Another enzyme proposed to contribute to demethylation is a glycosylase that has activity toward the 5' methylcytosine oxidation products 5-formylcytosine and 5-carboxycytosine.

In Figure 3.4, the modification of cytosine by DNMT activity and the modification of 5-formylcytosine by TET enzymes are illustrated.

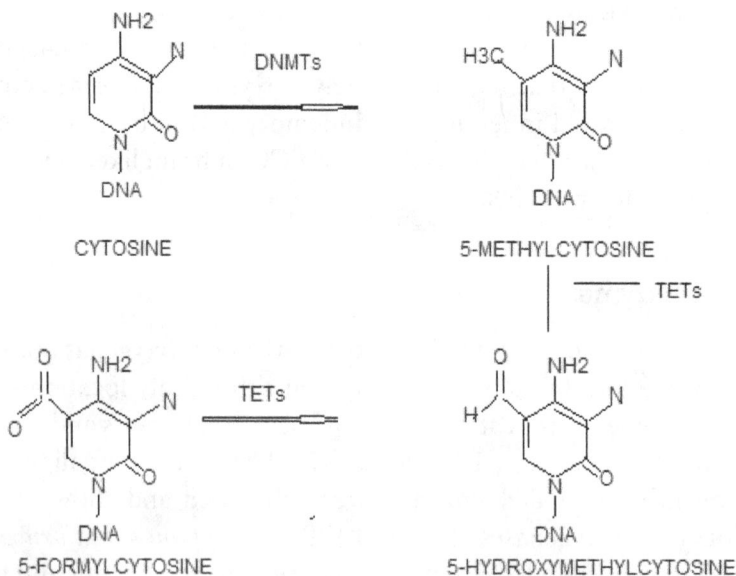

Figure 3.4. Functions of DNA methyltransferase (DNMT) and of TET enzymes.

3.6 MECP2

The MECP2 protein encoded on the X chromosome has been intensely studied since defective function of this protein occurs in Rett syndrome, a disease associated with neurological and cognitive deficits. The protein was initially characterized as binding to methyl CpG dinucleotides. There is now evidence that it also binds to CA (cytosine adenine) dinucleotides and that it preferentially binds to CA dinucleotides present in long genes. It is particularly important in binding to methylated CA dinucleotides in the brain (Gabel *et al.*, 2015).

There is evidence that the enzyme DNMT3A is particularly important in methylating CA dinucleotides in the brain. With knockdown of MECP2 in mouse models of Rett syndrome, genes that contain high levels of CA were found to be overexpressed. Furthermore, there is evidence that knockdown of Dnmt3a in mice leads to upregulation of genes similar in identity to those upregulated with defective MECP2 function.

MECP2 binds preferentially to methylated CA (cytosine adenine) dinucleotides. Chen *et al.* (2015) reported that there is a gradual accumulation of methylated CA dinucleotides in the brain between birth and adolescence. They postulated that this likely explains the later onset of manifestations of Rett syndrome.

Zylka *et al.* (2015) noted that the precise mechanisms through which MECP2 impacts gene transcription and gene transcription were not clear. There is now evidence that the transcriptional repressor function of MECP2 particularly impacts long genes. Many long genes are expressed in the brain and are involved in cell adhesion, and axon guidance. Zylka *et al.* and other investigators have noted that increased expression of specific genes, e.g. long genes that encode products involved in synaptic functions, alters the balance between excitatory and inhibitory functions. Zylka *et al.* proposed that long genes may be particularly sensitive to alterations in MECP2 function.

Genes repressed by normal MECP2 function include neuronal function genes, post-synaptic density (PSD) genes, and voltage gated channel activity genes. Gabel *et al.* noted recent reports that exposure of MECP2 deficient neurons to the topoisomerase inhibitor topotecan led to reversal of gene expression misregulation.

Topoisomerases resolve DNA supercoiling and therefore facilitate transcription, particularly of very long genes (King *et al.*, 2013). Topoisomerase inhibitors reduce expression of very long genes.

3.7 Chromatin Remodeling Complexes and Functions

Hargreaves and Crabtree (2011) reviewed the roles of macromolecular assemblies that are involved in chromatin remodeling. They noted that one important aspect of chromatin remodeling involves the movement of nucleosomes and the exchange of nucleosomes. ATP hydrolysis is utilized to provide energy to alter nucleosome position.

Chromatin remodelers regulate nucleosome positioning and DNA accessibility. The different remodeling complexes have different protein domains that target specific histone modification signals and interacting proteins. In addition specific non-protein coding RNA sequences can impact the targeting of chromatin modifiers. Längst and Manelyte (2015) reviewed chromatin remodeling complexes. They noted that the ATPase subunit in chromatin remodeling complexes facilitates use of ATP as energy source to carry out remodeling. Each of the four different families of chromatin remodelers that they reviewed has a DEXX DNA targeting domain and a HELICc domain for hydrolysis. In addition to these common domains shared by components of all four families, each family has family specific domains. Actin binding and bromo domains are present in SWI/SNF (BAF) family members. Chromo domains are present in CHD family members. INO80 family members have an actin binding domain and the DEX DNA binding domain and HELIC domains are separated. The ISWI family contains additional domains HAND, SANT and SLIDE involved in nucleosome recognition.

It is important to note that the subunits that contribute to each of the four families of chromatin remodelers are also often present in other complexes that may have different functions.

The mammalian SWI/SNF family of chromatin remodelers is also referred to as the BAF family. In humans there are 29 different genes present that encode BAF subunits. Different BAF complexes occur and each BAF complex is composed of 15 different subunits. For the different BAF subunits SMARC terminology is often used. Son and Crabtree (2014) emphasized the importance of BAF complexes in neural development and specific roles of these complexes in neural progenitor cell division and in maturation of neurons. They also noted the importance of BAF complexes in memory formation. MECP2 and ATRX also act as chromatin remodelers.

3.7.1 *Corepressor complexes*

In addition to coactivator complexes, there are also corepressor complexes with chromatin and DNA binding and histone modification properties. Schoch and Abel (2014) described the structures and functions of four co-repressor complexes NCOR, NURD, SIN3A and COREST. These complexes are bound to epigenetic modifier effector proteins and are recruited to chromatin by DNA binding or histone binding proteins.

Schoch and Abel noted further that gene silencing is associated with removal of activating histone modifications and addition of repressive DNA and histone modifications including methylation at H3K9, H3K27 amd H3K36.

They focused on the importance of co-repressor function in the brain, on signal transduction, plasticity and cellular memory. They emphasized the neuron specific expression of corepressor complexes. They also found that defects in specific components of corepressor complexes have been described in neurodevelopmental and neurological disorders. Their review focused on corepressor complexes implicated in memory and cognition; these included NCOR, (nuclear receptor corepressor), NURD (nucleosome remodeling and deacetylase complex), SIN3A (switch insensitive 3A) and COREST (RE1 element silencing transcriptional factor corepressor that binds to the C-terminal of REST repressor).

Schoch and Abel reported that repression of expression of specific genes is necessary for neural development and memory storage, and that MECP2 associates with multiple corepressors including SIN3A and COREST. This binding is influenced by phosphorylation of specific residues on MECP2. MECP2 has affinity with methyl cytosine in DNA and can interact with promoter sequences. Phosphorylation of MECP2 releases MECP2 from binding.

Activity of the NURD corepressor complex regulates nucleosome positioning and Schoch and Abel noted that this is linked to heterochromatin formation. They reported that the role of NURD1 in brain is not well documented. They noted that there is however evidence that NURD1 binds to MBD (methyl binding domain) proteins and there is evidence that MBD genes are mutated in some forms of autism.

SIN3A is a scaffold protein that interacts with many regulatory proteins. Schoch and Abel reported evidence that SIN3A plays critical roles in brain development and cognition. SIN3A brings about transcription silencing through interaction with histone deacetylase. It also interacts with the H3K9 methyltransferase SETDB1 and with other histone methyltransferases. There is evidence that loss of SIN3A activity leads to under-expression of some genes and over-expression of other genes. There is also evidence that SIN3A is a binding partner of TET1 enzymes that bring about the hydroxylation of 5-methyl cytosine.

3.7.2 *COREST*

COREST interacts with REST transcriptional repressor and with HDAC2 and it functions is to deacetylate chromatin. Schoch and Abel noted that there is also evidence that the lysine specific demethylase KDM1 (LSD1) interacts with COREST. Interactions of COREST with BRAF2 chromatin remodeling sub-unit plays roles in short term and long term neuronal gene repression. There is evidence that in brains dynamic regulation of SIN3A and COREST complexes occurs in response to calcium influx.

Schoch and Abel concluded that multiple subtypes of repressor complexes may exist and that they may have different functions. Furthermore components of COREST complexes are also present in other complexes. For example, SETDB1 and SET1 also occur in subtypes of the SIN3A complex. They noted that in brain dynamic regulation of SIN3A and COREST complexes occurs in response to calcium influx.

3.7.3 *Chromatin remodeling at double stranded DNA breaks*

Price and D'Andrea (2013) reviewed the importance of nucleosome repackaging and chromatin modification in promoting access of factors that repair double stranded DNA breaks. They noted that six different DNA repair pathways exist to restore DNA integrity following damage induced by ionizing radiation, ultra-violet radiation, environmental toxins, and endogenous reactive oxygen species. Repair of errors induced during DNA replication must also occur.

They reported that a critical step in the double stranded DNA repair pathway includes phosphorylation of the histone variant H2AX by the ATM kinase. This phosphorylation creates a binding site for the protein MDC1 (mediator of DNA damage checkpoint 1). The MDC1 protein then provides a binding site for additional proteins, including RNF ubiquitin ligase that ubiquitinates chromatin and facilitates recruitment of additional proteins. Insertion of the histone variant H2AZ into nucleosomes near the break also promotes repair.

Other chromatin changes that impact DNA repair include modifications in histone tails, for e.g. rapid acetylation in histone H2 and H4 that promotes development of an open chromatin structure.

Brownlee *et al.* (2015) reviewed the role of BAF subunit containing complexes (mSWI/SNF) in maintaining genomic stability. They reported that BRG1 (SMARCA4) and other BAF subunits played important roles in promoting repairs by homologous recombination and in repair by non-homologous end-joining.

3.7.4 *Maintenance of cell lineage epigenetic changes through cell division*

Cellular memory is needed to propagate specific patterns of gene expression in cell lineages. There is evidence that the Trithorax and Polycomb gene complexes play important roles in such cellular memory systems (Schuettengruber *et al.*, 2011).

Polycomb complexes serve to repress gene expression, while the Trithorax complex serves to activate gene expression. Schuettengruber *et al.* reported that Trithorax complex proteins are evolutionarily conserved chromatin regulators that can be divided into three classes. One class includes SET domain factors involved in histone methylation. The second class of Trithorax proteins includes ATP dependent chromatin remodelers that read histone modifications, particularly methylation modifications. The third group includes other modifiers and readers.

Set domain proteins are histone methyltransferases that catalyze mono, di, or trimethylation of lysine 4 in histone H3 (H3K4). These include the enzymes MLL1 (KMT2A), MLL3 (KMT2C) and MLL4 (KMT2B). A number of different subunits from ATP dependent remodeling systems are included in Trithorax complexes. These include SWI/SNF (SMARCA4), ISWI (SMARCA5) and chromodomain helicases CHD7 and CHD8 subunits.

Schuettengruber *et al.* reported that Trithorax complexes can also be recruited to specific genomic regions through non-coding RNAs. They noted that Trithorax complexes also play roles in regulating gene expression during cell division. Translocations that occur in tumor cells sometimes lead to abnormal expression of Trithorax components, particularly MLL subunits and are associated with aberrant cell cycle control.

Recruitment of Polycomb repressive complexes to DNA is dependent of specific elements PREs (PGC response elements). PRC2 has histone methyl transferase activity and following its recruitment to DNA it specifically induces trimethylation of lysine 27 on histone H3 (H3K27). This reaction is catalyzed by the enzyme EZH2 (enhancer of zeste) in the PRC2 complex. H3K27 methylation is recognized by a subunit in the PRC1 complex. Following coupling of PRC1 to H3K27 mono-ubiquitination of lysine119 on histone H2 occurs. This reaction is catalyzed by a ring finger protein RING1 and this reaction leads to gene silencing.

Over-expression of EZH2 occurs in certain cancers and this over-expression contributes to malignant transformation. Specific small molecule inhibitors of EZH2 are currently being investigated for cancer treatment (Xu *et al.*, 2015).

3.8 Epigenetic changes in Differentiation and Cell Lineage Determination

Analysis of factors involved in differentiation processes has been greatly enhanced through studies on transition of pluripotent stem cells to differentiated cells. Ziller *et al.* (2015) analyzed stages of differentiation from pluripotent stem cells to neuron cell types. They reported that selected enhancer regions were remodeled during differentiation and that regions exhibited gain of H3K4me1 and loss of DNA methylation. In addition they reported that specific transcription factors bound to core neural enhancers while other transcription factors bound only to enhancers expressed in specific neural lineages.

3.8.1 *Epigenomic signatures during development*

Romanoski *et al.* (2015) presented data on epigenomic signatures that determine differential gene expression during developmental transitions. Though some genes are required for general functions and are expressed in most cells, other genes are only expressed in subsets of cells. Genomic elements that enhance gene expression, enhancers, may be located upstream or downstream of genes even at some distance from genes and they may be located within genes. Most enhancers recruit transcription factors and co-regulators that modify chromatin. Romanoski *et al.* noted that enhancer elements that impact differentiation bind lineage determining transcription factors. During development, binding of specific factors can be facilitated or inhibited. Specific methylation patterns enhance transcription factor binding while other factors repress binding.

In the 2015 Epigenome Roadmap update, Farh *et al.* reported that through studies of 21 auto-immune diseases, they determined that 90% of the associated variants identified in genome wide association studies (GWASs) mapped in non-coding regions of the genome and 60% mapped in enhancer regions that were immune regulation sites. However they determined that these variants did not map at transcription factor binding motifs within enhancers. Romanski *et al.* (2015) noted that this suggests that enhancer elements are complex and that important function within enhancers may be located in sequences other than transcription binding sites.

3.8.2 *Pioneer transcription factors*

Pioneer transcription factors can bind specific sites in chromatin and they can open chromatin to facilitate the subsequent binding of other transcription factors. Zaret and Carroll (2011) reported that the transcription factors in the FOXA and GATA families are important initial chromatin binding factors and that these factors can bind to condensed chromatin. Stable binding of pioneer transcription factors occurred for a period and then binding of other transcription factors followed. FOXA transcription factors have DNA binding domains and histone binding domains. The pioneer transcription factors help organize chromatin through facilitating the nucleosome binding of chromatin modifiers.

3.9 Chromatin Mediated Control of Neural Development and Defects in Cognitive Disorders

A number of components of the BAF complexes have been implicated in syndromes associated with impaired cognitive development (Ronan *et al.*, 2013).

Other genes that impact chromatin, including writers, erasers and readers have also been found to harbor mutations that impair cognitive development (Kleefstra *et al.*, 2014). These include histone methyltransferase, histone acetyl transferase, histone lysine demethylases, histone deacetylases and histone ubiquitinases. In addition methylated DNA binding protein mutations may give rise to cognitive impairment.

Table 3.1.   Chromatin modifying enzymes mutated in cognitive disorders.

Histone methyl transferases MLL (KMT2A) MLL2 (KMT2D), MLL3 (KMT2C) EHMT1 (eukaryotic histone lysine N methyltransferases)
Histone demethylase KDM5C (JARID1C)
PHF8; histone acetyl transferases EP300, KAT6B
Histone deacetylase HDAC4
Histone ubiquitinase CUL4B

3.9.1 *Chromatin modifying enzymes mutated in specific cognitive disorders*

The different classes of chromatin modifying enzymes that have been found to be mutated in cognitive disorders are listed in Table 3.1.

3.9.2 *Chromatin remodeler defects and developmental delay: Coffin–Siris syndrome*

Coffin–Siris syndrome was first described in 1970 as a specific syndrome complex associated with developmental delay, intellectual disability, unusual facial features, feeding difficulties, hypoplasia or aplasia of fifth finger nails and fifth distal phalanges. Defects in a number of different BAF complex genes have subsequently been found in individuals with this syndrome (Kosho *et al.*, 2014). These include heterozygous mutations in the following genes listed in Table 3.2.

In addition SMARCB1 mutations occur in patients with a related syndrome referred to as DOORS syndrome and in Kleefstra syndrome. Additional features in patients with SMARCB1 mutations are hypertrichiosis and hearing impairment.

Other non-BAF complex genes have been found to be mutated in some patients with Coffin–Siris syndrome features. These include mutations in ADNP (activity dependent neuroprotector homeobox) and TBC1D24 (TBC1 domain containing; TBC1 domains interact with GTPase [guanosine-triphosphate hydrolase] proteins and impact signaling).

Table 3.2.   Genes mutated in Coffin–Siris syndrome.

Gene	Alias	Chromosome
BAF47	SMARCB1	22q11.23
BRG1	SMARC A2/A4	9q24.3
BAF190	SMARCA2	9p22.2
BAF190	SMARCA4	19p13.2
BAF57	SMARCE1	17q21.2
BAF250A	ARID1A	1p36.11
BAF250B	ARID1B	6q25.3

3.10 X Chromosome Inactivation

Since females carry two chromosomes and the males carry one
X chromosome and a smaller Y chromosome with relatively few
genes, inactivation of one X chromosome in the female, in a process
that silences expression, has evolved as a dosage compensation
mechanism.

3.10.1 *Early discoveries of X chromosome inactivation*

In 1949 Barr and Bertram reported the presence of a darkly staining
body associated with the nuclear membrane in interphase nuclei in
female cats. Subsequent studies revealed that this structure, referred
to as a Barr body, was presented in cells of females of many species
(Moore and Barr, 1954).

Ohno *et al.* (1961) suggested that the Barr body represented a
darkly staining single X chromosome.

Mary Lyon (1961) proposed a hypothesis that consisted of three
parts:

1. The heteropycnotic X is genetically inactive
2. The inactivated X in a specific cell may be maternal or paternal
    in origin and different cells in the organism differ with respect to
    which X is inactivated
3. X inactivation occurs early in embryonic life

Subsequent studies by Ferguson-Smith (1965) revealed that the inactivated X was not inactivated along its whole length; the uppermost segment of the short arm does not undergo inactivation.

It became clear that in microscopic studies that the sex chromosome mass could not readily be identified through staining in all nuclei possibly because of its position within the nucleus. In normal females approximately 25% of buccal smear cells contained a readily distinguishable Barr body. However one rarely finds more than one sex chromatin mass per cell nucleus.

In 1964 Polani reported that in majority of cases of triple X syndrome females showed the presence of two Barr bodies per cell.

3.10.2 *X chromosome inactivation center*

In 1983 Rastan proposed that inactivation of the X chromosome initiates from a specific site, the inactivation site (Xic). Rastan and Robertson (1985) reported that 2 copies of X bearing sites per cell are required to be present for X inactivation to occur.

Brown *et al.* (1991) reported that the Xic region carries a specific gene Xist and that this gene is only expressed from the inactive X chromosome. Wutz *et al.* (2002) demonstrated that RNA encoded by the Xist gene spreads along the chromosome and triggers inactivation of the specific chromosome from which it is expressed (i.e. cis inactivation).

The XIC region on the Xq13.2 is almost 1 Mb in length. It contains protein coding genes and non-protein coding RNA genes that include XIST and TSIX. The transcript derived from XIST is 17kb in length. The transcript contains a series of specific tandem repeats. There is evidence that the TSIX gene in humans is a pseudogene and is not expressed. In mice, the TSIX transcript is expressed on the active X chromosome and on the active X is represses XIST expression (Lee and Bartolomei, 2013).

Regulation of Xist expression and random X inactivation in a specific time period in development has been studied intensively. A number of non-protein coding transcripts map close to XIST. These loci produce non-coding RNAs JPX (ENOX) and FTX and are activators of XIST. Another activator of XIST maps upstream this activator is RNF12 (RLIM), a ring finger ubiquitin ligase. The histone variant macroH2A (2AFY) is incorporated in the inactivated X chromosome in a XIST dependent manner.

Vallot *et al.* (2013) reported that a specific transcript coats the X chromosome in pluripotent stem cells. This transcript is XACT (X active specific transcript). Vallot *et al.* (2015) reported that in pluripotent stem cells XACT coating leads to erosion of X chromosome inactivation in a percentage of pluripotent stem cells.

Gendrel and Heard (2014) reported that there is definite evidence that Xist expression is controlled at a transcriptional level. Regulation of choice, i.e. whether the maternal derived or the paternal derived X chromosome will be inactivated, is complex and likely dependent on long-range regulatory factors. They reported that there is some evidence based on X chromosome deletion patients, that additional regulatory elements are encoded on the X chromosome but at some distance from Xist.

3.10.3   *Maintaining the inactive status*

Maintenance of X inactivation is dependent upon methylation of DNA. Gendrel and Heard reported that impaired function of the DNA methyltransferase Dnmt3a and Dnmt3b impairs maintenance of X inactivation.

Regulatory elements that ensure that Xist is only expressed from one X chromosome per cell nucleus are still poorly understood. There is some evidence for pairing of the Xic regions of the two chromosomes.

Following Xist RNA coating of the X chromosome there is loss of the histone modifications that are usually associated with expression, e.g. loss of histone H3 lysine 4 methylation, (H3K4me2 and H3K4me3) and loss of histone H3 and H4 acetylation. In addition, coating leads the polycomb repressive complexes PRC1 and PRC2 to be recruited to DNA. Methyltransferases then induce H3K27 methylation. These factors work together to promote inactivation.

Da Rocha *et al.* (2014) reported that the protein Jarid 2 played an important in binding PRC2 to the inactive X chromosome.

3.10.4 *Analysis of the topography of X chromosome inactivation*

Through the use of mice into which X chromosome reporter genes were cloned Wu *et al.* (2014) analyzed topography of X chromosome inactivation. They demonstrated that the degree of mosaicism for maternal derived X inactivation versus paternal derived X inactivation varied among neighboring cells and between left and right sides of the body. In some tissues there was marked asymmetry so that in blocks of tissue one specific X chromosome was more active than the other.

3.10.5 *Non-random X inactivation*

It is important to emphasize that non-random X inactivation in specific tissues may have clinical implications. For example non-random X inactivation in muscle cells may lead female carriers of the Duchenne muscular dystrophy gene to manifest symptoms.

3.11 Genes on the X Chromosome and Ovarian Development

At approximately 50–55 days of human gestation in 46XX females, ovaries develop. Simpson (2014) reported that the early ovary contains up to seven million cells; however large numbers of these subsequently undergo atresia. In 45X embryos atresia occurs more rapidly, indicating that two X chromosomes are necessary for ovary development. Simpson noted that ovarian differentiation and ductal genitalia differentiation are separately determined.

Several regions of the X chromosome carry genes important for maturation of the ovary and follicle development. Regions on the short arm that are most important include Xp11 primarily and the centromeric portion of Xp21. Simpson reported that women with deletions in these regions are frequently infertile and have amenorrhea. Key regions for ovary development also include Xp12. The XIST locus maps to Xq13. Premature ovarian failure has also been reported in women with deletions in the terminal region of the long arm of Xq27-q28.

Simpson reported that mutation in a number of different autosomal genes also impair gonadal development and maintenance and lead to hypogonadism. A key gene involved in ovarian maintenance include genes that encode enzymes in the steroid synthesis pathway, e.g. aromatase (CYP19P and 17lapha hydroxylase (CYP17)) and the Inhibin A gene. Inhibins are glycoproteins synthesized by granulosa cells in the ovary.

Forkhead transcription factors are also important for ovarian follicular maintenance. Mutations in FOXL2, FOXO1A and FOXO3A mutations impair ovarian differentiation and development.

3.12 Imprinting

Through imprinting, large clusters of genes in specific genomic regions are controlled so that they are expressed from only one member of specific pairs of homologous chromosome, i.e. from the maternally or the paternally derived chromosome. Imprinting requires co-ordinated control in cis over long periods of time and lncRNAs play important roles in imprinting control (Lee and Bartolomei, 2013). LncRNAs play important roles in recruiting chromatin modifying factors. These lncRNAs are frequently derived from loci close to the imprint control region. Lee and Bartolomei emphasized that all of the genes that map within imprinted clusters were not yet defined.

In addition to genes that give rise to non-coding RNAs, imprinted genomic regions include protein coding genes and genes that encode small nucleolar RNAs (Sno RNAs) and micro RNAs. In addition each imprinted locus has an imprint control region that undergoes parent specific epigenetic modifications. Lee and Bartolomei emphasized the importance of DNA methylation of imprint control regions. Deletion of an imprint control region leads to loss of imprinting of the related gene cluster.

3.12.1 *Imprinting mechanisms*

Imprinting is dependent on DNA methylation, histone modifications and on factors, including lncRNAs that bind to modified chromatin. The modifications that determine either maternal or paternal expression of a specific gene within an imprinted region are mitotically stable and are maintained throughout life. However, Horsthemke (2010) reported evidence that some silenced alleles are not completely silenced and retain some residual activity. In addition there is evidence that mono-allelic expression of specific genes in imprinted clusters may not extend to all tissues.

Imprinting is usually erased in primordial germ cells and it is then established later in their maturation so that gametes are transmitted with the imprint corresponding to the sex of the transmitting parent.

3.12.2 *Imprinting disorders*

Imprinting disorders may arise through genomic defects, through epigenetic specific defects or through gene defects. Genomic defects leading to imprinting disorders include chromosome defects such as deletions or duplications, and also through uniparental disomy. In uniparental disomy both copies of imprinted regions on a specific chromosome pair are derived from a one parent. In paternal uniparental isodisomy both copies of a specific imprinted region are derived from the father. In maternal uniparental isodisomy both copies of a specific imprinted region are derived from the mother. Horsthemke (2010) reported that there are apparently 60 known imprinted genes in humans but there may be many more. Soellner *et al.* (2015) reviewed imprinting disorders.

In Table 3.3, imprinted chromosome regions, genes in imprinted regions and most common human imprinting disorders are listed.

Soellner *et al.* (2015) listed the shared clinical characteristics of imprinting disorders; the specific characteristics varied depending on the type of imprinting disorder. General characteristics include pre and post-natal growth defects, growth retardation or overgrowth, hypoglycemia or hyperglycemia, abnormal feeding behavior

Table 3.3. Chromosome regions and genes most frequently impacted in imprinting disorders.

Chromosome region	Key genes	Imprinting disorder
15q11-q13	SNRPN, AS, MKRN3	Prader–Willi syndrome, Angelman syndrome, Precocious puberty
11p15.5	ICR1/2, IGF2, CDNK1C	Beckwith Wiedemann syndrome, UPD11pat, Russell–Silver syndrome, UPD11mat
14q32	MEG3, DLK1	Temple syndrome UPD14mat, Kagami–Ogata syndrome UPD14pat
20q13	GNAS	Pseudohypoparathyroidism CHECK UPDpat, Albright hereditary osteodystrophy
20		Intra-uterine growth retardation, UPD20mat
6q24	PLAGL1, ZFP57	Neonatal transient diabetes, Intra-uterine growth retardation
7		Russell–Silver syndrome, UPD7mat

in childhood or later, behavioral difficulties, and deviations in timing of puberty.

There are also a number of more recently described imprinting disorders. Birk Barel mental retardation arises due to maternally inherited KCNK9 mutations. KCNK9 is a potassium ion channel encoded on chromosome 8q24.3. Maternal mutations in the genes NLRP2 and NLRP7 lead to reproduction losses, or to hydatiform mole pregnancies. In addition mothers with specific mutations in the genes NLRP2, NLRP7 or NLRP5 may have offspring with methylation defects in multiple imprinted genes (Docherty *et al.*, 2015).

NLRP2, NLRP7 and NLRP5 all map to chromosome 19q13. There is evidence that NLRP7 function is particularly important for methylation in early pregnancy (Mahadevan *et al.*, 2014).

Eggerman *et al.* (2014) reported evidence that a locus on chromosome 11p15.2 apparently plays a central role in methylation in multiple genomic regions. It is also important to note that multilocus imprinting disorders occur.

3.13 Epigenomics GWAS Signals and Disease

GWAS signals identify DNA variations associated with specific diseases. It is likely that disease related signals in regulatory regions impact specific cell types associated with a specific disease. Mouse models of human brain diseases such as Alzheimer's disease were studied in the Epigenome Roadmap project.

Gjoneska *et al.* (2015) carried out studies of the hippocampus in a mouse model of Alzheimer's disease and determined that the pathologic changes in the mouse model corresponded to the pathologic changes found in post-mortem derived samples of hippocampi from Alzheimer's disease patients. Their studies revealed upregulation of expression of immune response genes.

Lunnon *et al.* 2014 studied methylomic variations in brain samples from 122 patients with Alzheimer's disease. They identified a region in the vicinity of the ANK1 gene that was substantially hypermethylated in the entorhinal cortex region of Alzheimer's disease patients. The entorhinal cortex is known to be a site of early neuropathology in Alzheimer's disease. Lunnon *et al.* reported that ANK1 is a transcriptionally complex gene that has alternative promoters and gives rise to multiple isoforms. The ANK1 encoded protein ankyrin links cell membranes to underlying spectrin and actin cytoskeletal proteins.

De Jager *et al.* (2014) carried out studies on 708 brain samples obtained at autopsy from Alzheimer's disease patients. They identified genes with differential methylation and altered expression. These genes included ANK1, CDH23 (cadherin related 23), DIP2A (disco involved in axon patterning), SERPINF1 and F2 (neurotrophic factor encoding).

4. Proteins, Proteomics, Protein Folding and Proteostasis

The functions of proteins are determined by their 3-dimensional structures that result from folding and maintenance of folding is essential for continued function. The term proteostasis refers to processes that control the abundance and the folding of proteins. These processes involve regulation of gene expression, signaling pathways, secondary modifications, action of chaperones and protein degradation processes. Disruption of proteastasis leads to the formation of protein aggregates which eventually lead to disease manifestations.

4.1 Proteomics

Significant advances in proteomics have been facilitated by the implementation of Fourier transform mass spectrometry to protein studies. Various steps are used for separation of proteins prior to analysis of proteins by this technology. Kim *et al.* (2014) reported use of SDS polyacrylamide gels (SDS PAGE) to separate proteins prior to trypsin digestion. Materials derived from trypsin digestion of SDS PAGE separated fragments were then separated on micro-capillary based reverse phase liquid chromatography. Fractions derived from that procedure were then analyzed using nanoflow electrospray ionization and then analyzed on mass spectrometry. Peaks derived from mass spectrometry were analyzed. Sequences not matched to documented protein references sequences were then translated to nucleotide sequences to determine if the translated sequences matched sequences in the human genome.

Kim *et al.* carried out proteomic analyses on 17 different adult tissues, on 7 fetal tissues and on 6 hematopoietic cell types. Their studies led to tandem mass spectrometry profiles corresponding to 17,294 genes; approximately 84% of these matched annotated human protein encoding genes. Importantly they also obtained evidence for 2,535 protein products not observed in proteomic databases.

They reported that some of these novel protein products corresponded to translation of genes annotated as pseudogenes or to sequences annotated as non-protein coding RNAs. In addition Kim *et al.* determined that novel transcription start sites, novel coding exons and novel gene exon extensions led to generation of these novel proteins. They identified 28 genes with previously undocumented upstream frames. One example was the solute carrier gene SLC35A4. A novel protein encoded by genomic region C11orf48 was identified only in the retina. A novel 524 amino acid encoding open reading frame expressed only in fetal ovary and adult testes was identified for CHTF8 (chromosome transmission fidelity factor 8).

Kim *et al.* generated an interactive web based resource of the Human Proteome Map: http://www.humanproteomemap.org. They emphasized that development of the human proteome map complements human genome and human transcriptome efforts.

4.1.1 *Multi-dimensional proteomics*

Larance and Lamond (2015) emphasized the multi-dimensional analyses of the proteome and their importance in understanding cellular and physiological processes. Multi-dimensional analyses include analysis of protein abundance, tissue distribution of proteins, their sub-cellular localization, isoform expression, post-translational modification, activities, and turnover including synthesis and degradation rates and protein interactions.

They emphasized the importance of mass spectrometry in analysis of the population of proteins present in a specific tissue or cell type. They noted that differential centrifugation based separation methods are valuable for prior separation of subcellular fractions. Larance and Lamond emphasized that cellular localizations have major impacts on protein functions. Furthermore cellular localizations may be influenced by post-translational modifications.

They noted that the unphosphorylated transcription factor FOXO1 localizes to chromatin; however following phosphorylation, FOXO1 accumulates in the cytoplasm and interacts with other proteins including proteins in the 14-3-3 family.

Functions impacted by post-translational processes include folding and protein interactions. Protein-protein interactions are frequently studied by immunoprecipitation methods and analyses of co-precipitated proteins. In some cases modified proteins differ in chromatographic properties. Modification-specific antibodies are also used to isolate modified proteins. Isoform expression may be analyzed through targeted mass spectrometry.

Larance and Lamond noted that most studies concentrate on relatively few post-translational modifications including phosphorylation, acetylation, ubiquitination, and glycosylation. They emphasized that the less well characterized protein modifications are also important. These include citrullination of arginine residues and hydroxylation of protein residues.

A number of on-line resources facilitate data analysis. These include proteomic DB, Human Proteome Map, phosphorylation database, and Encyclopedia of Proteome dynamics.

4.2 Post-translational Modifications of Proteins

Recent expansions in proteomic data include accumulation of details relating to post-translational modifications of protein. Databases have been developed to document types of post-translational modifications of proteins, e.g. Swiss-Prot database.

It is important to note that a specific protein may undergo a number of different modifications. Post-translational modifications broaden the range of protein functions.

Khoury *et al.* (2011) carried out analyses of frequencies of experimentally observed post-translational modifications. They listed the range of observed post-translational modifications in order of frequency with the most frequent first: phosphorylation, acetylation, N-linked glycosylation, amidation, hydroxylation, O-linked glycosylation, ubiquitination, pyrrolidone carboxylic acid addition, sulfation, palmitoylation, C-linked glycosylation, adenosine di-phosphate (ADP)-ribosylation, myristoylation, farnesylation, nitration, formylation, geranyl-geranylation, deamidation, S-nitrosylation, citrullination, S-diacylglycerol modification of cysteine, modification by glucose phosphate isomerase, bromination, and flavine adenine dinucleotide binding.

Phosphorylation and particularly tyrosine phosphorylation has emerged as an important mechanism for signal transduction and regulation in eukaryotic cells. Hunter (2009) reviewed growth of knowledge concerning tyrosine phosphorylation and demonstration of its importance in human diseases. He emphasized that availability of complete genome sequences were key in facilitating identification of the full complements of tyrosine kinases and of serine and threonine kinases.

Hunter noted that in 2009 the catalog of phosphorylated proteins was incomplete. The two major classes of tyrosine kinases were identified. The receptor tyrosine kinases are transmembrane proteins composed of an N-terminal extra-cellular domain and a C terminal cytoplasmic domain with a catalytic site. Hunter reported that 58 of the 90 tyrosine kinases encoded in the human genome were receptor tyrosine kinase. Non-receptor tyrosine kinases are primarily intracellular proteins; however they may undergo post-translational modification, such as myristoylation that leads them to be targeted to membranes.

Blume-Jensen and Hunter first reported in 2001 that 51 of the 90 tyrosine kinases were implicated in cancers through mutation, under- or over-expression.

Hunter noted that natural product tyrosine kinase inhibitors were first identified in the 1980s and that thereafter synthetic small molecule tyrosine kinase inhibitors were identified. He predicted that there would be continuous development of second generation tyrosine kinase inhibitors to circumvent resistance problems and to increase the specificity of tyrosine kinase inhibitors.

4.2.1 *Example of a protein that undergoes multiple post-translational modifications*

Protein phosphorylation involves addition of a phosphate group to amino acids, particularly serine, threonine and tyrosine in proteins. Other amino acids that are phosphorylated in some proteins are arginine, lysine and cysteine. Depending on the position of the amino acid that undergoes phosphorylation, this modification may activate, deactivate or otherwise modify protein function. Enzymes involved in phosphorylation are kinases. Amino acids in proteins are also readily dephosphorylated by phosphatases.

P53 is a protein that undergoes multiple post-translational modifications. These include phosphorylation, acetylation, methylation and ubiquitination. These modifications can occur at any one of 60 amino acids in the protein that contains 393 amino acids. The p53 modifications impact the ability of p53 to interact with DNA, chromatin, cofactors and other proteins.

Nguyen *et al.* (2014) reported that p53 mutations found in cancer cells very seldom involve the sites that are known to undergo post-translational modifications. They noted that the precise effects of specific modifications on p53 function and the relationship of these modifications to cancer were unclear.

4.2.2 *Post-translational modification by prenylation*

Prenyl moieties utilized in post-translational modifications include farnesyl and geranylgeranyl isoprenoids. Wolda and Glomset reported the first prenylated protein identified in mammals in 1988. This was farnesylated lamin B. They noted that protein prenylation is an irreversible reaction.

Palsuledesai and Distefano (2015) reviewed protein prenylation. Specific transferases carry out reactions. Farnesyl transferases and geranylgeranyl transferase transfer isoprenoid groups to cysteine residues that are located within a specific group of amino acids known as the CAAX box. This box contains cysteine, aliphatic amino acids and X amino acids that differ depending on whether geranylgeranyl or farnesyl moieties are to be transferred. Following attachment of these moieties to cysteine additional processing of the protein takes place, including cleavage of the AAX by endopeptidase activity and carboxymethylation of cysteine. That reaction requires activity of the enzyme isoprenylcysteine carboxymethyl transferase. The processing reactions are often followed by localization of the modified protein to membranes.

Palsuledesai and Distefano reported that signaling proteins including G proteins and RAS proteins require prenylation for activity.

4.2.3 *Lamin A farnesylation and Hutchinson–Gilford syndrome progeria (HGPS)*

HGPS is a sporadic dominant disorder characterized by growth retardation, premature aging, and alopecia, presence of minimal subcutaneous fat, skeletal abnormalities, osteoporosis, and manifestations of arteriosclerosis in childhood, premature aging and early death, often due to myocardial infarction.

In 2003, Eriksson *et al.* and De Sandre-Giovannoli *et al.* reported that a single base change in codon 608 in exon 11 of the lamin A/C (LMNA) gene led to HGPS. LMNA encodes proteins lamin A and lamin C. The LMNA encoded products play roles in membrane scaffolding and also in chromatin organization, DNA transcription and DNA replication. Lamin A forms the inner layer of the nuclear membrane.

Gruenbaum *et al.* 2005 reported that more than 180 mutations occurred in LMNA and that these different mutations led to a number of different disorders. In addition to aging syndromes such as HGPS and atypical Werner syndrome, specific LMNA mutations lead to forms of muscular dystrophy, mandibulofacial dysplasia, lipodystrophy and dermopathy. Dittmer *et al.* (2014) identified 337 proteins that normally bind to lamin A. The diversity of manifestations in different laminopathies is largely due to how different mutations in these disorders impact the interactions with different proteins.

The specific LMNA mutation that leads to HGPS generates a cryptic splice donor site and this leads to presence of an mRNA that lacks 150 nucleotides and 50 amino acids. The deleted protein is designated progerin. The truncated lamin A protein, progerin, is permanently farnesylated. The progerin protein has a dominant negative effect on the function of the normal lamin A. The dominant negative effect interrupts the normal function of the wild type lamin A in formation of the inner membrane of the nuclear membrane. Abnormalities in nuclear shape therefore occur in HGPS and are referred to as nuclear blebbing. In nuclear blebbing lamin A is located in nucleoplasmic aggregates.

Treatments that inhibit farnesylation relieve the dominant negative effect of progerin and lead the wild type lamin A to be normally distributed. Farnesyl inhibitors have been utilized in treatment of HGPS patients. Gordon *et al.* (2014) reported evidence that treatment with farnesylation inhibitors increases survival of patients with progeria.

Capell *et al.* (2005) reported that a second modification of wild type farnesylated lamin A protein occurs. This reaction involves removal of the carboxyl terminal 15 amino acids. This second cleavage requires the zinc metallopeptidase ZMPSTE24. Mutations in the zinc metallopeptidase ZMPSTE24 are present in patients with mandibuloacral dysplasia, a condition that has some features in common with HGPS.

4.2.4 *Protein modification through lipidation*

The hedgehog family of proteins sonic hedgehog (SHH) desert hedgehog (DHH) and Indian hedgehog (IHH) play important roles in cell growth and during development. Mutations in SHH and IHH are associated with specific developmental defects in humans. SHH is the most intensely studied hedgehog protein in humans.

Porter *et al.* (1996) reported that SHH protein undergoes a series of post-translational modifications that are critical to its function. After removal of an N-terminal signal sequence, auto-cleavage occurs at the C-terminal end of the protein between cysteine and glycine. This leads to carboxyl ester group generation, followed by linkage of cholesterol to the N-terminal segment which leads it to be functional, while the C-terminal segment is non-functional. The

N terminal fragments with bound cholesterol are then bound to cell membranes. Further modification of the functional SHH protein then occurs at the N-terminal. Varjosalo and Taipale (2008) reported that this modification involved linkage of palmitic acid to the N-terminal through activity of the enzyme hedgehog acetyl transferase (HHAT). In 2015, Long *et al.* reported that SHH bound to a diverse spectrum of fatty acids at the N-amino terminal and that the diverse lipidation species of SHH likely have different physiological functions.

4.2.5 *Palmitoylation of signaling proteins*

There is growing evidence that palmitoylation, i.e. the attachment of palmitic acid to one or more cysteine residues in the protein, plays important roles in regulation the function of signaling proteins. Song *et al.* (2013) reported that NRAS and HRAS palmitoylation play critical roles in determining signaling activity and trafficking of these proteins. They demonstrated that NRAS that had not undergone palmitoylation did not sustain signaling following activation of epidermal growth factor signaling.

Gorinski and Ponimaskin (2013) reported that palmitoylation regulates multiple functions of G protein coupled receptors. These authors also reviewed the important role of palmitoylation in determining the function of 5-hydroxytryptamine (serotonin) receptors.

4.2.6 *Adenosine di-phosphate (ADP) ribosylation of proteins*

ADP ribosylation is a protein modification in which one or more units of ADP ribose are transferred to a specific amino acid in a protein (Scarpa *et al.*, 2013). In mammals two families of transferases are involved in ADP ribosylation reactions. The ARTD (ADP ribosyl transferases) family includes at least 15 different poly-ADP ribose polymerases, PARP1-PARP15. ADP ribose units are linked together through O-glycosidic bonds; transfer of these units to proteins occurs primarily in the nucleus. PARP1 is known to be activated by DNA breaks. PARP1 inhibitors are in use in anti-cancer therapeutics.

Scarpa *et al.* noted that mono-ADP ribosylation has been less well studied in mammals. Mono-ADP-ribose units are transferred to proteins by enzymes in the ARTC family, ARTC1-5. Monoribosylation plays important roles in immune regulation in part through ADP monoribosylation of T cell co-receptors.

4.3 Protein Modification through Glycosylation and Congenital Disorders of Glycosylation

Glycosylation modifies the three dimensional structures of proteins, promotes their stability, impacts their interactions with other molecules and influences their locations within the cell. Freeze *et al.* (2014) reported that 2% of the genome encodes proteins involved in glycosylation related processes. Glycosylation involves the secondary modifications of proteins and lipids through addition of glycans composed of monosaccharides that are linked together by glycosidic bonds to form homopolysaccharides or heteropolysaccharides. The key types of glycosylation of proteins in humans are N- and O-glycosylation (Jaeken, 2013).

In N-glycosylation, glycans are attached to the nitrogen residues in amino acids, primarily asparagine. Jaeken reported that the majority of N-glycosylation defects were assembly defects. A few of the N-glycosylation defects are processing defects. Processing involves trimming of glucose or mannose from glycans and this trimming occurs primarily in the endoplasmic reticulum (ER) prior to transfer of glycosylated proteins to the Golgi or to the endosome lysosome system. N-glycosylation syndromes are frequently associated with neurological symptoms, psychomotor disabilities, ataxia, and hypotonia.

In O-glycosylation, glycans are attached to the hydroxyl groups of amino acids, primarily threonine and serine. O-glycosylation takes place mainly in the Golgi and involves O-mannosylation or O-fucosylation, i.e. the addition of these monosaccharides to specific proteins through activity of specific glycosyl transferases.

It is important to note that defects in the early stages of synthesis of glycosamines and in the synthesis of dolichol may lead to combined N- and O-glycosylation defects. In addition there is evidence that the glycolipid glycosyl phosphatidylinositol (GPI) attaches to proteins and promotes their transport and attachment to membranes. Specific defects in the enzymes and proteins involved in GPI protein formation and functions have been delineated in recent years (Fujita and Kinoshita, 2012).

In recent years there has been a significant increase in the delineation of congenital disorders of glycosylation. This increase has come about primarily as a result of exome and genome sequencing in patients and discovery of defects in genes that encode enzymes and proteins involved in glycosylation pathways. Freeze *et al.* (2014) reported that over 100 human glycosylation disorders were known. They emphasized that in addition to disorders that involved the known glycan biosynthesis pathway, unanticipated defects in glycosylation were discovered through exome and genome sequencing with follow-up glycosylation analyses.

4.3.1 *Degradation of misfolded glycoproteins*

In 2001 Park *et al.* identified a hydrolase, N-glycanase that reacted with proteins with N-glycan linkages. Misfolded N-glycan proteins undergo hydrolysis in the ER to remove the N-glycan linkages and the released proteins undergo proteolysis.

Enns *et al.* (2014) reviewed the phenotypes of 8 patients with defects in the N-glycanase encoding gene NGLY1 (N-glycanase 1). They note that N-glycanase cleaves the bond between N-glycan and the aspartic acid and specifically the beta- aspartyl-glycosylamine linkage.

Clinical features in patients with N-glycanase deficiency include cortical vision impairment, corneal ulceration, and absence of tear formation, hypotonia, seizures and developmental delay. Patients often manifest abnormal choreoathetotic movements and lip smacking. Laboratory findings include elevated liver enzymes.

The first patient reported with N-glycanase deficiency manifested compound heterozygous loss of function mutations in NGLY1. Five of the eight patients reported by Enns *et al.* had c.1201 A > T mutations leading to protein p.R401X (X indicates a stop codon). This mutation was homozygous in four patients and in one patient it was present in compound heterozygous form. Enns *et al.* reported that even in patients with the same mutations the degree of disease severity varied.

Misfolded glycoproteins are transported out of the ER and into the cytoplasm where they undergo cleavage by N-glycanase. In the absence of N-glycanase abnormal substances accumulate in liver cells, and these substances are the unfolded and undegraded glyco-proteins.

4.4 Intrinsically Disordered Proteins and Protein Regions

Early studies on protein structure indicated that specific folding and 3 dimensional (3D) structure was necessary for protein functioning. Levinthal (1968) first proposed that a polypeptide chain might assume a large number of possible conformations. In some cases the entire protein does not fold into a stable globular 3D structure. Such proteins are referred to as intrinsically disordered proteins. There is evidence that proteins may have regions that are intrinsically disordered.

Berlow *et al.* (2015) reported that intrinsically disordered proteins participate in many important regulatory processes in the cell. Wright and Dyson (2015) reported that intrinsically disordered proteins are characterized by a biased amino-acid composition and low sequence complexity. They have low proportions of bulky hydrophobic amino-acids and high proportions of charged hydrophilic amino acids. These proteins fluctuate through a range of conformations. They noted further that sometimes the whole protein has a disordered structure while in other cases only specific regions within the protein are disordered.

Wright and Dyson emphasized that the degree of flexibility of folding of disordered proteins facilitates their interactions with a range of different partners. Intrinsically disordered proteins and proteins with intrinsically disordered regions frequently constitute multi-functional proteins and macromolecular complexes, examples of which include ribosomes and nuclear pores.

The p53 protein is an example of protein in which disordered regions occur at the N-terminus and the C-terminus of the protein and impact protein function.

There is also evidence that the conformational plasticity of disordered proteins facilitates post-translational modification and that different modifications may impact different functions of the protein.

4.5 Conformational Changes in Peptide Bonds Present in Proteins

Peptide bonds between amino-acids in proteins are usually in the trans conformation. Proline is an outlier since bonds between proline and adjacent amino acids are frequently altered to be in the cis configuration. Proline residues are particularly abundant in disordered proteins. Prolyl isomerases interconvert the cis and trans isomers of peptide bonds of proline. There are at least three families of prolyl isomerases; cyclophilins, F506 binding proteins and parvulins. PIN1 prolyl isomerase is a member of the parvulin gene family. The change in shape of proteins induced by the cis to trans orientation has profound effects on protein function.

Tau protein is an intrinsically disordered protein that becomes hyperphosphorylated in a number of neurodegenerative disorders. PIN1 prolyl isomerase specifically acts to convert phosphorylated Thr231-Pro in Tau from cis to trans. Kondo *et al.* (2015) reported evidence that PIN1 inhibits tauopathy by converting phosphorylated Thr231 from cis to trans. In the absence of PIN1 activity the kinase GSK3 hyperphosphorylates tau and cis-p-tau was shown to self-aggregate.

They reported that following traumatic brain injury neurons produce phosphorylated cis tau and that this disrupts microtubule networks and mitochondrial transport and leads to apoptosis. They reported that treatment of brain injured mice with an antibody toward cis tau prevented tauopathy.

Phosphorylation of the amyloid protein is directly induced at threonine 668-Proline by kinase activity. Dysfunctional cis proteins including tau and beta amyloid accumulate in the brain in Alzheimer's disease. Driver *et al.* (2014) reported that there is evidence that the cis isomers of these proteins are resistant to dephosphorylation while the trans isomerase can be more readily dephosphorylated. There is evidence that antibodies or vaccines against the dysfunctional cis proteins may be useful in the diagnosis and therapy of Alzheimer's disease.

4.6 Protein Folding

Kim *et al.* (2013) reviewed folding of proteins synthesized on cytoplasmic ribosomes and the roles of chaperones and chaperonins in determining correct folding.

Proteins are synthesized as linear amino acid structures and they undergo some degree of folding based on their amino-acid sequence. Kim *et al.* noted that in the dense molecular environment in cells proteins fold less efficiently. Furthermore exposed hydrophobic amino acids on surfaces particularly impact correct folding. Chaperones assist in correct folding through prevention of undesirable intermolecular interaction. In their review Kim *et al.* defined and illustrated states of folding and aggregation.

Linear proteins initially form folding intermediates. Chaperones facilitate folding of the protein to its functional native state.

Partially folded proteins may transform to amorphous aggregates, to oligomers and to amyloid fibrils driven by hydrophobic forces. Oligomeric intermediates and fibrillar aggregates frequently co-occur. Fibrillar aggregates often contain proteins that have parallel beta sheet structures.

There is evidence that many proteins are transported to specific sub-cellular compartments prior to undergoing folding. However multi-domain protein may undergo folding during translation.

Kim *et al.* reported that the heat shock families of chaperone proteins play important roles in protein folding and in protein homeostasis. These chaperones include Hsp70, Hsp70L, Hsp40 (DNAJ) and Hsp70 chaperone bound proteins are frequently transferred to the ATP dependent Hsp90 chaperone system.

Chaperonins are large double ring complexes and proteins enter the central cavity of these ring structure to undergo folding. Kim *et al.* noted that chaperonins provide an isolated environment to promote optimal folding.

The immunophilin FKBP10 is a chaperone that promotes collagen folding and triple helix formation. Mutations in FKBP10 lead to osteogenesis imperfecta. Hemizygous or compound heterozygous damaging mutations in FKBP14 lead to a specific form of Ehlers–Danlos syndrome with progressive kyphoscoliosis, myopathy and hearing loss.

Kim *et al.* (2013) defined interconnected pathways that protect functionally active forms of proteins and to eliminate damaged and aggregated forms of proteins. Increased expression of chaperones under stress conditions may prevent unfolding. Damaged non-functional proteins may be degraded in the 26S proteasome following ubiquitin tagging. Aggregated proteins may be removed by autophagy and degradation in the endosome lysosome system. Unfolded proteins in the ER may trigger the ER unfolded protein response and the endoplasmic degradation processes, such as ER-associated degradation (ERAD).

4.7 Small Heat Shock Proteins, Chaperones and Their Functions

Small heat shock proteins act as chaperones to prevent irreversible aggregation of denatured proteins. Alpha crystallins encoded by the CRYAA and CRYAB genes are among the best studied on small heat shock proteins. Alpha chain alpha crystalline CRYAA occurs primarily in the lens of the eye. CRYAB beta chain of alpha crystalline has a wider tissue distribution.

Haslbeck and Vierling (2015) reviewed structures and function of small heat shock proteins and their roles in protein homeostasis. They reviewed the binding of small heat shock proteins to destabilized substrate proteins. This binding facilitates stabilization of the substrate proteins and subsequent binding of the ATP dependent Hsp 70 chaperone system.

In humans, at least 13 different genes encode proteins in the Hsp70 heat shock family. The heat shock protein encoded by Hsp9

(Grp15) functions in mitochondria. The Hsp70 chaperones promote stabilization of partly folded proteins, promote transport of proteins and facilitate folding of unfolded proteins.

Hsp70 chaperones have a central substrate binding domain and an ATP binding domain. ATP hydrolysis is essential for the function of Hsp70 proteins (Mayer, 2013). The Hsp70 substrate binding domain recognizes a specific short motif in proteins. This motif is composed of a core of five hydrophobic amino acids followed by positively charged amino acids. Mayer noted that this core is common in proteins, and the same core amino acid sequences occur every 30 residues. However the core is only exposed when proteins are unfolded.

The N-terminal region of each Hsp70 chaperone protein is composed of four sub-domains that form two lobes. There is a crevice between the lobes and ATP binds in this crevice.

There is also evidence that Hsp70 chaperones facilitate degradation of proteins that cannot be folded. This function is facilitated through the interaction of Hsp70 protein with the product of a gene referred to as STUB1 (also known as CHIP, C-terminus of Hsp70 interacting protein). The product of this gene functions as an ubiquitin ligase. Following ubiquitination the damaged protein can be degraded in the proteasomal proteolysis system.

Hsp70 chaperone proteins often co-localize with co-chaperones. These include proteins in the Hsp40 family and NEF (nucleotide exchange factors) nucleotide exchange proteins that facilitate recycling of ADP generated through ATP hydrolysis during protein folding. At least 42 genes in humans encode the Hsp40 (DNAJ) chaperones.

There is evidence that Hsp proteins may have a protective role against neurodegenerative diseases that are characterized by accumulation of protein aggregates (Turturici *et al.*, 2011).

4.8 Protein Interaction Networks and Multiprotein Complexes

The functions of the protein products of genes are dependent upon their interactions with other gene products, including protein chaperones and interactions with cofactors. A number of investigators have proposed that reference protein interactome maps would be critical to understanding genotype phenotype interactions (Rolland *et al.*, 2014). Protein-protein interactions can be defined through *in silico* methods or through experimental procedures. *In silico* methods include analysis of protein domains and their structural properties. Experimental procedures include co-precipitation, affinity purification, and yeast-two hybrid isolation analyses.

Wodak *et al.* (2013) noted that there has been a remarkable proliferation in data on protein interaction networks but that the specific networks described were often dependent on the detection method used.

Greene *et al.* (2015) emphasized that the precise action of gene products depends on their tissue context. They noted that many existing networks which describe interaction lack tissue specificity. They emphasized too that multifunctional organisms often have multifunctional proteins. Tissue specific interactions of a protein may be dependent on which other proteins are present in that tissue.

It is interesting to consider the possibility that some variations in the function of a protein in different tissues may be related to the generation of different tissue specific isoforms of the protein.

Greene *et al.* integrated data relevant to 144 different tissue types and cell lineages. They determined for example that for the LEF1 (lymphoid enhancer binding factor 1) gene encoded product, a transcription factor involved in Wnt signaling, different protein interactions occurred in each of in four different tissues studied: B lymphocytes, hypothalamus, osteoblast and trachea.

Gross and Ideker (2015) emphasized that analyses of tissue specific network interaction are likely to improve possibilities for identification of new candidate genes for specific diseases that impact specific tissues.

4.8.1 *Genome-tissue Expression (GTEx) consortium studies*

GTEx consortium (2015) studies reported association of whole genome sequence data with RNA sequencing data across 50 tissues. They note that information on how variants in a specific gene impact transcription of that gene in a particular tissue will provide information relevant to genotype–phenotype analyses. The GTEx studies also provide information on how cis quantitative trait loci impact expression of a gene in a specific tissue.

Melé *et al.* (2015) analyzed RNA sequence data generated by the GTEx consortium. Their studies revealed that specific tissues exhibit characteristic expression patterns and noted that primary transcription was the major driver of tissue specific expression. They demonstrated age-related changes in tissue expression patterns in 1,993 genes. They also identified sex based differences in tissue expression in 753 genes.

4.8.2 *Multiprotein complexes*

It is important to note that in large multiprotein complexes such as mitochondrial complex 1, it is not yet clear to what extent the different components physically interact even though many are individually critical for function of the complex. Fassone and Rahman (2012) reported that the mitochondrial complex 1 is composed of at least 45 different sub-units, 14 of which are necessary for catalytic function. The remaining subunits include assembly factors and accessory factors.

4.9 Endoplasmic Reticulum (ER), Golgi and Proteins

4.9.1 *Transfer of protein to the ER*

Many membrane proteins and extracellular secrctcd proteins are transferred to the ER lumen following synthesis on ribosomes. These proteins have specific signal peptides in their amino acid sequence. Transfer may occur co-translationally or following completion of translation. Specific transport mechanism occurs in the ER membranes. Zimmerman *et al.* (2011) reported that the Sec61 complex present in ER membranes recognizes signal peptides on proteins destined for the ER. The protein BiP encoded in human by the HspA5 gene binds to incoming proteins and facilitates their transfer and completion of folding. BiP is a chaperone protein and a member of the family of heat shock protein chaperones. BiP is also involved in binding to the pre-active forms of key proteins in the unfolded protein response.

Proteins in the ER include secreted proteins destined for extracellular function and cell surface proteins and cell surface receptors. The ER is involved in protein modification, protein folding and homeostasis and also in lipid synthesis and calcium metabolism. Gidalevitz *et al.* (2013) reviewed mechanisms of folding in the ER. Hsp70 chaperones also expedite folding in the ER and N-glycosylation facilitates folding of ER proteins. Furthermore many ER proteins are rich in disulfide bonds. Protein disulfide isomerases and FAD (flavin adenine dinucleotide) dependent enzymes promote oxidative folding of proteins. Other ER specific enzymes involved in protein folding include peptidyl-prolyl isomerases and cyclophilin B. Calcium and calcium binding proteins are abundant in the ER. Calreticulin and calnexin are ER unique chaperones.

The ER is involved in calcium storage and release. Calreticulin has a globular N terminal domain, a central proline rich domain and a C terminal domain rich in negatively charged amino-acids that bind calcium.

Many proteins destined for secretion, cell membranes, lysosomes and endosomes pass from the ER to the Golgi complex with its systems of membranes and cisternae. There, proteins undergo further modification through the addition of polysaccharides, oligosaccharides, glycosaminoglycans and lipids. Glycosylation processes include N-glycosylation and O-glycosylation through activity of glycol transferases. Transport from the ER to the Golgi takes place in vesicles.

ER homeostasis is negatively impacted by depletion of calcium ions (Ca^{2+}), by oxidative stress and by accumulation of misfolded proteins. Placido *et al.* (2014) reported that calcium ion pumps play active roles in storage and release of calcium from the ER. Furthermore in regions of the ER that are associated with mitochondrial membranes, specific ion channels are present that permit the passage of calcium ions between ER and mitochondria. Impaired calcium balance predisposes cells to apoptosis.

Gurel *et al.* 2014 reported that interactions between the ER and microtubules are important in maintaining accurate ER structure.

Senft and Ronai (2015) reported that diverse cellular stresses lead unfolded or misfolded proteins to accumulate in the ER. These stresses include redox imbalance and perturbations in calcium homeostasis.

4.9.2 *Tagging of newly synthesized polypeptides with glycans*

Tagging of newly synthesized polypeptides with glucose, mannose and N-acetyl glucosamine (Gly3-Mann3-GlcNac2) takes place in the ER membrane through activity of oligosaccharyl transferase. Kamiya *et al.* (2012) reported that N glycosylated protein folding further involves molecular chaperones and thiodisulfide oxidoreductases in the ER lumen. Correctly folded and glycan tagged molecules then pass from the ER to the Golgi. They emphasized that N-glycan functions as quality control tags.

4.10 Unfolded Protein Response

Accumulation of unfolded protein in excessive quantities in the ER activates the three arms of the unfolded protein response. Collectively the components of this response function to reduce mRNA translation, to enhance expression of genes involved in protein folding and genes involved in the synthesis of ER components.

Walter and Ron (2011) reported that each of the three arms utilizes specific signal components to trigger expression of specific transcription factors. The ER resident signaling components included IRE (inositol requiring enzyme), PERK (RNA activated protein kinase) and ATF6 (activating transcription factor 6).

They also reported that ATF6 in the resting state is bound to BiP. ATF6 dissociates from BiP under condition of ER stress and this dissociation unmasks a Golgi localization signal on ATF6. Following transport of ATF6 to the Golgi, it undergoes proteolysis through activity of proteases S1P and S2P. This proteolysis generates a fragment ATF6N that migrates to the nucleus where it acts as a transcription factor. ATF6N in the nucleus enhances expression of genes that produce products which reduce ER stress. These include BiP and Derlin 3, both of which play roles in destruction of misfolded proteins.

Activation of the PERK arm of the unfolded protein response requires that PERK be released from BiP protein. On activation PERK protein undergoes oligomerization and phosphorylation and it then phosphorylates EIF2A (eukaryotic translation initiation factor 2A), a subunit of EIF2. EIF2A phosphorylation leads to inactivation of EIF2 and this then inhibits mRNA translation. However not all mRNAs undergo translation inhibition. One uninhibited mRNA is translated to transcription factor ATF. ATF enters the nucleus and enhances expression of CHOP, a protein that controls apoptosis. Another uninhibited protein that is translated is GADD34 (growth arrest protein). GADD34 expression can inhibit PERK phosphorylation of EIF2A. Walter and Ron reported that balanced regulation of EIF2A phosphorylation is required to achieve homeostasis.

The IRE arm of the unfolded protein response was the first arm described. IRE1 (inositol-requiring enzyme 1) is bound to BiP and under ER stress conditions it is released. IRE1 has kinase and endoribonuclease domains. The ribonuclease domain of IRE1 is activated following oligomerization of the IRE1 protein. This ribonuclease then cleaves the mRNA that encodes the XBP1 protein (X-box binding protein) at an unusual splice site. Following cleavage, the two fragments of XBP1 mRNA are united through activity of a tRNA ligase to generate XBP1 (X-box binding protein 1) that acts as a transcription factor. XBP1*(XBP1S) enters the nucleus and enhances expression of genes involved in lipid synthesis and genes that encoded proteins involved in the development of ER. There is evidence that IRE1 ribonuclease participates in the degradation of mRNAs in a process referred to as RIDD (regulated IRE1 dependent degradation) (Brewer, 2014).

4.10.1 *ER and overload response*

Smith *et al.* (2011) reported that many misfolded proteins in the ER are terminally misfolded and in some cases these proteins are returned to the cytosol where they may undergo degradation in the ubiquitin ligase proteasome system.

Roussel *et al.* (2013) described an additional form of ER stress referred to as ER overload response. In this situation the ER is distended with misfolded proteins. They emphasized that cells that undergo cell divisions can tolerate accumulation of unfolded proteins since these become diluted on cell division. Neurons however are post-mitotic cells and the absence of cell division results in neurons being intolerant of such accumulations.

These investigators noted that ER overload occurs in familial encephalopathy where unfolded forms of the protein neuroserpin accumulate. Neuroserpin is a serine protease inhibitor. In familial encephalopathy the neuroserpin aggregates do not activate the UPR and instead they lead to ER overload and degeneration.

They noted that post-mortem studies of the brain in cases of Alzheimer's disease, Parkinson's disease, amyotrophic lateral sclerosis, multiple sclerosis and cerebral hypoxia, there was evidence of ER dysfunction. Roussel *et al.* reported that in multiple sclerosis, impaired regulation of EIF2A phosphorylation occurs. In cerebral hypoxia the UPR is impaired through energy depletion.

4.10.2 *Unfolded protein response in neurodegeneration*

Halliday and Mallucci (2014) emphasized that recent studies indicate that the unfolded protein response is activated in neurodegenerative diseases including Alzheimer's, Parkinson's, Huntington's and rare prion diseases. Modulation of this response is being considered in the development of therapies for these diseases.

4.11 Ubiquitination and its Consequences, Proteasomes

Ubiquitination was initially discovered as a form of post-translation modification that targeted protein substrates for degradation. In recent years it has become evident that ubiquitin modification takes place in diverse forms and that it plays important roles in signaling. Three classes of enzymes are involved in the transfer of ubiquitin molecules to substrates. These include E1 ubiquitin activating enzymes (two in humans), E2 ubiquitin conjugating enzymes (35 in humans) and hundreds of E3 ubiquitin ligase enzymes (Vittal *et al.*, 2015).

Ubiquitin is a 76 amino acid protein encoded by 4 different genes in humans. UBB and UBC encode polypeptide precursors with 4 tandem repeats of ubiquitin. UBA52 (ubiquitin A-52 residue riboso-mal protein fusion product 1) and RPS27A (ribosomal protein S27a) are fusions between ubiquitin genes and ribosomal protein encoding genes.

E1 ubiquitin activating enzyme carries out a two-step reaction. In the first step there is activation of E1 enzyme with ATP. In the second step a thioester bond is formed between the E1 active site cysteine and the ubiquitin C terminus and AMP is released. In the second reaction the ubiquitin charged E1 interacts with E2 ubiquitin conjugating enzyme and ubiquitin is transferred to the active cysteine in the E2 enzyme. The third reaction involves activity of E3 ubiquitin ligase that transfers ubiquitin from the E2 active site cysteine to lysine side chains in the substrate. Substrates can be modified with one ubiquitin molecule or by an ubiquitin chain. The ubiquitin mol-ecule itself has seven lysines and the lysine side chains can be ubiq-uitinated to give rise to polyubiquitin chains.

There is now evidence that three distinct classes of E3 ubiquitin ligases carry out the transfer of ubiquitin from the E2 bound form to the substrate lysine. These include HECT (homologous to E6-AP carboxyl terminus) ubiquitin ligase, RING (Really Interesting New Gene) E3 ubiquitin ligases and a newly discovered class, RING-BetweenRING-RING (RBR) E3 ligase (Berndsen and Wolberger, 2014). They reported that RING E3 ligases transfer ubiquitin directly from E2 to substrate lysine. HECT E3 ubiquitin ligases carry out a two-step reaction including a thioesterification reaction in which ubiquitin is transferred to a cysteine in HECT and in a subse-quent reaction ubiquitin is transferred between HECT and substrate.

Galligan *et al.* (2015) identified 300 different substrates that interact with HERC2 ubiquitin ligase. Key interactions included proteins associated with metabolism and growth e.g. EIF eukaryotic transcription initiator complex, GSK glycogen regulator, phosphory-lase kinase, phosphatide inositol kinases and proteins involved in fatty acid transport. HERC2 (HECT and RLD domain containing E3 ubiquitin protein ligase 2) was also associated with proteins involved in gene repair.

There are approximately 600 different genes in the human that encode components of the RING E3 ligases. RING E3 ligases are frequently multi-subunit complexes. Cullin E3 ubiquitin ligases each contain a single cullin (CUL) sub-unit, either CUL1, CUL2, CUL3, CUL4A, CUL4B or CUL5 subunits. The CUL subunit acts as a scaffold. The N-terminal domain in the CUL subunit interacts via linkers with the substrate to be ubiquitinated. The C-terminal of the CUL subunit binds ROC (regulator of CULs) as a linker that binds E2 ubiquitin ligase. Ubiquitin is transferred from the E2 bound ROC to the substrate. An ubiquitin like modifier NEDD8 (neural precursor cell expressed, developmentally down-regulated 8) binds to the 3' end of CUL and serves as an activator (Jackson and Xiong, 2009).

RBR E3 ligases have two RING domains separated by an in-between domain, and RBR E3 ligases have features of RING E3 ligase functions and features of HECT function. RBR E3 ligases include Parkin and human homolog of Ariadne (HHARI).

Ubiquitin itself is a target of ubiquitination. There are different types of ubiquitin linkages that involve different amino acids in the ubiquitin molecule. The best known linkage is to lysine 48 and proteins with chains comprised of ubiquitin 48 linkages are usually targeted to proteasomes for degradation. Linkage between lysine 63 (K63) in ubiquitin have been implicated in non-proteolytic processes.

When M1, the first methionine in ubiquitin is involved in linkages, this leads to a head to tail linkage where the C-terminal of one ubiquitin conjugates with M1 in the next ubiquitin. This linkage predisposes to linear ubiquitin linkages. There is evidence that linear ubiquitin linkages are important in the activity of signaling complexes.

The degree of ubiquitination of specific proteins can vary and a number of different types of ubiquitin chains are generated. Furthermore, ubiquitin-like molecules can also be attached to proteins. Studies of specific disease have provided insight into ubiquitin related mechanisms of proteolysis.

There is evidence that targeting of proteins with ubiquitin has a number of different physiological functions (Rieser *et al.* 2013). Parkin a RBR E3 ligase has at least 30 different substrates. Müller-Rischart *et al.* (2013) reported that Parkin enhances linear ubiquitination of the protein NEMO (NF-κB essential modulator) and this leads to increased expression of the mitochondrial GTPase OPA1 (optic atrophy 1) that is a key regulator of mitochondrial cristae integrity and of mitochondrial fusion.

4.11.1 *Ubiquitin modification and proteosomal degradation of damaged proteins*

Cellular protein concentrations are regulated by synthesis and degradation. The proteasome is involved in the degradation of defective and misfolded proteins and in degradation of regulatory proteins. Proteasomes are now known to also play roles in the immune system by producing peptides that serve as antigens and are presented bound to major histocompatibility complexes. Key elements in the proteasomal system include attachment of ubiquitinated linkages to protein.

Inobe and Matouschek (2014) reviewed proteosomal structure and mechanisms of protein degradation in the proteosome. Through the control of degradation of intra-cellular proteins, including regulatory proteins, the proteosome impacts a number of important processes including cell cycle regulation, cellular differentiation and signaling. In addition this system plays critical roles in the degradation of damaged and unfolded proteins in the cytoplasm. The proteosome recognizes ubiquitin tagged molecules.

4.11.2 *Proteosome structure*

The core of the proteosome is a cylindrical structure formed by four rings, two composed of alpha subunits and two composed of beta subunits. The rings are arranged in the order alpha, beta, beta, and alpha. There are seven different types of alpha sub-units and seven different beta subunits. Genes encoding the alpha subunits are designated PSMA1 (proteosome subunit alpha 1)-7; genes encoding the beta subunits are designated PSMB1 (proteosome subunit beta 1)-7.

Gomes (2013) reported that the beta 1 subunit encoded by PSMB6 has caspase like activity. The beta 2 subunit encoded by PSMB7 has trypsin like activity and the beta 5 subunit encoded by PSMB5 has chymotrypsin like activity.

Four different forms of proteosome cap structures occur. Caps are located at either end of the core structure. The best-characterized cap structure is the regulatory 19S complex that is involved in recognizing ubiquitin tagged substrates and transporting these to the core. Regulatory subunits (RPT1-5) with ATPase function occur in the region of the 19S complex located adjacent to the alpha rings of the 20S core. PSMC genes encode RPT subunits. The 19S cap also contains non-ATPase subunits that cleave ubiquitin from substrates. RPN1 and RPN2 sub-units form a clamp that attaches a lid-like cap structure to the proteosome. PSMD genes encode different RPN subunits. Gomes (2013) reported that some proteosomes have a single lid while others have two caps. Together the 20S core and the 19S complex form the 26S proteasome.

There are different types of proteosomes. In some proteosomes a chaperone VCP (valosin containing protein) is bound to the 20S proteosome core. This form of proteosome was reported to be particularly involved in degradation of unfolded proteins that are translocated from the ER.

4.11.3 *ER and degradation processes (ERAD)*

Proteins that cannot be correctly folded are degraded in the ERAD system. Olzmann *et al.* (2013) reviewed the ERAD system. They noted that there are four key processes in the system. The first involves substrate selection and identification of misfolded proteins. The second step involves dislocation across the ER membrane. In a third step the dislocated substrate is polyubiquitinated. In the fourth step the dislocated polyubiquitinated proteins are degraded in proteasomes. The ERAD system therefore spans the ER membrane bilayer.

Olzmann *et al.* (2013) reported that post-translational modifications of proteins in the ER include addition of high mannose core glycans. These are composed of glucose, mannose and N-acetylglucosamine. The ERAD system identifies substrates for dislocation depending on conformation and glycan status. Specific adapters in the ER membrane recognize as substrates inadequately folded proteins. These protein adapters include ERLIN1 and 2 lipid raft associated proteins.

Specific proteins are also required for substrate dislocation. These include derlins 1–3 encoded by DERL1, DERL2 and DERL3 genes.

Olzmann *et al.* reported that VCP is an enzyme that utilizes ATP hydrolysis to pull targeted proteins out of the ER. The process of dislocation likely utilizes additional enzymes, including rhomboid proteases (RHBDL2, RHBDF1/F2, and RHBDD1).

The enzyme NGLY1 catalyzes deglycosylation of N-linked glycoproteins that have been retrotranslocated to the cytosol. Mutations in NGLY1 cause the loss of enzyme activity and lead to a condition characterized by developmental delay, seizures, peripheral neuropathy, lack of tears sometimes leading to corneal ulceration, and impaired liver function (Enns *et al.*, 2014).

The final steps in the ERAD process include ubiquitination of the dislocated substrate and degradation. Ubiquitination is dependent on the activity of E3A ubiquitin ligases.

4.11.4 *Ubiquitin linkages*

At least eight different types of ubiquitin linkages can occur and give rise to different types of chains (Komander, 2009). The chain type influences the outcome of protein ubiquitination.

During the past decade experiments have revealed that the processes of ubiquitination are much more complicated than initially proposed. Proteins may undergo mono-ubiquitination where a single ubiquitin is attached to lysine in the protein. They may also undergo multi-mono-ubiquitination where a single ubiquitin is added to multiple lysines within the protein. Polyubiquitination also occurs that involves linkage of a series of ubiquitin to a specific lysine residue in the protein. The ubiquitins within the ubiquitin chain may be linear or non-linear.

Komander (2009) reported that mono-ubiquitination of cell surface receptors promotes their internalization and degradation in the endosomal lysosomal system. He also noted that mono-ubiquitination is involved in the DNA damage response and that histones and PCNA (proliferating cell nuclear antigen) undergo mono-ubiquitination.

Early studies revealed that protein degradation in the proteosome utilizes polyubiquitinated proteins and that linkages through lysine 48 in ubiquitin promote delivery to the proteosome.

LUBAC is a dedicated ubiquitin ligase involved in assembly of linear ubiquitin complexes (Rieser *et al.*, 2013). There is evidence that linear ubiquitin complexes play significant roles in signaling. Mass spectroscopy has proved to be particularly valuable in analysis of ubiquitin linkages. In addition to ubiquitin, a number of ubiquitin-like molecules are also involved in protein modification.

4.11.5 *Deubiquitinases*

It is important to emphasize that ubiquitination is a reversible process. Eletr and Wilkinson (2014) reported that there are five different families of deubiquitinating enzymes. These include hydrolases and ubiquitin-specific protease, including metalloproteinase. They reported that deubiquitinases often remain inactive until recruited to specific substrates of substrate adaptors. In addition to deubiquitinating specific proteins, these enzymes play roles in recycling ubiquitin.

4.12 Proteostasis Network

Amm *et al.* (2013) reviewed aspects of protein quality control and the role of the ubiquitin proteosomal system. They noted that misfolded proteins might arise as a result of mutations or as a result of cellular stress. If correction of folding is not achieved through action of chaperones, misfolded proteins tend to be non-functional and aggregate. Defective proteins and aggregates may be removed through autophagy. Misfolded protein may be removed in the ubiquitin proteosomal system.

Following ribosomal synthesis proteins adopt a specific three-dimensional (3D) structure. The precise 3D structure may also depend on post-translational secondary modifications.

Amm *et al.* noted that a complex system of chaperones exists to facilitate 3D structure maintenance. Stress factors that impair structure maintenance include metal ions, and oxygen radicals. In addition to facilitating protein folding, chaperones serve to guide proteins to degradation sites.

Defective nascent proteins may be generated on ribosomes through transcription defects, stop codon defects and these may undergo ubiquitination. Newly synthesized proteins that do not have appropriate signal sequences to target them to specific sites within the cells bind to chaperones and are targeted for proteosomal degradation. Amm *et al.* reported that specific chaperones, e.g. in the Hsp70 and Hsp40 systems, can act to correct misfolding and can inhibit aggregate formation. The protein STUB1 (CHIP) functions as a co-chaperone and as ubiquitin ligase.

The proteostasis network acts to ensure protein integrity and to prevent aberrant protein interaction. Key activities in this network are molecular chaperones that assist with protein folding and systems that target denatured proteins for destruction. Hipp *et al.* (2014) reviewed the extensive network that a cell possesses to safeguard proteome integrity and to maintain homeostasis. Breakdown in the molecular chaperone systems or in the proteolytic degradation machinery lead to accumulation of aberrant protein conformation and protein aggregates.

They noted that the tendency of a specific protein to aggregate is dependent on the amino acid composition, the intrinsic stability of its folded structure and the cellular concentration of the protein. They noted further that approximately 30% of proteins have intrinsically unstructured, disordered regions and that those proteins tend to form particularly toxic aggregates.

Aggregation of proteins occurs when the chaperone and degradation systems are overwhelmed. Chaperones also play roles in the initial folding of proteins and facilitate the refolding of proteins. Refolding of proteins is achieved through the co-operative activities of small heat shock proteins and the Hsp70 chaperones that have ATP dependent functions.

Proteins that cannot be refolded are frequently recognized by ubiquitin ligases and subsequently undergo proteasomal degradation. Larger aggregates and complexes may also be removed by autophagy and lysosomal degradation.

Hipp *et al.* reported that mechanisms involved in proteostasis apparently decline during aging. This decline promotes accumulations of toxic aggregates that may lead to cell death. The proteostasis mechanism may however simply be overwhelmed by the chronic production of aggregated proteins. There is evidence that a specific transcription factor NFY that stimulates Hsp70 production is abnormally sequestered by specific aggregates, e.g. polyglutamine aggregates. Ubiquitin and Hsp70 chaperones may also be abnormally sequestered in aggregates.

4.13 Autophagy

Autophagy is a process that serves to maintain quality control through breakdown of damaged cellular organelles and complex molecules and to recycle the breakdown products. Autophagy serves to replenish nutrient in situations of nutrient deprivation. The possible importance of autophagy in prevention of neurodegenerative diseases has fueled intense investigations into mechanisms of autophagy (Nixon, 2013).

Many different components of the autophagic machinery have been characterized in recent years. In humans many of the genes that encode autophagic components have been shown to be homologous to yeast autophagy genes. At least 101 genes in humans have been designated as autophagy (ATG) genes. In addition many different autophagy regulatory proteins and complexes have been characterized.

The autophagy processes are triggered by starvation and also by the accumulation of cellular debris, protein aggregates lipid molecules and pathogens. Levine *et al.* (2015) reported that key steps in autophagy involve initiation, with generations of an isolating membrane the phagophore, elongation of the membrane and ingestion of material, maturation of the membrane and full vesicle formation. The autophagosome with ingested material then fuses with a lysosome and digestion of the engulfed material follows. Released degraded material may be re-utilized as nutrients.

The autophagy process is stimulated by starvation. Nixon noted that autophagosome biogenesis is initiated when MTORC1 activity is inhibited and when the ULK1, ATG13, ATG101 autophagy complex is phosphorylated by AMPK (adenylate monophosphate kinase). The ULK autophagy complex then phosphorylates a downstream complex composed of ATG14 and the phosphatidyl inositol complex PI3CIII. This complex along with lipids and additional ATG proteins promotes membrane elongation and generation of the phagophore.

Russell *et al.* (2013) reported a mechanism through which ULK1 promotes autophagy. They reported that activated ULK1 phosphorylates Beclin-1 on serine 14 and this enhances activity of the pro-autophagic lipid kinase VPS34 and this promotes autophagic initiation. Binding to BCL2 protein inhibits the Beclin complex.

A number of investigators have described different forms of autophagy. Nixon (2013) described three forms of autophagy particularly important in the brain. In macroautophagy ubiquitinated proteins, including misfolded proteins and protein aggregates are recognized by phagophore adaptor receptor proteins p62 (sequestosome SQSTM1), NBR1 and NIX and these react with LC2 (MAP1LC3A) on the phagophore. They are then taken up into the autophagosome that forms from the phagophore.

In mitoautophagy, the ubiquitin ligase, Parkin, interacts with the mitochondria protein PARKIN and with the VDAC1 protein on the outer membrane of damaged mitochondria. These proteins along with the damaged mitochondria are taken up on the phagophore and autophagosome. The autophagosome then fuses with a lysosome.

In chaperone mediated autophagy an unfolded protein that expresses a specific amino acid motif KFERQ on its surface interacts with the Hsp70 chaperone. These are then delivered to the lysosome and taken up through activity of LAMP2 (lysosomal membrane protein 2).

Schneider and Cuervo (2014) described three types of autophagy. Macroautophagy involves the encapsulation of damaged structure in double membrane vesicles, autophagosomes, and their delivery to lysosomes through vesicular fusion. Microautophagy involved capture of damaged material in single membrane structures that are invaginations of lysosomes or endosomes. Chaperone mediated autophagy enables damaged material to enter lysosomes directly facilitated through the action of chaperones including Hsp70. The lysosome associated membrane protein LAM2A is important for transfer into lysosomes.

Schneider and Cuervo (2014) described the steps and effectors of autophagy. The first step is cargo recognition through action of adaptor molecules that include LC3, also known as MAP1LC3A, microtubule associated light chain. Kinase complexes that include ULK1 and Beclin activate ATGs that form autophagy membranes. Phosphorylation of lipids also occurs and contributes to phagosome membrane formation. Sealing of phagophore membranes and their subsequent fusion with lysosomes is facilitated through activity of small GTPases. Dynein proteins facilitate microtubule trafficking of autophagosomes. Proteins involved in fusion of phagosomes with endosomes and lysosomes include SNARE proteins, LAMP2, ubiquilin, RAB 11, VCP and MCOLN1 the lysosomal receptor.

The protein TFEB (transcription factor EB) enhances degradation in the lysosomes. Degraded materials are released from the lysosomes following permease digestion of lysosome membranes.

4.13.1 *Autophagy in disease states*

In neurodegenerative diseases, dysfunctional mitochondria and protein aggregates are present. Their removal from cells may be impaired through dysfunctional autophagy. This was demonstrated in Huntington's disease where autophagy was impaired through sequestration of ATG proteins in huntingtin protein aggregates (Choi *et al.* 2013). These authors also noted that in Alzheimer's disease, accumulated beta amyloid protein and phosphorylated tau protein might impair lysosomal function. In some cases Paget's disease of bone associated with fronto-temporal dementia are due to mutations in the autophagy related protein sequestosome (SQSTM).

4.13.2 *Autophagy and aging*

Cuervo and Macian (2014) reported that malfunction of autophagy occurs in many organs and tissues in aging. They noted further that most interventions that improve life span activate autophagy.

Impaired autophagy results from decreased transcription of ATG genes and decreases in chaperone-mediated uptake of degraded materials. Impaired degradation of cargo can be visualized as lipofuscin deposits in lysosomes.

5. Genotype–Phenotype Correlations and Complexities

5.1 Phenotype Description

As genomic studies and next generation sequencing (NGS) become increasingly available, there is growing emphasis on the relevance of DNA sequence changes in the causation of phenotypic abnormalities. It is important that comprehensive databases are developed with accurate information on genetic variants and on phenotypic manifestations in patients. Correlations of genotype information with phenotypic features are essential in developing insights into disease pathogenesis through knowledge of functions of gene products, and ultimately in enhancing patient management and therapy.

Documentations of phenotypic manifestations and genotype–phenotype correlations in different patients are facilitated by the use of standardized terms to define phenotype. The human phenotype ontology project (HPO) was designed to provide comprehensive terms for phenotypic abnormalities. Other databases designed to present standardized terms for medical conditions include the medical usage system (UMLS) and other language systems designed for specific structures, e.g. elements of morphology databases that specifically describe features of the face, hands and feet.

Templates designed to capture phenotypic features include Phenotips and PhenoDB. Phenotips templates include categories such as demographic information, family history, medical history, measurements (including prenatal and post-natal growth and growth curves), phenotypic information using HPO terms and diagnosis.

Specific phenotypic features are documented as observed, not observed and relevant, not investigated, or not relevant. Each phenotypic feature is also documented in terms of age of onset, pace of progression, clinical symptoms and physical findings. In addition, as phenotypic features are documented searches are done automatically in Online Mendelian Inheritance in Man (OMIM) to find matching disorders.

The Phenotips system requires recording of patient phenotypes using a standardized vocabulary. It is also important that standardized abbreviations be used so that phenotypic information is collected in a format suited for automated analysis (Girdea *et al.*, 2013). Furthermore the system requires comprehensive information and precision in descriptions of phenotypic features. An example given is the term developmental delay where the entry in the template is expanded to report whether this is global, whether it relates to gross motor skills, fine motor skills and/or language.

Girdea *et al.* noted that Phenotips follows the patterns of clinical examination work flows. The pattern followed includes patient and demographic information; prenatal, perinatal and post-natal history; medical and developmental history. For family history Phenotips also has the capacity to upload pedigrees. Clinical symptoms and physical findings are included. Additional files include genetic testing.

Phenotypic information and clinical symptoms are organized in sub-categories. An example presented involves patients with seizures. If seizures are present more extensive information on seizure type is collected.

Pheno DB is another web based system for collection and storage of phenotypic features. The template includes patient and family information, physical features, special studies (including imaging studies), digitized pathology slides and samples collected.

5.2 Classification of Mutations in Protein Coding Genes through DNA Sequencing

The use of next generation sequencing techniques in clinical settings prompted the publication of several reviews and guidelines for interpreting and documenting nucleotide variants.

5.2.1 *American college of medical genetics (ACMG) and association for molecular pathology recommendations*

Processes utilized for establishing classification of variants found in next-generation sequencing include population data computational data, functional data, and segregation of variants in affected individuals.

ACMG (Richards *et al.*, 2015) defined guidelines for DNA sequence evaluation in cases of rare diseases likely to be of genetic origin. ACMG recommends that tests be performed in a CLIA laboratory (a diagnostic testing laboratory regulated under the Clinical Laboratory Improvement Amendment) and that results be interpreted by a board certified clinical molecular geneticist, a genetic pathologist or equivalent.

Sequence polymorphisms are described as changes in the nucleotide sequences that occur with a frequency greater than 1% in the specific population. It is important to note that the frequencies of variants differ in different populations. ACMG recommended that the terms mutation and polymorphism be replaced by the term variant and that variants then be further qualified as pathogenic, likely pathogenic, uncertain significance, likely benign, or benign. Furthermore pathogenicity should be reported relative to a disease condition and inheritance pattern.

The term, likely pathogenic, should be used if there is a greater than 90% certainty of pathogenicity. Similarly the term, likely benign, should be used if there is a greater than 90% chance that the variant is benign. Variant description should include sequence, g for gene position, c for position in coding sequence, p for position in the protein sequence, m for position in the mitochondrial sequence.

An example is the pathogenic variant in phenyl alanine hydroxylase, c.1521-1523 del CTT, p. Phe 508del. For coding sequences position p1 is the A in ATG translation initiation codon. The transcript used should be the longest transcript. ACMG recommends however that the impact of the variant on all transcripts of a particular gene should be noted.

The sequence utilized should be the reference sequence in the NCBI (National Center for Biotechnology Information) database and the term Ref seq. and the sequence version number e.g. h19.

In silico tools are available for interpretation of the sequence variant in terms of the amino acid level and the impact of the variant on the primary transcript and alternative transcripts. Important questions are whether or not the nucleotide variant alters splicing and whether it alters the protein.

Criteria used to assess the impact of missense changes include evolutionary conservation of an amino acid, location in the protein, and biochemical consequences of the amino acid change. Various prediction software programs exist, including PolyPhen-2, SIFT, MutationTaster, Provean and the program Condel that combines SIFT and PolyPhen-2, a mutation assessor. Nucleotide conservation prediction programs include GERP (genomic evolutionary rate profiling) and PhyloP.

ACMG criteria include subcategories for reporting variants that take supporting evidence into account; these include very strong (PVS1), strong (PS1–4) and moderate (PM1–6). PVS1 variants include variants that are likely to disrupt function, e.g. nonsense mutations, frameshift mutations, canonical splice site variants, initiation codon variants, single exon deletions and multi-exon deletions. In addition to pathogenicity inheritance recessive or dominant, the relation of protein to disease must also be considered. Variants are defined as *de novo* if they are absent from biological parents.

Truncating variants when the terminal exon in the gene is involved must be functionally evaluated since such variants may be consistent with the production of a full length protein. Splice variants may lead to exon skipping or to inclusion of intronic material.

Functional assay of RNA or protein is recommended since exon skipping may involve non- critical regions of the protein.

A moderate degree of certainty of pathogenicity can be applied to documented mutation hotspots and to function-altering mutations at very low frequency in the general population. A moderate degree of certainty of pathogenicity can also be assigned to deleterious variants in trans with a known pathogenic variant in cases of recessive disorders. It may be necessary to carry out family studies to prove that the two mutations are in trans (each inherited from a different parent) in cases of likely recessive disease. Reliable information on the occurrence of a specific variant associated with a particular disease and phenotype in other patients also represents supporting information.

ACMG noted that assays at the RNA level are useful in determining messenger RNA processing, stability and processing. Functional assays must reflect the biological function of the protein.

Data to be used to define variants as benign include lack of evidence that the variant impacts splicing or protein function. Variants that have a frequency higher than 5% in the general population are likely benign. Variants leading to an in-frame deletion or insertion in a repeat sequence region are likely benign. Variants in cis with a pathogenic variant are likely benign if the disease is suspected to be autosomal recessive. ACMG recommends caution in interpreting missense changes.

Mitochondrial sequence variants should be documented according to the Revised Cambridge Reference Sequence and the level of heteroplasmy should be recorded. Variants that impact gene products involved in drug metabolism should be compared with variants reported in the Pharmacogenomics database. ACMG emphasizes that common complex disease risk alleles typically confer low relative risk and their value for patient care is currently unclear.

Databases with information on mutation frequencies in populations are very valuable in determining the likely disease relevance of a specific mutation; these databases include the 1,000 genomes data bases at NCBI and the ExAC (Exome Aggregation Consortium) database available through the Broad Institute. ExAC has information on at least 60,000 (control) individuals.

5.2.2 *Additional guidelines for investigating the causality of sequence variants in human disease*

MacArthur *et al.* (2014) emphasized that false assignments of pathogenicity to sequence variants can have severe consequences, and lead to incorrect advice to patients. Also false assignment of pathogenicity impacts research efforts. The MacArthur review focused on the impact of variants in monogenic diseases and they also focused on rare large effect risk variants in complex diseases. They adopted five terms to describe sequence variants. These include:

(i) Pathogenic variants defined as contributing to disease though not necessarily causing disease
(ii) Implicated variants with a pathogenic role and a defined level of confidence
(iii) Associated variants enriched in disease cases versus controls
(iv) Damaging variants that alter the normal levels of a protein or its function
(v) Deleterious variants that reduce the reproductive fitness of carriers

They noted that unambiguous assignment of the level of pathogenicity of a variant was often impossible. They note that it is important to not assume that a variant is fully penetrant and that it explains all the features of the disease.

Key steps in assigning pathogenicity include deriving information about variants from public databases, including information on functional data; experimental data on model organisms could be considered. Statistical evidence about the distribution of the variant in patient and control populations should be considered. For monogenic disorders, the search for information regarding variants of the specific gene in other cases with the same disease is important.

They emphasized that in compiling statistics evidence about the relevance of variants for a disease the gene size and mutation rate should be taken into consideration.

5.2.3 *Mutation constrained genes*

Availability of genome sequencing data on large numbers of healthy individuals has revealed that genes vary in their mutation rate. The background mutation rate for nucleotides was calculated to be 1.6×10^{-16} per base (Samocha *et al.*, 2014). Examination of gene sequences and documentation of specific mutations in healthy individuals (e.g. the ExAC database) revealed that there are genes that have fewer nucleotide mutations than expected. These genes were labeled as "intolerant" genes by Petrovski *et al.* (2013) who developed intolerance metrics. Of particular relevance is the fact that there are genes that have fewer missense mutations and fewer loss of function mutations than would be expected based on the background mutation rate. These genes were labeled as "constrained" by Samocha *et al.* (2014) who identified a list that included approximately 5% of all genes in the human genome. They determined that mutations in the constrained genes occurred in cases of intellectual disability.

In a study of 151 families, Samocha *et al.* determined that loss of function (LOF) mutation per exome occurred with a frequency of 0.09 in control. In cases of intellectual disability (ID) they occurred with a frequency of 0.24. The significance value of the difference in LOF frequency in control and ID cases was 6.49×10^{-7}. Specific genes that were identified as significantly constrained in normal populations but were found to be mutated with higher than background frequency in cases of intellectual disability included SYNGAP1 (synaptic GTPase activating protein), SCN2A (voltage gated sodium channel) and STXBP1 (syntaxin 1 binding protein involved in neurotransmitter release).

Samocha *et al.* noted that another way to identify constrained genes is to compare the frequency in a specific gene of synonymous mutations (that do not change the amino acid) to the frequency of mutations that change the amino acid code (their ratio of synonymous to nonsynonymous sites deviated from the genome-wide average at $p < 0.001$). They identified 377 such genes and this list included genes that encode RYR2 (calcium channel component), KMT2A, KMT2D (lysine methyltransferase), and SYNGAP1.

The availability of the ExAC database with frequency of mutations in more than 60,000 control individuals will facilitate identification of constrained genes.

5.2.4 *Functional analyses of mutations*

As noted above functional studies on model organisms can be used to investigate pathogenicity of specific gene variants. Increasingly, cell systems, including studies of pluripotent stem cells and of tissue type specific cells, are used to investigate pathogenicity of variants. Chakravarti *et al.* (2013) proposed that the equivalent of Koch's postulates (designed for confirmation of pathogenicity of putative infectious agents) be applied to determination of functional significance of variants. They proposed the following postulates:

(i) Candidate gene variants are enriched in patients
(ii) Disruption of the gene in a model system gives rise to a model phenotype that is accepted as relevant and "equivalent" to the human phenotype
(iii) The model phenotype can be rescued with the wild-type human alleles
(iv) The model phenotype cannot be rescued with the mutant human alleles

5.2.5 *Exome sequencing leading to diagnosis and effective therapy*

There are growing numbers of reports in the literature of cases where exome sequencing led to diagnosis and then to successful treatment. One example includes the case reported by Petrovski *et al.* (2015). They described a case of a 20 month old female with a rapidly deteriorating neurological condition. The patient was initially diagnosed with a possible auto-immune condition; however her condition did not respond to treatment with steroids. Exome sequencing revealed that the patient had compound heterozygous deleterious mutations in the gene SLC25A2, a riboflavin transporter. She had inherited a rare SLC25A2 mutation from her father and a rare nonsense mutation in SLC25A2 from her mother. Based on reports in the literature of other cases with pathogenic mutations in SLC25A2 and patients' responses to riboflavin therapy, high dose riboflavin therapy was initiated in the patient described by Petrovski *et al.* and steroid therapy was discontinued. That therapy led to rapid and sustained improvement in the patient's condition.

5.3 Impact of Mutations in the Non-protein Coding Genome

As discussed above, interpretations of non-synonymous nucleotide variants in protein coding regions of the genome are facilitated by a number of different computational tools, including SIFT, PolyPhen-2, MutationTaster and GERP. In addition, bioinformatic resources are available that provide information on the frequency of specific nucleotide variants in the general population. Resources and databases are progressively being developed that document occurrence of specific mutations associated with medical conditions; e.g. ClinVar and GeneTests. However there are few resources currently available to interpret the significance of single nucleotide variants in non-coding (non-protein coding) regions of the genome.

The fact that nucleotide variants identified as disease risk variants in genome-wide association studies (GWASs) most frequently map in the non-coding regions of the genome has spurred effects to functionally assess these variants. There is growing evidence that GWASs identified disease-associated variants map in enhancer sequences and in DNAse I hypersensitivity regions of the genome, i.e. regions where chromatin has lost its compacted structure and is more accessible to transcription factors.

Zhang and Lupski (2015) reviewed functional annotations of non-coding single nucleotide variants (SNPs) mapped as disease associated through GWASs. They noted that a specific SNP rs7539120 was associated with a cardiac phenotype and abnormal

QT interval on electrocardiogram. This SNP maps in an enhancer element that impacts expression of the NOS1AP gene that encodes nitric oxide synthase adapter protein.

There is also evidence that a non-coding variant that interrupts a specific enhancer sequence leads to diminished BCLIIA expression and to increased expression of HBF (fetal hemoglobin) (Bauer and Orkin, 2015).

Zhang and Lupski (2015) emphasized that nucleotide variants in long non-coding RNAs (lncRNAs) and in microRNAs also play roles in disease. Variants in microRNAs MIR137 and MIR2682 have been implicated in schizophrenia. In addition variants in the 5' and 3' untranslated gene regions (UTR) also play roles in disease. The risk allele in SNP variants rs11603334 located near the promoter of the gene ARAP1 led to increased expression of ARAP1 that impacts cell specific trafficking. The increased expression of ARAP1 led to decreased expression of pro-insulin in pancreatic B cells and to susceptibility to type 2 diabetes.

Zhang and Lupski noted that there are examples of cases where nucleotide variants in the 3' UTR of a gene created a novel microRNA binding site that impacted transcript translation. The mutant allele at rs2266788 increases microRNA binding and downregulated APOA5 apolipoprotein expression and promoted hypertriglyceridemia.

They also emphasized the importance of non-coding quantitative trait loci in regulating gene expression and noted that computational programs for prediction of pathogenic non-coding variants are being developed. These include genome-wide annotation of variants (GWAVA) and Combined Annotation Dependent Depletion (CADD) that can be applied to short insertions and deletions.

5.3.1 *Copy number variants (CNVs) in non-coding regions*

Zhang and Lupski noted that CNVs that impact protein coding variants are usually considered to exert their effects through alterations of gene dosage. They noted that there are also CNVs in non-coding regions that apparently have functional effects. Examples include duplications or deletions either upstream or downstream of the SOX9 gene that lead to sex reversal. Deletion of a 380 kilobase region upstream of SOX9 occurred in a patient with male to female sex reversal. Duplication upstream of SOX9 led to female to male sex reversal.

Deletion of a genomic region on chromosome 16q24.1 that contained a lncRNA (long non-coding) gene led to a lung developmental disorder. A patient with the same phenotype was found to have deletions in the FOXF1 gene that maps in vicinity in 16q24.1 (Szafranski *et al.*, 2014).

Zhang and Lupski proposed that contributions from a CNV in a particular chromosome region, together with SNP variants in the same region on the homologous chromosome, may account for variable phenotypic features associated with specific CNVs.

5.4 Somatic Mutations and Mosaicism

Somatic mutations can arise due to DNA replication errors; they can also be due to errors in cell division and abnormal chromosome segregation. It is however important to consider that mutations can arise in non-replicating DNA. These may be due to activity of mobile elements in DNA; mutations can also occur during transcription. Mutations and DNA breakage can also arise due to the action of harmful endogenous factors including reactive oxygen species and may occur due to the action of harmful environmental factors, including ultra-violet action, radiation and harmful chemicals, including nicotine.

The majority of *de novo* germline mutations in children are due to replicative defects inherited from fathers. Shendure and Akey (2015) reported that *de novo* mutations found in autistic children were inherited with a 4:1 paternal:maternal ratio, and the *de novo* mutations in paternal genomes increase at the rate of 0.26% per paternal year. There is also a paternal bias age effect for the origin of *de novo* CNVs found in children.

Somatic mutations due to DNA replication errors occur in early development and lead, in some cases, to monogenic disorders. Shendure and Akey noted that some of these disorders are exclusively associated with somatic mutations.

Deleterious mutations reduce reproductive fitness. Shendure and Akey emphasized that it is useful to think of pathogenicity of mutations as a continuous property. Some mutations are highly penetrant and lead to problems while other mutations have limited effects. Mutations in protein coding genes are responsible for most monogenic disorders. However in many cases a specifically named monogenic Mendelian disorder may be due to mutations in more than one gene. However in each particular individual with a monogenic disorder only one gene is impacted. We are becoming increasingly aware of mutations that are pathogenic but have limited effects. In addition modifiers that impact the severity of Mendelian diseases are being identified.

In complex common disease, mutations are often present in regulatory regions of the genome. In addition each mutation has a modest effect.

Shendure and Akey reported that rates of mutation at CpG dinucleotides are 10 times higher than the average mutation rate due to deamination of cytosine to thymine. They also noted that non-synonymous mutations in protein coding genes involve changes in arginine and glycine in 30% of cases due to higher rates of mutation in arginine and glycine. The nucleotide codons for arginine are CGT, CGC, CGA, CGG, AGA, and AGG. The nucleotide codons for glycine are GGT, GGC, GGA, and GGG.

5.4.1 *Somatic mutations due to replicative or mitotic errors*

Campbell *et al.* (2015) noted that there is significant evidence that the occurrences of mutation events during mitosis are not uncommon. Following mutation in a specific cell, that cell may undergo apoptosis or may be retained. Mutant cells in low numbers may not impact functions. Mutations that occur early in development when there are relatively few cells may impact significant regions of the body. However mutations that occur later in life may not be noticed.

5.4.2 *Hemimegalencephaly (HME)*

HME is a condition associated with overgrowth of one hemisphere of the brain. Abnormalities of cell growth, cell morphology and synaptogenesis occur in the affected hemisphere. This condition is associated with severe epilepsy, intellectual disability and hemiparesis. In a number of cases, studies on tissue derived from the hemisphere with abnormal growth have revealed mutations in gene products that function in the PI3K-AKT pathway (phosphatidyl inositol kinase and AKT serine threonine kinase). In some cases mutations in mTOR occur and there is evidence of mTORC1 activation (Saxena and Sampson, 2014). The PI3K-AKT and mTOR pathways are related to cell growth and proliferation.

5.4.3 *Segmental overgrowth syndromes*

Somatic mutations in genes such as AKT1 lead to developmental defects in parts of the body other than the brain. In Proteus syndrome, segmental overgrowth may impact one limb and is associated with overgrowth, lipomatosis and arteriovenous malformations. Lindhurst *et al.* (2014) reported that sequencing of affected tissues in 29 patients revealed that 26 patients had somatic activating mutations in AKT1 at position c.406A; p.Glu17Lys. Lindhurst *et al.* (2012) also reported that somatic mutations in PIK3CA lead to overgrowth of specific tissues.

5.4.4 *Somatic mutations in specific genes that lead to abnormal vasculogenesis*

The Sturge–Weber syndrome is a neurocutaneous syndrome associated with capillary malformation that occurs commonly on the face as a port-wine stain and may also impact leptomeningeal brain membranes. The facial capillary malformation occurs in the distribution of the ophthalmological branch of the trigeminal nerve. There is evidence that the Sturge–Weber syndrome is due to mutations in the GNAQ gene that encodes a subunit of the guanine nucleotide binding signaling protein. Shirley *et al.* (2013) examined paired samples of blood and affected tissues from 50 persons with this syndrome. They identified a specific non-synonymous variant leading to a protein substitution GNAQ p.Arg183Gln in affected tissues of participants. Nakashima *et al.* (2014) identified this same mutation GNAQ p.Arg183Gln in affected tissue in 12 of 15 patients with Sturge–Weber syndrome.

5.4.5 *Somatic mutations in mature neurons that are non-replicative cells*

Lodato *et al.* (2015) reported on results of their studies on the landscape of single nucleotide variants (SNVs) in the human brain. They emphasized that most neuronal mutations are not caused by DNA replicative errors since following their terminal differentiation neurons do not divide. Lodato *et al.* proposed that post-replicative neuronal mutations are in part due to errors during transcription. There is growing evidence that R loops that arise during transcription are a threat to genome stability. RNA–DNA hybrids form during transcription and activity of RNA polymerase. An R loop is a 3-stranded structure composed of a RNA–DNA hybrid and a displaced non-template DNA strand. It is important that R loops be resolved to allow complementary DNA strands to re-anneal. Specific proteins play important roles in R loop resolution. These include topo-isomerases and helicases such as senataxin encoded by the SETX gene.

Mutations in neurons may also arise due to activity of oxygen free radicals, radiation, mutagens and due to retroviral activity and insertions.

Lodato *et al.* demonstrated that somatic mutations are abundant in brain neurons and that they likely modify the penetrance of germline mutations. This may contribute to phenotypic variability in different family members, even phenotypic variability in identical twins with the same specific germline mutations.

5.5 Germline Mutations that only Manifest in a Specific Tissue Following a Second Hit

In patients with germline mutation in the TSC1 or TSC2 genes in tuberous sclerosis there is evidence that second hit mutations commonly give rise to the hamartomas (tumors) that occur in that condition. Hamartomas are defined as benign masses composed of disorganized elements of the tissue in which they arise. Loss of heterozygosity in TSC1 or TSC2 genes is a common phenomenon in a range of hamartomas including angiomyolipomas of the kidney, rhabdomyomas of the heart, in brain lesions (including tubers and giant cell astrocytomas) and in angiofibromas of the face. The second hit occurs in the normal gene that is the homolog of the gene that carries the germline mutations (Green *et al.*, 1994; 1996; Crino *et al.*, 2010; Tyburczy *et al.*, 2014).

Cerebral cavernous malformations are vascular lesions characterized by enlarged capillary cavities. In patients with cerebral cavernous malformations (CCM), capillary malformations may also impact skin and eye. Riant *et al.* (2010) reported that CCM may occur sporadically or they may occur in families. In familial cases germ-line mutations in any one of three genes have been identified in different forms of cerebral cavernous hemangiomas. CCM type 1 (CCM1) involves defects in KRIT1; in CCM type 2 (CCM2) mutations occur in MGC4607; in CCM type 3 (CCM3) mutations occur in PDCD10. KRIT1 has an ankyrin repeat domain; MGC4607 is a scaffold protein important in the MAPK (mitogen-activated protein kinase) signaling pathway and PDCD10 is described as a programmed cell death protein. Riant *et al.* reported that affected patients were heterozygous for mutations in these genes. There is evidence that the vascular lesions arise due to second hit mutations in endothelial cells.

More than 150 distinct mutations have been identified in CCM1, CCM2 and CCM3 patients. The mutations were loss of function mutations. Akers *et al.* (2009) cloned cells from vascular lesions derived from patients with germline CCM1, CCM2 or CCM3 and they established that somatic mutations were present in the lesions. Somatic mutations occurred in patients in these genes in patients with sporadic disease. Their findings suggested that there is complete loss of function of the relevant CCM determining gene in the endothelial cells of the vascular lesions.

5.6 Phenotypic Variation due to Different Mutations in the Same Gene

5.6.1 *Mutations in the ARX gene*

Structural brain malformations of different types occur in cases with different mutations in ARX (aristaless related homeobox gene). In addition mutations at some sites in ARX are associated with functional brain impairment without evidence of structural abnormalities detectable on magnetic resonance imaging (MRI). The ARX gene is located on the X chromosome and mutations lead to pathology almost exclusively in males. However some female carriers of specific ARX mutations were found to have agenesis of the corpus callosum (Kato *et al.*, 2003).

ARX is a homeobox gene that encodes a transcription factor aristaless. The gene is expressed in the fetal and adult brain, in testes, muscle and pancreas. The gene has five exons and four alternative polyadenylation sites. The ARX protein has an octapeptide domain, a DNA binding domain, a homeodomain and an aristaless domain. The octapeptide domain encoded by exon 1 has nuclear localization sequences. The homeodomain is encoded by exons 2, 3, and 4; the aristaless domain is encoded by exon 5 (Olivetti and Noebels, 2012). ARX is expressed during development in proliferative zones in the brain and it is expressed in interneurons during development and in adult life. There is evidence that ARX plays roles in radial and tangential neuronal migration and in Gabaergic interneurons.

The spectrum of clinical phenotypes due to the ARX mutations includes X linked intellectual disability in males, and seizures. Seizure types include seizures with a defined electroencephalogram (EEG) pattern referred to as Ohtahara syndrome, seizures of the infantile spasm type, tonic–clonic seizures, and episodes of dystonia with abnormal limb posturing. Abnormalities outside the central nervous system that sometimes occur include ambiguous genitalia.

The range of structural brain abnormalities found included lissencephaly (smooth brain with diminished numbers of sulci and gyri), hydrancephaly, agenesis of the corpus callosum, polymicrogyria and periventricular nodular heterotopia.

Shoubridge *et al.* (2010) reported data on the location of ARX mutations and phenotype in 97 cases. Mutations in a specific exonic polyalanine tract, the second polyalanine tract ARX c.429-452, occurred in 40% of their cases. Shoubridge *et al.*, noted that interfamilial and intra-familial variation occurred in patients with this specific mutation. Most of these cases presented with moderate to severe non-syndromic intellectual disabilities. However six of the patients with this mutation had intellectual disabilities and dystonic movements of the hands and speech difficulties. Patients in three families had seizures of the infantile spasm type. Other mutations in this ARX second polyalanine tract have also been described (e.g. ARX c.430–465) and these patients presented with infantile spasms.

Shoubridge *et al.* reported that brain malformations occurred particularly in cases with complete loss of function mutations including nonsense mutation and splice site mutations. Missense mutations in the homeodomain region also led to malformation syndromes. Seven of nine missense mutations in the homeodomain were associated with the syndrome X-linked lissencephaly and abnormal genitalia. The ARX protein homeodomain is a DNA binding region.

They also reported that in families where ARX mutations lead to severe malformation in the male patients, some carrier females manifested mild learning disabilities.

5.6.2 *Other examples of variations in phenotype depending on type of mutations present in a specific gene*

Genes in which loss of function mutations lead to widespread systemic effects and/or specific syndromes sometimes also have hypomorphic mutations that lead to more limited effects. These limited effects often lead to neurological or neurobehavioral deficits. Hu *et al.* (2014) postulated that these results likely reflect the increased sensitivity of the central nervous system to mild perturbations in protein levels or functions. Examples of variations in phenotype depending on type of mutations presented by Hu *et al.* include the spectrum of findings associated with mutations in PEX7 (peroxisomal biogenesis factor 7), VPS13B (vacuolar protein sorting 13B), ARX and AMT (aminomethyltransferase).

5.7 Diverse Phenotypes that Result from Mutations in Different Members of the Same Gene Family

5.7.1 *Tubulinopathies*

The spectrum of brain defects due to mutations in the tubulin genes are referred to as tubulinopathies. Alpha and beta tubulins in human are encoded by a series of genes and mutations in several of these genes are associated with structural brain defects, these include TUBA1A, TUBA8, TUBB2A, TUBB4A, TUBB2B, TUBB3 and TUBB5. Tubulin proteins are components of microtubules. Microtubules are composed of heterodimers of tubulin and in addition they contain a number of microtubule associated proteins.

5.7.2 *Structure of tubulins*

Romaniello *et al.* (2015) reviewed the structure of tubulins and reported that the tubulins encoded by different genes are highly homologous and have three distinct domains, including an N-terminal GTP binding domain, an intermediate domain and a C-terminal domain that binds the microtubule associated proteins. Different isotypes of microtubules form and these are composed of different combinations of tubulin proteins. It is likely that different microtubule isoforms perform different functions. The centrosome serves as the microtubule nucleating center and is critical to the formation of the mitotic spindle. Higginbotham and Gleeson (2007) reported that the centrosome position within differentiating neurons is important in determining neuronal migration, cell positioning and cortical layering.

Bahi-Buisson *et al.* (2014) reported that the phenotypes that result from mutations in TUBA1A, TUBB2B, TUBB3 and TUBB5 tend to overlap; however there are some phenotypic manifestations that are more prominent in patients with mutations in a specific tubulin gene. They noted further that disease-causing mutations in these genes were heterozygous and predominantly sporadic in origin (i.e. not inherited).

Bahi-Buisson *et al.* undertook mutation analyses and genotype phenotype correlations and detailed brain magnetic resonance imaging analyses in 74 tubulinopathy cases. TUBA1A protein mutations were found in 45 of the 78 patients. Three hotspot mutation regions were found and these all involved changes to arginine residues, Arg264, Arg402 and Arg 422.

Mutations in TUBB2B proteins were identified in 12 cases and in 6 fetuses. Three recurrent TUBB2B mutations occurred: p.I202T, p.S172P. p.F265L. TUBB3 mutations were found in eight cases.

Analyses of MRI brain images in the patients in the study revealed major malformations: microlissencephaly in 12 cases, agyria-pachygyria in 19 cases, central pachygyria with diffuse polymicrogyria in 6 cases, and simple gyral pattern with focal polymicrogyria in 19 cases. Key structural brain abnormalities that occurred in 75% of cases were dysmorphism or unusual orientation of the basal ganglia. Corpus callosum abnormalities, including complete agenesis or hypoplasia occurred in 40% of cases.

Bahi-Buisson *et al.* noted that different individuals with the same mutation at a specific site in a particular tubulin protein have similar phenotypes. Different patients with the R264C mutation in the TUBA1A protein had central pachygyria while cases with the R402F mutation in TUBA1A had classic lissencephaly. They also determined that the specific amino acid change at a specific site influenced the phenotype, patients with the TUBA1A p.R264C mutation had central pachygyria and patients with the TUBA1A p.R264H patients had microlissencephaly.

They noted that the impact of specific mutations on the tertiary structure of tubulin depended on the mutation position within the protein. N-terminal domain mutations impacted the GTP binding pocket while C-terminal domain mutations impacted interactions with microtubule associated proteins such as kinesin and dynein.

Bahi-Buisson *et al.* noted that there are published reports of cases with TUBB3 mutations that have ophthalmological and peripheral nerve pathologies.

Structural brain malformations similar to those that occur in tubulinopathies arise due to defects in genes that encode molecular motor proteins that promote microtubule associated movement. These include mutations in dyneins e.g. DYN1H1 and kinesins KIF5C, motor proteins that convert ATP contained energy to mechanical energy.

5.8 Clinical and Metabolic Phenotypes due to Mutations in Different Components of a Multi-subunit Complex

5.8.1 *Non-ketotic hyperglycinemia classical and variant forms*

Non-ketotic hyperglycinemia results from defects in the glycine cleavage pathway. The glycine cleavage system is composed of four proteins: P protein pyridoxal phosphate dependent glycine decarboxylase; H protein lipoic acid hydrogen carrier protein; L protein lipoamide dehydrogenase; T protein tetrahydrofolate dependent aminomethyltransferase.

The classical form of hyperglycinemia is associated with encephalopathy, hypotonia intractable seizures and possible early death. Dinopoulos *et al.* (2005) reported that there are non-classical forms of hyperglycinemia including infantile and juvenile forms that lead to intellectual impairment and behavioral abnormalities. In addition later onset forms occur that may manifest with mild cognitive impairment, behavioral problems, peripheral neuropathies or spinocerebellar manifestations.

Patients with classical non-ketotic hyperglycinemia have deficiency of P protein (GLDC gene defects) in 80% of cases and T protein defects (AMT gene mutations) in 15% of cases. Dinopoulos et al., reported that most patients with non-ketotic hyperglycinemia have compound heterozygous mutations. They reported that a few patients with non-classical forms of non-ketotic hyperglycinemia have AMT gene mutations; other patients with atypical forms have missense mutations in GLDC that encodes the P protein. The missense mutations in the patients with atypical forms tended to be associated with higher levels of the corresponding enzymes than were found in patients with the classical form. Dinopoulos *et al.* concluded that additional studies were required to define the cause of late onset adult forms of non-ketotic hyperglycinemia.

5.9 Similarities and Differences in Phenotypes in Patients with Mutations in Different Proteins that Function in a Specific Biochemical Pathway

5.9.1 *Monoamine neurotransmitter disorders and tetrahydrobiopterin*

A range of neurological symptoms occur in disorders due to defects in the pathway involved in the synthesis of monoamine neurotrans-mitters (Kurian *et al.*, 2011). The monoamines serotonin, dopamine and epinephrine all require tetrahydrobiopterin (BH4) for their synthesis. Dopamine norepinephrine and epinephrine share origins from tyrosine and the reactions that convert tyrosine to L-dopa. Furthermore conversion of L-dopa to dopamine and conversion of the tryptophan derivative 5-hydroxytrophan to serotonin are both dependent on the same enzyme and cofactor. This enzyme is aromatic L-amino hydroxylase (AADC) and its cofactor is pyridoxal phosphate (PLP). Dopamine is converted to norepinephrine through the activity of dopamine beta hydroxylase. Norepinephrine generates epinephrine through activity of phenylethanolamine-N-methyl transferase. The pathways for synthesis of serotonin, dopamine and noradrenaline are illustrated in Figure 5.1.

Indirectly related to the generation of neurotransmitter amines is the reaction which converts phenylalanine to tyrosine. This reaction also utilizes tetrahydrobiopterin (BH4).

5.9.2 *Tetrahydrobiopterin (BH4) synthesis*

The first step in BH4 synthesis involves conversion of guanosine-5-triphosphate to dihydroneopterin triphosphate through activity of the enzyme GTP cyclohydrolase. Dihydroneopterin is then converted

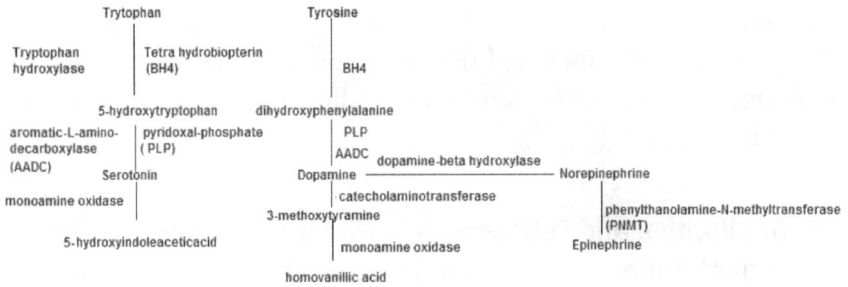

Figure 5.1. Neurotransmitter biosynthesis.

to 6-pyruvoyl-tetrahydrobiopterin through activity of 6-pyruvoyltet-rahydrobiopterin synthase. 6-pyruvoyl-tetrahydrobiopterin is then converted to sepiapterin through the enzyme aldose reductase. The final step in biopterin synthesis involves the conversion of sepi-apterin to BH4 through the activity of sepiapterin reductase.

BH4 can also be derived through a regeneration pathway also referred to as the pterin reductase pathway. In this pathway sepi-apterin can be stored and later converted to 7-8-dihydrobiopterin and this can then be modified through dihydrofolate reductase (DHFR) to generate tetrahydrobiopterin.

Werner *et al.* (2011) reported that enzymes that use BH4 as cofactor function are mixed function mono-oxygenases; one oxygen atom is incorporated into the substrate to yield a product and one oxygen atom is reduced to H_2O.

5.9.3 *Monoamine neurotransmitters breakdown*

Breakdown of the four monoamines serotonin, dopamine, norepi-nephrine and epinephrine all require monoamine oxidase. In addi-tion dopamine, norepinephrine and epinephrine require catechol-O-methyltransferase (COMT) and aldehyde dehydrogenase for breakdown. Monoamine breakdown products in urine can be quantified and assessed to facilitate diagnosis of specific monoamine defects. Assays of monoamine neurotransmitter levels in CSF are sometimes carried to establish diagnosis.

5.9.4 *Monoamine neurotransmitter tissue distribution and functions*

Dopamine is present at high concentrations in the human striatum. Dopamine neurons project to the midbrain areas and have connections to the nucleus accumbens, hippocampus and to corticolimbic structures and also to the pituitary. Dopamine impacts a number of functions including cognitive processes, emotions, locomotion and neuroendocrine functions (Kurian *et al.*, 2011).

Serotoninergic neurons occur in the midbrain raphe nuclei and have widespread projections to the supratentorial cortical areas including motor cortex, cerebellum and spinal cord. Serotonin plays key roles in motor control and it also impacts mood and specific autonomic functions. The norepinephrine and serotonin systems impact emotion and stress response. High affinity transporters for norepinephrine and serotonin occur in brain and ensure clearance of these substances from synapses.

Kurian *et al.* noted that monoamine neurotransmitters play important roles in brain development.

5.9.5 *Clinical manifestations of monoamine neurotransmitter disorders*

In children, dystonia and abnormal posturing of limbs is an early and important manifestation of deficiency of GTP cyclohydrolase (GTPCH), the product of the GCH1 gene and the first enzyme in the biopterin biosynthesis pathway. Heterozygous mutations in GCH1 lead to GTPCH deficiency. Kurian *et al.* reported that more than 100 different mutations have been described in GTPCH deficiency. They emphasized that early correct diagnosis is important since excellent clinical response can be achieved with levodopa and cardidopa. Supplementary therapy with tetrahydrobiopterin is also sometimes beneficial.

Deficiency of sepiapterin reductase that converts sepiapterin to BH4 also present with dystonia. Patients may also manifest with choreoathetosis, spastic ataxia, tremor, irritability, psychomotor retardation and oculogyric crises prolonged involuntary upward deviation of the eyes. Deficiency of 6-pyruvolyltetrahydrobiopterin synthase (PTPS) and deficiency dihydropeteridine reductase (DHPR) may present with similar symptoms.

Oculogyric crises (dystonia associated with prolonged upward deviation of the eyes), choreoathetosis and dystonia are also features of aromatic-L-amino-decarboxylase (AADC). Kurian *et al.* noted that patients with this deficiency may sometimes present with metabolic encephalopathy.

Patients with loss of function homozygous or heterozygous mutations in the SLC6A3 dopamine transporter protein manifest with dystonia, chorea, hyperkinesia and Parkinson-like tremor. They may also manifest with eye movement disorders and oculogyric crises.

5.9.6 *GTP cyclohydrolase variants in families with adult onset Parkinson's disease*

Mencacci *et al.* (2014) reported that detailed family history in the case of a child with dystonia due to GTP cyclohydrolase deficiency revealed that the father was diagnosed with dystonia at the age of 42 years and that the grandfather had developed Parkinson's manifestation at 59 years of age. All three affected individuals were heterozygous for a splice site mutation c.343+5 G>C. in GCH1. Transcript analyses revealed that this mutations led to intron retention in a proportion of the transcripts. Premature stop codons were present in the retained intron.

Mencacci *et al.* identified GCH1 gene mutations in several families where dystonia was present in a child and where an older relative manifested signs of late onset dystonia or symptoms of Parkinson's disease. Based on these findings they undertook GCH1 sequencing in a large cohort of 1,318 patients with early onset Parkinson's disease and 1,635 controls. Heterozygous damaging mutations were found in 10 out of 1,318 Parkinson's patients and in 1 out of 1,645 controls. They determined that in most cases with GCH1 mutations that manifested Parkinson's disease had nigro-striatal dopamine deficiency.

Selective serotonin deficiency disorders have been identified in patients with adult onset dystonia.

5.10 Example of a Clinical Syndrome with a Specific Name due to Defects in Different Genes that Encode Components of a Specific Structure e.g. Cilia

Joubert syndrome is an example of a specifically named syndrome due to defects in any one of 27 proteins that are important in the function of cilia. Bachman-Gagescu *et al.* (2015) reported results of clinical studies, targeted gene capture and next generation sequencing carried out on 532 individuals with Joubert syndrome. All patients studied had the core manifestations of Joubert syndrome. These include hypotonia, ataxia, cognitive dysfunction, oculomotor apraxia, abnormal breathing and molar tooth phenomenon on brain MRI. The molar tooth phenomenon describes thickened horizontally oriented cerebellar peduncles and a deep interpeduncular fossa.

Bachman-Gagescu *et al.* noted that subsets of the individuals with core Joubert syndrome manifestation had additional clinical findings. The range of additional manifestations included chorioretinal colobomas, retinal dystrophy, kidney defects, liver defects, and skeletal defects including polydactyly. They noted that the care of these individuals is complex and that multiple medical specialists are often involved.

In their study of Joubert syndrome patients, Bachman-Gagescu *et al.* identified 253 different mutations in 23 genes. It is important to note that in 38% of Joubert syndrome patients they did not identify gene mutations. They carried out detailed studies to determine the frequencies with which different genes were mutated and also the incidence of additional phenotypic manifestation in patients with mutations in a specific gene.

Mutations in five different genes each accounted for between 6 and 9% of all Joubert syndrome cases. These genes and the protein functions included C5ORF42 (transmembrane protein), CC2D2A (coiled-coiled domain), AHI (required for cerebral and cerebellar development), CEP290 (centrosomal protein), and TMEM67 (transmembrane protein). Three genes each accounted for approximately 3% of cases: CSPP1 (centrosome and spindle protein), TMEM216 (transmembrane protein), and INPP5E (inositol polyphosphate phosphatase). When they examined the additional features in patients, Bachman-Gagescu *et al.* determined that retinal abnormalities were significantly more common in patients with CEP290 mutations. Colobomas and liver abnormalities were significantly more common in patients with TMEM67 abnormalities.

This study illustrates extreme genetic and phenotypic heterogeneity in a specifically named syndrome with defined core features.

5.11 Examples of Clinical Syndromes each with a Specific Name and each due to Defects in Multiple Different Genes

5.11.1 *Charcot-Marie-Tooth (CMT) disease*

The basic constellation of clinical manifestations of CMT disease include progressive muscle weakness of distal limb muscles and atrophy of those muscles, depressed tendon reflexes and distal sensory loss (Rossor *et al.*, 2013). This constellation of manifestations can result from mutation in any one of at least 60 different genes. The basic constellation of symptoms may be inherited as autosomal dominant, autosomal recessive or as X linked traits. There are additional manifestations that can occur; these include hearing loss, ophthalmoplegia, optic atrophy, cataracts, facial defects, vocal cord paralysis, tongue defects including fungiform papillae, kidney defects including glomerular sclerosis, anhydrosis and autonomic dysfunction. Other names associated with specific CMT forms include hereditary motor neuropathy (HMN) and hereditary sensory neuropathy (HSN).

Functional defects occur in gene products that function in the axon, in neuronal cell bodies and in Schwann cells. In addition affected proteins may function in different cellular structures, including mitochondria, cell membranes, channels, endosomal sorting systems, endoplasmic reticulum, Golgi, proteasome, cytoskeleton and nuclear membranes. Other specific functions that may be impacted include cell signaling, RNA processing, myelin synthesis and synaptic transmission.

5.11.2 *Rett (RTT) syndrome and Rett-like syndrome*

The diagnosis of RTT syndrome is based on clinical features (Neul *et al.*, 2010). The key features include initial normal development and onset of symptoms around one year of age. These symptoms include impaired walking, loss of purposeful hand movements. Patients may also start to manifest social withdrawal; however eye-based contact continues. Later manifestations include stereotypic hand movements (hand wringing, squeezing, tapping, mouthing), loss of speech, sleep disturbances and breathing abnormalities. In classical RTT syndrome there is often a period of regression followed by a period of stabilization.

Suter *et al.* (2015) reported that in the majority of RTT cases, mutations occur in the MECP2 gene (methyl CpG binding protein 2). These investigators noted that more than two hundred different MECP2 mutations have been documented in RTT syndrome patients. In 2010 Neul *et al.* reported that in 70% of cases the mutations were sporadic (not inherited). They noted that C to T mutation occurred at eight specific sites. Four of these changes led to missense mutations (p. R016W, p.R133C, p.R158M and p.R306C). Four other C to T changes led to nonsense mutations (p.R168X, p.R255X, p.R270X and p.R294X).

Other common MECP2 mutations in RTT syndrome patients include small insertions or deletions in the C-terminal domain leading to frame shifts or to truncations. However there is evidence that at least 5% of patients with typical RTT syndrome do not have MECP2 mutations (Suter *et al.*, 2015). These investigators also described patients who did not meet diagnostic criteria for RTT syndrome or for atypical RTT syndrome but had MECP2 mutations. They described four such patients. Two of the four female patients met diagnostic criteria for autism; one patient met diagnostic criteria for PDD NOS (pervasive developmental disorder non-specific); one patient met diagnostic criteria for attention deficit hyper-activity disorder and obsessive compulsive disorder. One of these four patients had a p.R306H mutation that also occurs in approximately 9% of typical RTT syndrome cases. Another of these four patients had a MECP2 C-terminal truncation c. 1164-1207. MECP2 mutation chromosome inactivation studies on blood samples from each of the four patients did not reveal evidence of increased inactivation of the chromosome that carried the MECP2 mutation.

It is interesting to consider the possibility that the relative ratio of inactivation of the MECP2 mutation carrying X chromosome and the normal X chromosome may be different in brain than in peripheral blood samples.

5.11.3 *Molecular mechanisms in RTT syndrome with MECP2 mutations*

Pohodich and Zoghbi (2015) emphasized that the clinical picture produced by MECP2 mutations can be influenced by favorable X inactivation patterns and by specific MECP2 mutations. They noted that disease manifestations and neurological manifestations occurred in patients with reduced levels of MECP2 protein and in patients with excess levels of MECP2 protein that occur in patients with MECP2 gene duplications. These findings indicate that an appropriate level of MECP2 protein is required for normal neurological function.

Pohodich and Zoghbi reported that brain size is reduced in RTT syndrome and that neurons are smaller and densely packed, and dendritic complexity is reduced; however gross brain malformations do not occur. MECP2 is widely expressed in the body however expression is particularly high in the brain. These investigators noted that the early normal development in RTT syndrome patients is puzzling.

There is evidence that MECP2 protein binds to methylated CpG (cytosine-guanine) nucleotides, it also binds to methylated CpH dinucleotides, mCH (H=A, C or T), and to hydroxymethylcytosine hMC. Pohodich and Zoghbi noted that loss of MECP2 has effects on chromatin beyond its effects on methylated DNA; it is able to bind to and impact A-T (adenine thymidine) rich DNA and to impact the binding of proteins to chromatin. However binding of MECP2 to

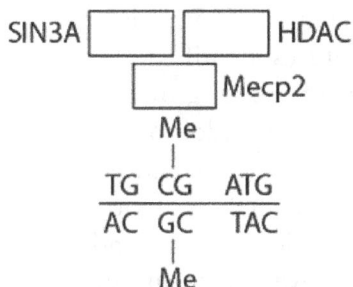

SIN3A [ ] [ ] HDAC

[ ] Mecp2

Me
|
TG CG ATG
AC GC TAC
|
Me

MECP2 binds to methylated CpG dinucleotides
in DNA. Following this
HDAC and other chromatin modifiers are
recruited by DNA bound MECP2

Figure 5.2. MECP2 binding to DNA and recruitment of chromatin modifiers.

chromatin likely follows its binding to methylated DNA via the MECP2 methyl binding domain (MBD). Figure 5.2 illustrates the binding of MECP2 to CpG dinucleotide.

These authors noted that decreased dendritic arborizations that results from MECP2 mutations likely leads to alterations in neuronal circuits. In addition there is evidence that neurotransmitter profiles are altered in MECP2 deficient patients. They noted further that the electroencephalogram (EEG) profiles are altered in these patients and they manifest cortical hyper-excitability and decreased inhibitory synaptic control. Network hyper-excitability likely plays roles in the increased seizure incidence in RTT syndrome patients and in the abnormal breathing that occurs in these patients. Increased excitability occurred in the nucleus tractus solarius of the brain stem in MECP2-deficient mice. Studies in mouse models of RTT syndrome revealed imbalance between neuronal excitability and inhibition.

5.11.4 *Other genes implicated in RTT syndrome and atypical RTT syndrome*

Bahi-Buisson *et al.* (2008) reported that some patients with mutations in the X-linked cyclin dependent kinase like 5 gene (CDKL5) have seizure disorders characterized by infantile spasms while other patients with mutations in CDKL5 have features of RTT syndrome. The RTT syndrome features identified in CDKL5 mutation patients included deceleration of head growth, hand stereotypies, and impaired walking abilities. In addition 17 of the 20 patients they identified with CDKL5 mutations had autistic features and all 20 patients manifested intellectual disabilities. Brain MRI studies in these patients revealed cortical atrophy and in 65% of patients white matter hyper-intensities were present. Eighteen different CDKL5 gene mutations were identified in the 20 patients.

Copy number variation and mutations in the Forkhead box G1 transcription factor gene (FOXG1) have been described in patients with neurodevelopmental disorders. Microdeletions in FOXG1 were identified in patients with RTT syndrome features. Allou *et al.* (2012) identified patients with severe Rett-like neurodevelopmental disorders and FOXG1 point mutations. The mutations they identified in cases clinically diagnosed as RTT syndrome included indels that led to frame shifts c.256dupC, and C.460dupG. These patients did not have mutations in MECP2 or CDKL5.

A SHANK3 (SH3 and multiple ankyrin repeat domains 3) mutation and a Rett-like syndrome occurred in a female patient described by Hara *et al.* (2013). This patient had normal early development but delayed walking and speech development. She began to regress between two and three years of age and lost verbal and motor skills. She developed stereotypic hand movements and exhibited autistic behaviors. Mutations were not detected in the MECP2, CDKL5 or FOXG1 genes. Exome sequencing revealed a frameshift mutation in the SHANK3 gene. Hara *et al.* noted that SHANK3 is one of the most CG rich genes in the genome; it has a CpG island in exon 21. They noted further that there is evidence that MECP2 protein binds to the CpG in SHANK3.

5.12 Identification of Genetic Factors that Modulate Phenotype in a Monogenic Disorder

Huntington's disease (HD) is a dominantly inherited Mendelian neurodegenerative disorder due to expansion of a CAG nucleotide repeat in the huntingtin gene HTT. Manifestations of the disease include motor impairments, movement disorders and cognitive difficulties that may include psychiatric manifestations. The age of onset of symptoms is largely, but not entirely, dependent on the degree of expansion of the CAG repeat. The Genetic Modifiers of Huntington's Disease Consortium (GeM-HD consortium, 2015) reported results of studies carried out in large cohorts of HD patients. They determined that the age of onset of symptoms was not entirely dependent on the length of the CAG repeat. They identified a phenotypic characteristic defined as the residual age of onset, determined as the difference in years between observed age of onset and that expected based on the length of the CAG repeat determined in prior studies.

They then carried out genome wide association studies in which the single nucleotide polymorphism (SNP) minor allele frequencies were determined in individuals whose age of HD onset was at the 20% extremes respectively, representing earlier or later than expected age of onset. These GWASs were carried out in three separate groups of HD patients.

Genome wide significance of association was achieved for three loci, two loci on chromosome 15 and one locus on chromosome 8. On chromosome 15 SNP rs 146353869 at 31,126,401 base pairs (bp) achieved significance $p = 4.3 \times 10^{-20}$ and snp rs 2140734 at 31,243,792 bp achieved significance $p = 7.2 \times 10^{-14}$. The chromosome 8 locus that achieved significance was rs1037699 at position 103,250,930bp, $p = 2.7 \times 10^{-8}$.

The closest genes to the GWAS signals on chromosome 15 include MTMR10 (myotubularin related protein 10) and FAN1 (Fanconi associated nuclease 1); several long non-coding RNAs, lincRNAs, also map in the region. Genes close to the chromosome 8 associated signal include RRM2B (p53 inducible ribonucleotide reductase) and UBR5 (ubiquitin ligase); the region also included micro RNA encoding loci. Top associated loci that did not achieve gene wide significance levels but were significantly associated at the $p = 2.2 \times 10^{-7}$ level included a locus on chromosome 3 in the MLH1 mismatch repair gene.

GeM-HD consortium (2015) then carried out pathway analyses of GWASs associated markers in HD patients using three different programs that included SETscreen, ALIGATOR and GSEA (gene set enrichment analysis). The pathways that reached highest levels of significance with all three programs included DNA repair pathways, mitochondrial fission and oxidoreductase activity. The authors noted that FAN1 is a member of the nucleotide excision repair pathway. They noted that the proteins encoded by loci in chromosome 15 associated regions FAN1 and MTMR10 have functions previously implicated in HD pathogenesis. FAN1 is involved in structure specific DNA handling, repair of DNA interstrand complexes. MTMR10 and other myotubularin act on phosphatidyl inositol phosphates.

Products of genes closest to the chromosome GWAS signal RRM2B and UBR5 impact mitochondrial energetics, oxidative stress and proteastasis. Products of RRM2B also impact DNA synthesis and repair.

The GeM-HD consortium concluded that the study of genetic modifiers of HD identified DNA handling as an important pathway impacted in HD and that manipulation of this pathway may provide therapeutic benefit.

The authors also emphasized that similar study methods might open the way to identification of modifiers in other Mendelian diseases.

5.13 Conclusion

It is likely that as information accumulates, having patients' fully sequenced genome information available will enable more accurate diagnoses and prognosis determinations and will facilitate design of therapies.

6. Immune System §

"Immunology is a major force in medicine. It is needed to understand how prevalent diseases come about and how to develop preventions and treatments."

R.M. Steinman and J. Banchereau, 2007

Advances in knowledge about immunological mechanisms and the application of biotechnological advances over the past two decades have had considerable impact on diagnosis and therapy of a number of different diseases, including infectious disease, immune-deficiencies, auto-immune diseases and cancer. Advances in cancer immunology were selected as the breakthrough of the year by Science Magazine in 2013.

In this chapter aspects of immunological mechanisms and advances in immunology will be reviewed.

6.1 Innate System

Cells of the innate immune system include dendritic cells, macrophages, granulocytes, monocytes and innate lymphoid cells. There is collaboration between the innate and adaptive immune systems and with non-hematopoietic cells. Together the systems promote immunity, inflammatory response, repair and homeostasis (Artis and Spits, 2015).

Innate lymphoid cells have lymphocyte morphology but do not have the cell-surface markers that are characteristic of adaptive system lymphoid cells. The common lymphoid progenitor (CLP) cells give rise to sub-populations which include the innate lymphoid cells and the natural killer (NK) cells.

The innate system lymphoid cells include cytotoxic NK cells that produce specific interleukins (IL12, IL15m IL18), perforin and granzyme. Non-cytotoxic innate lymphoid cells are separated into three classes (ILC1, 2 and 3) and LTi cells (lymphoid tissue inducer cells). ILC1 cells produce gamma interferon and tumor necrosis factor. Artis and Spits reported that the activities of these cells are directed against intra-cellular microorganisms. Cells in the ILC2 group produce cytokines and interleukins IL4, 5, 9, 13 and amphiregulin. ILC2 cells are active in the inflammatory response, in the allergic response and in tissue repair.

Artis and Spits reported that NK cells are part of the innate immune system and are involved in killing tumor cells that have lost expression of class I major histocompatibility complex (MHC) antigens.

6.2 Dendritic Cells and Pattern Receptors

Ralph M. Steinman and coworkers described dendritic cells and delineated their functions over several decades, and Steinman's work was acknowledged when he received the Nobel Prize in 2011. He established that dendritic cells control a spectrum of innate and adaptive immune responses.

Steinman and Banchereau (2007) emphasized that response to disease is determined not only by antigens and lymphocytes but also by dendritic cells that control antigen presentations. They reported that dendritic cells are abundant on the body's surface, including skin and mucosal surfaces. They are also present in intact lymphoid tissues. Dendritic cells have specialized endocytic systems. Their key function is to capture and process antigens and to convert proteins to peptides that link to molecules of the MHC. The linked molecules act as antigens that are then recognized by T lymphocytes.

Specific T cells stimulated by antigens then undergo clonal proliferation. Antigen binding to the cells may stimulate specific T helper cells to produce cytokines such as interferon or to produce interleukins. Presentation of MHC bound antigens may also stimulate proliferation of killer T cells. Activated T cells then mobilize other cells of the innate immune system.

Steinman and Banchereau reported that dendritic cells are also present in tumors and that they can be utilized in cancer prevention and treatment since oncogenic viral antigens and tumor antigens presented to dendritic cells are immunogenic. They also emphasized that immune attack on cancer cells can be broad and can also encompass NK cells and alpha beta T cells and gamma delta T cells. They noted that dendritic cells loaded with tumor antigens could be injected or could be targeted specifically to lymphoid tissue.

Dendritic cells also play roles in immune tolerance. Dendritic cells drive differentiation of T regulatory (Treg) cells that express FOXP3 (forkhead box P3), a transcription factor. The breakdown of dendritic cell mediated tolerance can lead to disease. Steinman and Banchereau reported that over-production of a specific cytokine tumor necrosis factor alpha (TNFalpha) by subsets of dendritic cells can induce auto-immunity. In specific auto-immune diseases e.g. lupus erythromatosis, subsets of dendritic cells over-produce type 1 interferon and/or interleukin 23.

Cellular nucleoproteins may activate specific receptors (TCR7 and TCR9) on dendritic cells, and stimulate interferon production.

6.2.1 *Pattern receptors*

Bruce A. Beutler also received the 2011 Nobel Prize in Medicine for his work in innate immunity. Beutler and coworkers (2006) established that cells of the innate immune system expressed evolutionarily conserved receptors encoded in the germline and that these receptors, referred to as pattern recognition receptors, recognized pathogen derived ligands. The downstream effects of ligand recognition by these receptors is activation of the immune response. Wagner (2012) noted that Toll receptors were the first recognized pattern receptors; subsequently other pattern receptors were discovered.

Toll receptors are pattern recognition receptors located on specific cells in the innate immune system, including macrophages, dendritic cells and monocytes. There are at least 13 different Toll receptors. Toll receptors have extra-cellular, transmembrane and cytoplasmic domains. They react to a variety of different antigens, including lipoproteins, lipopolysaccharides, heparan sulfate, hyaluronic acid, RNA and DNA. Activation of these receptors leads primarily to cytokine production.

In addition to reacting to foreign organisms, Toll receptors react to damaged protein derived from the host. These responses may facilitate repair of damaged tissue but may also stimulate associated immune responses.

6.3 Adaptive Immune System

The adaptive system creates immunological memory and it includes humoral and cell mediated responses. The humoral response involves B cells, and the cellular adaptive response involves T cells. In this system there is high specificity to specific antigens and recognition of non-self-antigens. The immune response involves generation of specific receptors. Key features of receptor formation involve DNA rearrangements, recombination and somatic hyper-mutations that are then passed on to the progeny of a specific cell.

Three types of cells are defined: Naïve B and T cells that have had no antigen contact, effector cells that are antigen activated and memory cells.

Nucleated cells can act as antigen presenting cells when specific antigens are bound to molecules derived from the MHC. In addition co-stimulatory molecules may be present.

Dendritic cells in the innate immune system are frequently involved in ingesting pathogens and transporting these to lymph nodes.

6.4 T cells, Receptors, Antigen Recognition and MHC Region

6.4.1 *Thymus and T cell differentiation*

Hematopoietic cells from bone marrow seed the thymus. Differentiation to T cells occurs there and this differentiation includes rearrangement of T cell receptors. Selection then occurs and this includes apoptosis of T cells that develop receptors with high affinity to self-peptides (Labrecque *et al.*, 2011).

T cell receptors are keys to the function of T cells that respond to antigens coupled to MHC receptors and presented by antigen presenting cells (illustrated in Figure 6.1). In lymphoid organs dendritic cells are antigen presenting cells. T cell receptors differ based on gene rearrangements that take place during development in the thymus. Broere *et al.* (2011) noted that mature naïve T cells are present in lymphoid organs and in blood. Due to the great diversity of the receptors created, the T cell receptors have affinity for a range of different antigenic peptides. Binding of an antigenic peptide to a T cell receptor leads to activation of that specific T cell and clonal expansion. Binding of antigen to receptors on cytotoxic CD8 positive T cells leads to the production of cytokines that can lead to lysis of cells that carry that antigen. CD4 positive T cells are classified as helper T cells that act to promote cytokine production. CD4 positive helper cells may also facilitate antibody production by B cells.

CD8 T cells recognize antigens bound to molecules of MHC class 1. CD4 cells recognize antigens bound to MHC class II molecules.

T cell receptor engagement and binding of antigen MHC complexes then requires CD8 cell receptors and CD4 cell co-receptor stimulatory action to initiate activation.

When acute infection subsides many effector T cells die; however a sub-population remains and these constitute memory T cells. Broere *et al.* distinguished two types of memory T cells: CCR7 T memory cells with cytokine receptor and central memory T cells.

6.4.2 *Cytotoxity pathways*

Broere *et al.* reviewed two major pathways of T cell cytotoxity: the perforin granzyme pathway that is calcium dependent and the FAS ligand pathway of apoptosis that is calcium independent. Regulatory T cells play key roles in peripheral tolerance and down regulation of the immune response.

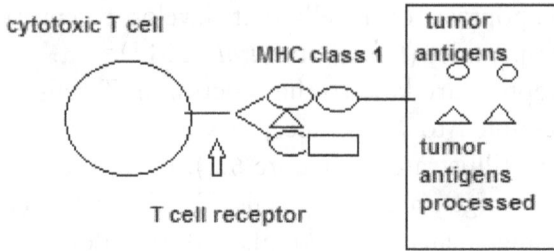

Figure 6.1. Tumor antigens bound to MHC class 1 antigens and T cell receptors.

6.4.3 *T cell receptors*

T cells with dimeric alpha and B chains are most common. Both alpha and beta chains have variable and constant regions. The T cell alpha chain maps to human chromosome 14, the beta chain maps to chromosome 7 and this locus is duplicated. The T cell alpha locus comprises 100 variable regions, 61 J regions and 1 constant region. On rearrangement one V region is combined with approximately 4 J regions and a single C region.

The germline T beta locus comprises 52 variable regions, a DB segment, 2 blocks of J segments each containing 6 segments and two C segments.

Transport of T cell receptors from the endoplasmic reticulum and their assembly on the cell surface requires the CD3 complex.

T cell receptors composed of delta and gamma chains are much less common. Broere *et al.* noted that T cells with gamma delta receptors are more common in skin and mucosa. The alpha and beta chains of the T cell receptor form a complex with CD3 co-receptor that is composed of three dimeric structures.

Following antigen binding, activation occurs. The first step in downstream signaling involves recruitment of a kinase LCK and phosphorylation of a tyrosine motif in CD3. This then leads to recruitment of the protein kinase ZAP70 that activates tyrosine residues on the LAT (linker for activation of T cells). LAT then recruits a number of different proteins to form a signalosome. Within this complex are molecules that act to trigger three signaling pathways. The activation of signaling and mobilization of NF kappa B lead to the passage of transcription factor to the nucleus and to expression of genes that promote differentiation growth and proliferation. Signals from the signalosome also trigger integrin release and actin reorganization (Brownlie and Zamayska, 2013).

6.4.4 *T cell receptor signaling network*

The T cell receptor signaling network was illustrated by Brownlie and Zamoyska. In this receptor, alpha and beta T cell chains are associated with a CD3 complex composed of delta, gamma and epsilon chain. The intra-cellular conserved domains of this complex are phosphorylated by a specific kinase LCK. This phosphorylation then leads to activation of ZAP70 and subsequently to phosphorylation of the LAT signalosome. The activated LAT signalosome then propagates signaling through the phospho-inositol, diacyl glycerol and RAS pathways to induce transcription activation and their passage to the nucleus activates gene expression.

6.4.5 *Antigen recognition and MHC region*

T cells recognize antigens bound to MHC complex. MHC class I molecules primarily present peptide antigens derived from within cells. They mainly present antigens to cytotoxic T cells that express CD8. MHC class II molecules can present extra-cellular products and products that were taken up by phagocytic cells. MHC molecules class II molecules present antigens to helper T cells. B cells may also present antigens to helper T cells. B cells with cell surface immunoglobulins recognize bacteria. These are then taken up into vesicles within the cell and are degraded. Antibodies bound to MHC class II and transferred to cell surfaces may activate T helper cells. Antibodies also eliminate extra-cellular pathogens.

Trowsdale and Knight (2013) reported that full sequencing of the MHC region on chromosome 6p21.3 revealed that it contains approximately 260 genes that span 4 megabases.

The MHC class I region contains class I genes in three categories: HLA-A, HLA-B and HLA-C. The class II region contains genes involved in antigen processing: HLA-DR, HLA-DQ, HLA-DP, TAP (transporter associated with antigen processing), LMP (peptidase) and BRD2 (bromodomain containing). The class III region contains genes involved in innate immunity, tumor necrosis factor (TNF) gene, complement genes and non-immune function related genes. Within the MHC region there are also genes not related to immune function. These include CYP21 congenital adrenal hyperplasia related gene in the HLAIII region. The recombination rate in the MHC region is much lower than in other regions of the genome.

Trowsdale and Knight reported that there are some micro-organisms that suppress expression of the MHC region genes and thereby reduce efficiency of antigen presentation and limit possibilities of their destruction.

A disease that is associated with a specific MHC haplotype is narcolepsy. Toxic reactions to specific small molecule drugs have been associated with specific HLA alleles. Abacivar toxicity in individuals with HLAB57.01 and HLAB15.02 is associated with sensitivity to carbamazepine.

6.4.6 *Treg cells*

FOXP3 expressing Treg cells are key regulators of immune tolerance. These cells function to antagonize immune responses. Liston and Gray (2014) reported that these cells are a subset of CD4 T lymphocytes and they protect against damaging immune responses, including auto-immune responses. They reported that Treg cells have distinct T cell receptors that have particular affinities for self-antigens. Binding of self-antigens to these receptors triggers transcription of FOXP3. They noted that the thymus is the major site of Treg cell induction. However emigrant cells from the thymus that are in the circulation or in other lymphoid tissue can also be induced to express FOXP3.

Liston and Gray reported that Treg cell are particularly sensitive to environmental influences. This is true of Treg cells in lymphoid tissue and in the circulation. These cells are derived from CD4+ CD25+ T cells that express FOXP3 transcription regulator.

Regulatory T cells also require interleukin 2 (IL2) expressions for their differentiation.

6.5 Interleukin and receptors

IL2 is a cytokine produced by CD4+ T cells and also by CD8+ T cells. It is also produced by activated dendritic cells of the innate immune system and by mast cells. The induction of expression of IL2 is activated by specific transcription factors including NFAT, NFKB, OCT1, JUN1 and AP1.

6.5.1 *Interleukin receptors and IL2R*

There are different classes of interleukin receptors that differ in their composition of alpha beta and gamma chains. The gamma sub-unit is a component of a number of different interleukin receptors ILR2, ILR4, ILR7, ILR15 and ILR21. Common gamma chain, sometimes referred to as the X linked gene IL2RG, encodes IL2R gamma.

The IL2 receptor is composed of a central long molecule, the IL2R beta subunit and lateral molecules, sub-unit IL2R alpha and the gamma sub-unit. The intra-cellular region of the alpha sub-unit associates with JAK1 and the intracellular portion of the gamma sub-unit associates with JAK3 (Malek and Bayer, 2004). Activation of the receptors leads to activation of the Jak 1 and Jak 3 kinases and to phosphorylation of tyrosine residues in the beta-subunit intracellular region and to phosphorylation of linked signal trans-ducers, including STAT, MAPK (mitogen-activated protein kinase) and PI3K-AKT signaling. Activation of signaling leads then to transcription.

6.6 NF kappa B (NFkB)

NFkB is a complex composed of either NFkB1 or NFkB2 bound to REL, RELA or RELB. RELA maps to chromosome 2p13, RELB to 19q13.32, NFkB1 to chromosome 4q24 and NFkB to 10q24. The NFkB complexes act as transcription factors that bind to specific sites in the genome and play roles in activating or inhibiting transcription (Oeckinghaus and Ghosh, 2011).

NFkB expression is activated during infection, following antigen receptor binding to receptor and through cytokine expression. Oeckinghaus and Ghosh noted that multiple regulatory elements control NFkB activity. When NFkB is not required it is sequestered in the cytoplasm bound to IkappaB (IkB). IkB binds specifically to REL within the NFkB complex. When NFkB is required the protein NEMO binds to and phosphorylates IkB that is then targeted for proteolytic degradation. The protein NEMO is also sometimes designated as IkBG and it is encoded on Xq28.

Oeckinghaus and Ghosh reviewed NFkB activating pathways. The pathway, designated as the physiological or canonical pathway, initiates with signaling through antigen binding to pattern recognition receptors (including NOD), and through cytokines, interleukins and the tumor necrosis receptor. Signaling through this pathway leads to release of NFkB from IKK and subsequently to translocations of REL heterodimers to the nucleus. In the non-canonical pathway, specific cytokines lead to the release of NFkB and generation of RELB complexes that enter the nucleus.

It is important to note that NFkB also has signaling functions in the nervous system and that it plays roles in dendritic growth and synaptic plasticity. There is evidence that NFkB can induce transcription of target genes in a variety of cell types (Smale, 2011). However regulatory factors determine expression in a specific cell type. Transcription factor binding of NFkB related components depends on chromatin structure. Specific barriers may impact binding. Repressive methylation signals on histone may need to be removed prior to signaling. Smale noted that the synergy between transcription factors that bind to gene promoters and those that bind to enhancers likely play important roles. In addition binding of co-regulatory molecules is important.

6.7 NK Cells

These cells represent a third type of lymphocyte. They differ from B and T cells in that they do not rearrange receptors. They function as cytotoxic lymphocytes and are particularly important when combating viral infections. Receptors on NK cells include activating receptors KIR1DS, KIR2DS and KIR3DS that bind class I MHC antigens and NKG2D, NKP30 and NKP44 that bind lectins. They also have inhibitory receptors KIR2DL.

NK cells are lymphoid cells that are part of the innate immune system. These cells differentiate in the bone-marrow, tonsils and thymus. NKT cells have receptors encoded by the leukocyte receptor complex that maps to human chromosome 19q13.4. This complex extends over one megabase and harbors genes that encode immunoglobulin-like extracellular domains. Genes in this complex include:

- LILR genes leukocyte immunoglobulin-like receptor
- LAIR genes leukocyte associated immunoglobulin-like genes
- KIR genes; these include 15 different genes that encode killer cell immunoglobulin-like receptors and 2 pseudogenes

NK cells release cytokines in response to antigen presentation.

6.8 Immunotyping

The Cluster of Differentiation (CD) system is used to classify cell surface molecules that distinguish different cells in the immune system. The CD molecules may constitute receptors but may also have different functions. As of November 2014, 364 different CD molecules were defined for immunotyping. Cells are classified as positive or negative for particular CD molecules (antigens). CD antigens are used for cell sorting. For example the CD4 antigen is expressed on helper T cells. The HIV Aids virus binds to CD4 and to a specific chemokine receptor on T cells. CD8 forms co-receptors for MHC class 1. CD25 is present on activated T cells and activated B cells; it is also present on many thymocytes and myeloid precursor cells.

6.9 Immune Checkpoints

The balance of co-stimulatory and inhibitory signals impacts the immune response. Inhibitory signals are often referred to immune checkpoints. Pardoll (2012) reported that checkpoint pathways are essential for maintenance of self-tolerance and to modulate the immune response.

He noted further that many genetic and epigenetic alterations occur in cancerous tumors and lead to generation of novel antigens that can be recognized by T cell receptors. However many cancers produce substances that inhibit the function of T lymphocytes that if they were not inhibited could lead to destruction of the tumors.

CD8 effector T cells (cytotoxic T cells) work together with CD4 helper cells to integrate innate and adaptive immune responses. Agonists of costimulatory receptors or antagonists and inhibitory signals can amplify T cell responses. He noted that agents that block checkpoint inhibitory signals are transforming cancer therapeutics. Complexity lies in the number of different inhibitory and co-stimulatory molecules.

Key factors to be considered are mechanisms of action of inhibitory molecules. These may be over-expressed in tumor cells or in their micro-environments.

6.10 B Cells and Antibodies

B cells develop in bone-marrow from multipotent progenitor cells. B cells then migrate to the spleen and to lymphoid tissues. B cells can be activated by T cell dependent antigens. They can also be activated by T cell independent antigens.

Parham (2005) defined 4 stages in the B cell life cycle. In the first stage, maturation takes place in the bone-marrow when B cells develop receptors through immunoglobulin gene rearrangements. During the second stage, selection takes place to remove those B cells with receptors that react against self (auto-reactive B cells). In the third stage, B cells that are not auto-reactive migrate in the blood stream to secondary lymphoid tissue. Contact of a B cell with cognate antigen then leads to the fourth stage when B cells differentiate into antibody producing plasma cells or memory B cells.

Mauri and Bosma (2012) reviewed the immune regulatory functions of B cells. They noted that B cells have previously been considered primarily in relation to their roles as antibody producers. However in recent decades additional B cell functions have been elucidated. These include production of cytokines including secretion of IL10 by regulatory B cells. These cells suppress specific autoimmune responses. B cells also express Toll receptors.

Antibodies have a Y shaped structure. The variable region of the antibody forms the V portion of the Y and constant regions form the perpendicular part of the Y. Duplicate light and heavy chain variable regions form the V portion of the Y and together form the antigen binding region. Only heavy chains form the perpendicular region of the Y.

There are 2 different forms of light chains, kappa encoded on chromosome 2 in humans and lambda encoded on chromosome 22. There are 5 different types of heavy chains and all are encoded on human chromosome 14. Light and heavy chain regions within the V portion of the antibody molecule each contain variable and constant regions.

The lambda light chain gene on chromosome 22 has at least 40 variable segments, 5 joining segments and 4 constant segments. The kappa light chain gene on chromosome 2 has at least 40 variable segments, a cluster of 4 joining segments and 1 constant region. The entire kappa light chain gene is duplicated and it extends to 3.2 megabases. The heavy chain has at least 150 segments including variable segments, diversity segments, 6 joining segments and a single constant region (Parham, 2005).

6.10.1 *Heavy chain immunoglobulin locus*

The constant (C) region cluster of the heavy chain Ig locus comprises 9 different segments: Cm, Cd, Cg3 Cg1, Ca1, Cg2, Cg4, Ce4 and Ca2.

Following rearrangement of the VDJ segments, transcription of the C segments commences, followed by splicing and ultimately translation. The m and d heavy chains are first transcribed giving rise to a B cell that expresses IgM and IgD. Parham noted that the two heavy chains are expressed in naïve B cells (prior to contact with antigen) and that these represent the only two immunoglobulins that are co-expressed on a single B cell. He also noted that immunoglobulins occur as membrane bound and secreted forms. The M heavy chain has the highest molecular weight. IgG forms have the lowest molecular weight and IGG1 is the most abundant immunoglobulin in serum.

6.10.2 *Recombination of the V(D)J regions*

This requires activity of the VDJ recombinase complex that contains a number of different enzymes including RAG1 and RAG2 (recombinase activating enzymes), Artemis nuclease, XRCC4 (XR cross-linking repair), NHEJ1 (non-homologous end-joining protein), DNA dependent protein kinase, DNA polymerases and DNTT (terminal nucleotide transferase). The VDJ recombinase complex recognizes recombination signal sequences located adjacent to the variable (V), diversity (D) and joining (J) segments.

VDJ recombination events take place during the development of B cells and T cells. Schatz and Ji (2011) reviewed recombination during development of B cell and T cell receptors. They noted that immune responses are dependent on the availability of a large repertoire of antigen receptors and that recombination occurs during the early stages of lymphocyte development.

The recombinase complex induces double stranded breaks in DNA strands. Following recombination these breaks must be repaired. Non-homologous end-joining processes usually carry out this process. Schatz and Ji reported that specific recombination signals (RSS) flank each V, D and J segment. Furthermore accessibility of the chromatin within which each RSS is located impacts recombination. Covalent histone modifications in these regions impact recombination. There is also evidence that the 3-dimensional chromatin architecture, chromosome looping and condensation impact the process. RSS signals are composed of conserved heptamer and nonamer elements separated by 12 to 23 base pairs. Recombination preferably occurs between 12RSS and 23RSS signals. Recombination signals include 5'CACAGTG 12bp and 23bp ACAAAAACC3'.

There is evidence that SWI/SNF or BRG1 chromatin remodeling complexes are recruited to specific regions undergoing remodeling. With respect to nuclear architecture, Schatz and Ji reported that the receptor complex genomic segments move away from the nuclear membrane when they undergo recombination and return to these regions following recombination.

6.10.3 *Somatic recombination and somatic hyper-mutation*

Somatic recombination involves cutting out and splicing gene segments. A DNA segment which is formed in this manner is composed of a single type of variable, joining and constant segments. The variable recombinants formed lead to antibody diversity.

Somatic hyper-mutation occurs following antigen binding. Single nucleotide substitutions take place within the rearranged variable regions of heavy and light chain genes. The mutation rate in these segments is enhanced by the presence of activation induced cytidine deaminase (AICD) and uracil DNA deaminase.

6.10.4 *Natural antibodies*

There is growing evidence that a sub-group of natural antibodies that target the body's own molecules (self-antigens) may have beneficial effects. This subgroup of natural antibodies likely acts to clear debris which results from normal cell degradation processes. Previously antibodies directed against self-antigens were considered to be uniformly detrimental and responsible for auto-immune diseases (Leslie, 2015).

6.10.5 *Pre-B receptor*

Conley *et al.* (2009) described and illustrated signaling through the pre-B cell receptor. Pre-B receptor components include IGM heavy chain (mu heavy chain), IGLL1 that forms the pre-B light chain, and CD79A and CD79B (pre-B cell associated molecules). CD79A and CD79B are bound to the intracellular conserved domains of the receptor (CD79A and B were previously referred to as Ig alpha and Ig beta). CD19 A and B intracellular motifs are then phosphorylated by a specific kinase LYN. Following LYN phosphorylation a specific hematopoietic cell kinase SYK then phosphorylates the CD19 linked scaffold protein BLINK (B cell linker) and proteins assemble on BLINK. These proteins include BTK Bruton kinase, VAV (guanine nucleotide exchange protein), phospholipase gamma 2, MAP4K1 and GRB2. These phosphorylated signaling factors then lead to transcription factor activation (NFAT and NFKB). Activated transcription factors enter the nucleus and promote gene expression.

6.11 B Cell Immunodeficiencies and Agammaglobulinemia

Conley *et al.* reviewed B cell immunodeficiencies. Several result from single gene disorders while others are multifactorial in origin. They emphasized that there is heterogeneity of manifestations even among patients with the same B cell immunodeficiency disorder. Environmental factors and exposures impact presentation. Patients commonly manifest increased susceptibility to infections with encapsulated bacteria such as *Streptococcus pneumoniae* and *Haemophilus influenzae*. These infections lead to bronchitis, pneumonia and otitis media. Giardia infections are also common in patients with antibody deficiencies.

Conley *et al.* defined three categories of antibody deficiencies:

(i)   Defects in early B cell development
(ii) Hyper-IgM syndromes due to class switch recombination defects
(iii) Common variable immunodeficiencies

6.11.1 *Defects in early B cell development*

These defects are most commonly due to a mutation in the X-linked Bruton tyrosine kinase (BTK) and they present in males. BTK is expressed in B cells and monocytes. Nomura *et al.* (2000) discovered that the X linked Bruton form of agammaglobulinemia was due to mutations or deletions in a specific kinase BTK.

Conley *et al.* reported that patients with BTK deficiency usually present between 3 and 18 months of age with recurrent bacterial infection. They defined this as a leaky disorder since in some children the mutations present lead them to have few B cells and low levels of immunoglobulin.

Bone marrow studies in these patients reveal that the block in pre-B cell differentiation or proliferation and IgM is usually absent from cells. BTB base is a database that includes information on at least 600 different mutations, deletions, inversions and DNA rearrangements that lead to BTK deficiency.

6.11.2 *Autosomal recessive forms of agammaglobulinemia*

Conley *et al.* reported that specific mutations or deletions in IGHM, the gene that encodes the constant region of immunoglobulin M, lead to autosomal recessive forms of agammaglobulinemia. This condition overlaps in manifestations with X-linked agammaglobulinemia. Autosomal recessive forms of agammaglobulinemia may also result from deficiency of other products involved in B cell maturation. These include deficiency of IGLL1, the pre-B cell light chain, and deficiency of CD79A or CD79B that constitutes the pre-B cell receptor.

In some patients the B cell developmental defects are due to mutation in the scaffold protein BLNK on which BTK and other signal transduction molecules assemble.

6.11.3 *Hyper IGM and class switch defects*

The CD40 receptor is a member of the TNF tumor necrosis family of proteins. A specific ligand CD40LG binds to this receptor and plays an important role in immunoglobulin class switching. The gene that encodes CD40LG is located on Xq26.

Mutations that impair the function of CD40LG therefore lead to disease in males. The CD40 receptor maps to chromosome 20q12-q13.2, and specific mutations lead to autosomal defects which in turn lead to hyper-IGM syndrome.

Conley *et al.* reported that autosomal recessive mutation in AICDA (activation induced cytidine deaminase) encoded by a gene on chromosome 12p13 occur in 10–15% of patients with class switch recombination defects. Class switch recombination defects may also result from mutations in uracil glycosylase (UNG) encoded by a gene on 12q23. This is one of a number of different DNA glycosylases. Patients with UNG mutations have elevated or normal levels of IGM and low levels of IgA, IgG and IgE.

Class switch isotype switching involves recombination events that enable the V region to be used with other heavy chain C regions. This switching takes place at highly repetitive sequences within switch regions. Class switching involves generation of DNA breaks that are then repaired.

Deficiency of the gamma sub-unit of the interleukin receptor leads to X-linked severe combined immunodeficiency (SCID) (Noguchi *et al.*, 1993). Mutations in the alpha sub-unit of the IL2 receptor lead to an autosomal recessive immunodeficiency. Mutation in the JAK3 (Janus kinase 3) that interacts with the gamma subunit of interleukin receptors also leads to autosomal recessive immunodeficiency (Mella *et al.*, 2001).

6.12 Immune Deficiencies Due to Defects in T Cell Differentiation

Signaling from IL2 encoded on chromosome 4q26, and activity of FOXP3 transcription factor that is encoded by a gene on Xp11.23 are required for differentiation of Treg cells. A specific form of X-linked immune dysregulation results from mutation in FOXP3. This immune deficiency is often associated with polyendocrinopathy. Mutations in other gene products related to Treg cell differentiation can also give rise to this syndrome. These genes include STAT5B, CD25 NEDD4 and ITCH ubiquitin ligases.

6.13 SCID

Cases of SCID differ with respect to the cell deficits they manifest. Specific patients may be deficient in B cells or T cell or NK cells.

Table 6.1. Gene defects in SCID.

IL2RG mutations, defects in the X linked gene that encodes the common gamma chain for interleukin receptors occurred in 48% of cases

ADA mutations in adenosine deaminase occurred in 16% of cases

IL7R mutations in the alpha subunit of the interleukin 7 receptor (IL7) occurred in 10% of cases

JAK3 mutations in the downstream signaling Janus kinase 3 occurred in 6% of cases

RAG1/RAG2 mutations in the recombinase activating enzymes occurred in 6% of cases

DCLRE1C mutations in a product required for DNA recombination occurred in 5% of cases

Other rare causes of SCID include:

STAT5 signal transduction molecule mutations

CORO1A mutations in coronin that mediates lymphocyte emigration from thymus

NHEIJ mutations in non-homologous end-joining enzymes

AK2 mutations in adenylate kinase a metabolic enzyme

Rivers and Gaspar (2015) reported that mutations in at least 18 different genes may lead to SCID and in 10% of cases a genetic diagnosis cannot be made. In these cases exon sequencing is indicated. Gene defects leading to SCID are listed in Table 6.1.

6.13.1 *T cell receptor rearrangement and functions and SCID*

During T cell differentiation, rearrangement takes place within the T cell receptor genes. This rearrangement involves the V, D and J gene segments and recombination signal sequences. Hazenburg *et al.* (2001) reported that during the course of rearrangement, V, D and J gene segments are deleted as circular DNA excision products, defined as TRECS (T cell receptor excision circles). They noted further that the levels of TRECS in the peripheral circulation are dependent on thymus output and rate of T cell emigration from the thymus. They proposed that TREC level assessment would serve as an elegant test for measuring thymic function and output.

Puck (2012) reported that the TREC test has proven useful as a newborn screening test SCID. TRECs are readily detectable in DNA extracted from dried blood spots obtained for routine screening of newborns for inborn errors of metabolism. The TREC test proved useful for a range of lymphopenic conditions in newborns. In infants who have low levels of TREC in blood spots follow-up testing includes quantitative determination of lymphocyte subsets.

Conditions with low levels or absence of TRECs and significant lymphopenia include:

- Typical SCID due to mutations in any one of 18 genes
- Leaky SCID due to hypomorphic mutations in SCID causing genes
- Variant SCID with no defects in currently known SCID causing genes
- Multi-system syndromes with variably affected cellular immunity (these include DiGeorge, CHARGE and Jacobsen syndromes)
- Secondary T lymphopenia associated with extreme prematurity, gastroschisis, or thymus removal during cardiac surgery

Puck reported the frequency of SCID gene mutations found in patients in the California newborn screening program. The latest results from newborn screening for SCID and SCID variants indicated that 1 in 58,000 newborns were affected (Kwan and Puck 2015). Newborn screening for SCID has been advocated since hematopoietic cell transfusion has been curative in many cases. SCID due to adenosine deaminase (ADA) deficiency has responded to enzyme infusion and gene therapy. Gaspar *et al.* (2013) distinguished four immunotypes in SCID. All had very low or absent T cells. In category 1, B cells were present. Subdivision occurred in that some patients were positive for NK cells and others were NK negative.

In category 2, B cells were negative. Subdivision occurred in that some patients were NK positive while others were NK negative.

6.14 Other Immune Mediated Diseases

6.14.1 *Common variable immunodeficiency (CVID)*

CVID includes several different disorders, as listed in Table 6.2. Conley *et al.* reported that patients have normal or low levels of immunoglobulins. Levels of B cells may be normal or low. Patients with CVID often have manifestations of auto-immune diseases. CVID is often inherited as an autosomal recessive disorder. However there is evidence the CVID may in many cases be multifactorial in origin.

Immunodeficiency may also occur as one manifestation of specific syndromes. Examples include syndromes associated with DNA damage repair defects; see Table 6.3.

6.14.2 *Immune dysregulation*

Immune dysregulation and manifestations of auto-immunity sometimes occur in the same patient. There are disorders with these manifestations that follow Mendelian patterns of inheritance, and they include IPEX (immunodysregulation polyendocrinopathy enteropathy X-linked) and IPEX-related disorders. Verbsky and Chatila (2013)

Table 6.2.   Genes defective in CVID.

CD19 cell surface molecule that assembles B cell receptors
CD81 tetraspanin mediates signal transduction and cell development
CR2 membrane receptor that participates in B cell activation
ICOS facilitates interactions between B cells and T cells
LRBA associates with protein kinase c aids in effector functions
MS4A1 surface molecule involved in B cell differentiation and development
NFKB2 sub-unit of NFkB transcription factor
PRKCD protein kinase family signaling molecule
TNFRSF13B molecule on the surface of B cells that promotes signaling
TNFRSF13C enhances B cell activation
Syndromic immunodeficiency

Table 6.3. Syndromic disorders in which Immune deficiency may be a manifestation.

Wiskott–Aldrich syndrome

Ataxia telangiectasia

Nijmegen breakage syndrome

Bloom syndrome

Centromere instability syndromes

Telomerase defects (TERC, TEL1 or TERT defects)

Immunodeficiencies may also arise due to metabolic defects or deficiencies in Folate or B12 metabolism:

TCN2 transcobalamin transport defect

SLC46A1 folate transporter defects

MTHFD1 folate reductase deficiency

reviewed these disorders. IPEX disorders include immune dysregulation, polyendocrinopathy and enteropathy that are X-linked. IPEX results from mutations in the X-linked gene FOXP3. The disorder primarily affects Treg cells. IPEX-related disorders are in some cases due to mutations in STAT5A, STAT5B and STAT3 (signal transducer and transcription activators) that are encoded by contiguous genes on chromosome 17q11.2. Defects in these genes also impact Treg cells. Mutations in CD25 and ITCH (and E3 type ubiquitin ligase that maps to chromosome 20) may also lead to IPEX-related disorders. Gain of function mutations in STAT1 lead to IPEX-related disorders.

IPEX manifestation appears early in life, leading to enteropathy, auto-immune endocrinopathy and eczematous dermatitis. IgE (allergen-specific immunoglobulin E) levels are often significantly elevated.

Treg cells are primarily generated in the thymus and have high affinity for self-antigens. Treg cells normally manifest high levels of FOXP3. Loss of FOXP3 leads to T cell activation, lymphoid and myeloid hyperplasia and intense inflammatory response. Loss of FOXP3 transcription factor is also characterized by loss of CD4 and CD25 expressing Treg cells and this leads to auto-immunity.

Liston and Gray (2014) reported that Treg cells are developed both in the thymus and in the periphery. They emphasized that dynamic homeostatic processes control the number of Treg cells. Cell numbers are controlled by rates of cell production and apoptosis. When the numbers of Treg cells are too low auto-immune reactions occur. When Treg cell numbers are too high immune responses are suppressed.

FOXP3 levels are important in controlling Treg cell numbers. Other important factors in control of Treg cell function include the IL2 receptor (IL2RA), CTLA4 (cytotoxic T lymphocyte antigen) and tumor necrosis factor receptor TNFRSF18.

There is evidence that impaired Treg response may play roles in non-hereditary immune dysfunction including systemic lupus erythromatosis, multiple sclerosis and type 1 diabetes mellitus.

Sorrentino (2014) reviewed the impact of genome-wide association studies (GWASs) on insights into the genetics of auto-immune diseases. She noted that each variant found through GWASs to be associated with a specific auto-immune disease usually alters risk to a very small degree. In addition a growing number of variants were found to be associated with a specific auto-immune disease. Furthermore specific GWAS variants are often associated with more than one auto-immune disease.

6.14.3 *Genetics and common auto-immune diseases*

Knight (2013) emphasized that the MHC region on chromosome 6p21 plays a predominant role in auto-immune and inflammatory disease and is also involved in some forms of drug sensitivity reactions.

More significant associations have been made between specific conditions and the MHC regions than with any other region of the genome. The associations include inflammatory diseases, auto-immune diseases, drug hypersensitivity and neuropsychiatric diseases. In initial MHC association studies serological MHC markers were used. More recently single nucleotide polymorphisms in the region have been used. Some groups have initiated high throughput sequencing to examine markers in highly polymorphic MHC regions. However given the extensive linkage disequilibrium that exists in the region it is difficult to progress from association to identification of specific causal variants. There is also evidence that epigenetic factors impact expression of genes in the MHC region. Sekar *et al.* (2016) reported that variants in the complement component 4 gene in the MHC region contribute to the risk of schizophrenia.

Concordance rates for diseases such as type 1 diabetes, multiple sclerosis, rheumatoid arthritis, and inflammatory bowel disease in monozygotic twins range between 25–50%. These concordance rates indicate that genetics plays an important role in disease pathogenesis but that non-genetic risk factors are also contributory. Concordance rates for these diseases in dizygotic twins are 2–12%.

Goris and Liston (2012) emphasized the importance of common genetic variation in susceptibility to auto-immune disease. Most of the risk alleles associated with these diseases are of small effect and alter risk on average by 1.2%.

6.14.4 *GWASs in immune mediated diseases*

Parkes *et al.* (2013) reported that immune-mediated diseases affect 3–5% of individuals. Recent studies of these disorders have utilized GWASs and immune-chip microarray analysis to investigate the genetic basis of factors involved in susceptibility. The immunochip interrogates 195,800 single nucleotide polymorphisms (SNPs) and small insertion deletions that have been reported to be associated with immune mediated disease. Parkes *et al.* reported that very few rare alleles, i.e. alleles with frequency lower than 0.5%, have been associated with common immune related diseases.

6.14.5 *Overlap between associated loci in different immune diseases*

Significant overlap of associated loci was found for inflammatory bowel disease, rheumatoid arthritis, psoriasis and type 1 diabetes. Partial overlap was found in loci associated with ankylosing spondylitis, type 1 diabetes, rheumatoid arthritis, psoriasis, coeliac disease and inflammatory bowel disease.

Although SNPs found associated with immune-mediated diseases frequently mapped close to or within a specific gene, it is important to note that many associated SNPs are mapped in gene deserts and in putative regulatory regions.

6.14.6 *Gene loci and pathways*

Parkes *et al.* reported that seropositive diseases (diseases associated with specific antibodies) were typically found to be associated with class II HLA alleles. These diseases include type 1 diabetes, systemic lupus erythromatosis associated with the HLA-DR3-DQ2 haplotype and rheumatoid arthritis associated with HLA-DRB1.

Seronegative disorders were found to be associated with disease specific HLA class I antigens; ankylosing spondylitis with HLA B27, psoriasis with HLA-Cw6, Behcet's disease with HLAB51. Crohn's disease was found to be associated with HLA class I but not with specific alleles.

6.14.7 *Other key pathways in immune mediated disease*

Parkes *et al.* reported that the following pathways were also important in the pathogenesis of immune mediated disease: Interleukin pathways (especially IL23, and IL2, IL21 and IL10), the NFkB pathway, and the IRF pathway involved in the production of interferon regulatory factors that control interferon release in response to viral infection.

Other important proteins linked to immune mediated diseases include the protein tyrosine phosphatase (PTPN2 and PTPN22), and tyrosine kinases including tyrosine kinase 2 (TYK2) and Janus kinase (JAK2).

Parkes *et al.* emphasized that a specific allele might constitute a risk factor for one type of immune associated disease and that the same allele may be protective against another. They concluded that the biggest remaining challenge relates to clinical translation of GWASs.

6.14.8 *Environmental factors*

In a review of mapping immune mediated genes Ricaño-Ponce and Wijmenga (2013) emphasized that immune mediated disease develops in individuals genetically at risk and in response to interactions between the immune system, the environment and microbial factors. The MHC is the strongest locus for association. There is evidence that 28% of genes expressed from the MHC locus impact immune function. They emphasized that when the HLA regions was excluded from analysis of immune mediated diseases, the associations found were primarily with common variants and the risk effect sizes were a modest 1.04 +/- 3.99.

6.15 Complement

Ricklin *et al.* (2010) reported that complement proteins play roles not only in the elimination of microbes but also in eliminating cellular debris and in immune surveillance. Furthermore these investigators noted that although complement pathways are frequently presented as linear, these pathways are in fact connected to a number of different systems.

The complement component C1q recognizes distinct molecular structures on microbial cells or apoptotic cells and it may recognize specific patterns and recognition molecules. Following the binding of C1q to surfaces, C1s and C1r are activated; C1s then cleaves C4 to C4a and C4b. C4b derivatives then generate activity in the C3 convertase and classical complement pathway.

In the lectin pathway mannose binding lectins bind to pathogen surfaces. Specific serine proteases (MASP1, 2, 3) that are similar to C1r and C1s and cleave C4 to C2. This then leads to activation of C3.

In the alternative pathway specific molecules on surfaces induce complement activation.

C1q functions in pattern recognition. C1s, C1r C2 and MASP1–3 function as proteases. C3–C9 are complement components that promote opsonization of pathogens to facilitate uptake by phagocytes. Activated complement also promotes perforation of pathogen membranes and promotes recruitment of inflammatory cells. Soluble and cell bound regulators control the impact of complement. These include CFH (factor H) and clusterin.

Complement functions not only in the active phase of infection and inflammation but also in the resolution phase to clear apoptotic cells and immune complexes.

Ricklin *et al.* (2010) reported that there is cross-talk between complement and coagulation pathways. Fibrinolysis is inhibited by complement component C5a. Coagulation and clotting likely act locally to limit the spread of infection.

6.15.1 *Complement deficiencies*

Grumach and Kirschfink (2014) reviewed complement deficiencies. These authors noted that these conditions are under-recognized. They reported that deficiencies of the early components of the complement pathway C1q, C1r, C12, C2 and C4 are often associated with manifestations of auto-immune disease.

Deficiency of C1q inhibitor (C1NH, also known as SERPING1) leads to hereditary angioedema associated with local sub-cutaneous edema and mucosal edema. Type III hereditary angioedema is also caused by defects in the coagulation factor F12.

Auto-immune disease manifestations associated with complement deficiency include rheumatic disorders, systemic lupus erythromatosis, membrano-proliferative glomerulo-nephritis and vasculititis.

Properdin complement factor (CFP) is a positive regulator of the alternate pathway of complement activation. Properdin binds to microbial surfaces and to the surfaces of apoptotic cells. Together with C3 and C5 it facilitates formation of membrane attack complexes. Deficiency of properdin leads to increased susceptibility to meningococcal infections.

6.15.2 *Complement regulation and complement factor H related proteins*

The regulation of complement activation gene cluster on chromosome 1q32 includes loci for factor H and 5 CFHR (complement factor H related) loci. Within this genomic region there are large segments of repeat sequences that predispose the region to rearrangements. Furthermore the 5 CFHR genes contain short repeat elements and high degrees of sequence homology occur in the N-terminal exons of the 5' CFHR genes. C terminal regions of these genes show high homology with the complement FH gene (Skerka *et al.*, 2013). The CFHR region is therefore is an unstable region. Deletions, duplication and variants that occur in this region are associated with human disorders. These include atypical hemolytic uremic syndrome, glomerulopathy, glomerulonephritis, dense deposit (kidney) disease, age related macular degeneration, and systemic lupus erythromatosis.

Hemolytic uremic syndrome (HUS) is associated with microangiopathy, fragmentation of erythrocytes, and thrombocytopenia. Skerka *et al.* reported that acute renal failure results from platelet rich microthrombi that form in small vessels. HUS may arise due to infection with Shiga toxin producing bacteria. It may arise due to mutations in complement H genes and in some cases it is due to mutation in C1 or C3. In some cases it is linked to the formation of auto antibodies.

Variants in complement genes C1, C2 and C3 and variants in complement factor H are also associated with age related macular degeneration. In addition this disorder may be due to defects in proteins outside the complement pathway.

Age related macular dystrophy (AMD) is a slowly progressive degenerative ophthalmologic condition that normally presents after the 6th decade of life. This condition leads to loss of central vision (Schramm *et al.*, 2014).

A number of different genetic variants have been associated with AMD and several of them impact the complement pathway. The associated variants include highly penetrant rare variants and common variants that increase risk to a low degree.

Environmental factors, particularly smoking, also increase risk. Characteristic lesions of AMD are drusen, which are composed of extra-cellular deposits of lipids and proteins that expand and separate the photo-receptor cells from the retinal pigment epithelium.

Schramm *et al.* reported that in some cases drusen formation leads to excessive blood vessel growth. Late stage AMD is subdivided into neo-vascular wet forms and dry forms with geographic atrophy.

Variants in complement factor H and in complement factor H-related genes alter risk for AMD. Three highly penetrant rare variants in complement factor that increase AMD risk are R1210C, R53C and D90G. Variants in other complement pathway genes that contribute to AMD risk occur in C3 and in complement factors B and I.

6.15.3 *Complement related diseases*

Ricklin *et al.* (2010) noted that unsuccessful removal of debris due to impaired activity of C1, C4 or C2 likely contributes to auto-immune diseases, including systemic lupus erythromatosis and age related neurodegeneration.

They noted that complement might play roles in Alzheimer's disease since amyloid fibers accumulate C1q and C3. Binding of complements might promote inflammation, release of cytokines that lead to neuronal dysfunction. Age related macular degeneration was found in some cases to be associated with specific polymorphisms in the complement regulatory proteins FH and FHL1. In these cases complement activation occurred due to aberrant stabilization of the regulator.

There is evidence that tumor cells have high levels of complement regulators bound to their surface and that this promotes tumor cells to evade clearance through complement activities.

6.16 Monoclonal Antibodies and Engineered Antibodies Used in Therapies

Production of monoclonal antibodies through hybridoma technologies initially involved fusion of splenic cells from a hyper-immunized mouse with mouse myeloma cells and then isolation of individual clones that produced antibodies against specific antigens or epitopes.

Since problems arose when mouse antibodies were used in the treatment of humans, different modifications of antibodies were used to humanize antibodies. These included using only the specific antigen binding portion of the antibody and fusing this to a human non-antigen binding immunoglobulin Fc region.

Later transgenic mice were produced to derive mouse strains that produced only human forms of immunoglobulin. More recently phage display platforms have been used to produce highly specific antibodies (Deantonio *et al.*, 2014).

By fusing the coding sequence of the antibody variable region to the gene encoding the phage minor coat protein, antigen binding regions of antibodies can be displayed on the surface of a bacteriophage. Bacteriophages that display effective antibodies can be selected and propagated. The variable region antibody segments can then be excised from phage DNA and recloned to generate a full-length antibody.

6.17 Harnessing the Immune System for Cancer Therapy

For a number of years, monoclonal antibodies against specific factors that are over-expressed in tumors have been used in therapy. These include antibodies against EGFR, ERBB2, and VEGF.

Savage (2014) reviewed approaches to harness the immune system for cancer therapy and included three methods. In the first, adoptive cell therapy was involved by isolating tumor infiltrating lymphocytes from a resected tumor, expansion of these cells in *in vitro* cultures, followed by infusion into the patient from whom the tumor was derived.

The second approach involves use of antibodies against inhibitory receptors on T cells (immune checkpoint antibodies). This approach will be discussed in detail later in this chapter.

The third approach discussed by Savage involves vaccination with high affinity tumor-derived mutant peptides to expand immunity to neoantigens on tumors. Savage described mechanisms through which tumor mutations might be immunogenic. He emphasized that the mutant amino acid produced in the tumor must be processed and loaded onto an MHC molecule and then be presented to T cells that could distinguish the neoantigens from self-antigens.

Robbins *et al.* (2013) analyzed samples from melanoma patients who had regression of tumors following adoptive T cell transfer. They carried out exome sequencing on the initially resected tumors and on the T cell cultures used for transfer. They determined that the autologous T cells reacted to 23 different neo-epitopes produced by the tumor.

Van Rooij *et al.* (2013) reported identification of specific tumor neoantigens in melanoma patients. They identified the neoantigens through exome sequencing and reported that there were T cells reactive to the tumor epitopes in cultures of lymphocytes that infiltrated the tumor.

Gubin *et al.* (2015) reviewed tumor antigens that might be derived from mutations within tumor genomes, from genomic rearrangements in tumors or from oncogenic viruses. These are defined as tumor-specific antigens (TSA). Two other categories of tumor antigens are distinguished. Tumor-associated antigens (TAA) are defined as normal proteins aberrantly expressed in tumors. TAA may also be normal proteins that have undergone unusual post-translational modification. Gubin *et al.* noted that TAAs have low T cell receptor affinity. Another category of tumor antigens includes specific proteins that have highly restricted expression and are referred to as cancer germiline or cancer testis antigens (CTAs).

Steps involved in the process of identifying tumor specific antigens include sequence analysis to define tumor specific mutations and bioinformatic analyses that include the use of epitope prediction algorithms. These algorithms determine the likelihood of specific abnormal peptides being appropriately processed, and binding to MHC class 1 molecules. They noted that suitable peptides are usually 8 to 11 amino acids long. These peptides are transferred to the endoplasmic reticulum by transfer proteins (TAP) and are then loaded onto MHC class 1 complexes. They reported that there are approximately 2,500 human MHC1 allelic sequences. Specific bioinformatic tools are available including the immune epitope database and analysis resources IEDB and NETMHC.

Carreno *et al.* (2015) utilized exome sequencing to identify somatic mutations in tumor samples from tumor patients. They determined the MHC binding capacity of the mutation containing peptides and established that vaccination of the patient with high affinity specific tumor-derived mutant peptides expanded the anti-tumor response in the patients.

6.17.1 *Immune checkpoints and cancer therapy*

Immune checkpoints serve to modulate the intensity of the immune response and to limit the response to self-antigens. T cell receptor activation following antigen binding is subject to stimulatory and antagonist inhibitory responses.

CTLA4 is an immune checkpoint receptor that down-modulates T cell activation (Pardoll, 2012). There are a number of different mechanisms through which CTLA4 inhibits T cell activation. CTLA4 out-competes the activated T cell receptor for binding to stimulatory ligands.

Specific tumor antigens stimulate CTLA4 and inhibit T cell activation. Antibodies to CTLA4 neutralize the down-modulatory effect of this receptor on T cell activation and have been particularly useful in the treatment of specific cases of melanoma.

PD1 (programmed cell death 1) is an immune checkpoint protein. Pardoll reported that this protein limits T cell effector function in peripheral tissues. Specific tumor ligands bind to the PD1 receptor and lead it to limit T cell effector function. Antibodies to PD1 have also proven useful in treatment of certain cancers.

Rodić *et al.* (2015) reported that PD-L1 ligand expressed in melanomas correlates with the response to PD1 antibodies. However they determined that PD-L1 expression did not correlate with expression of the melanoma driver mutation BRAF V600E.

There is evidence that CD4 helper T cells detect as antigens mutated peptides generated by tumor cells. Mutated peptides generated by cancer cells are referred to as somatic neo-epitopes. Snyder *et al.* (2014) carried out studies to define the neo-epitope landscape in tumors from 64 patients. They identified a specific neo-epitope signature in melanomas that proved responsive to treatment with anti-CTLA4 antibodies. They concluded that there is a defined genetic basis for response to CTLA4 antibodies.

Their study involved exome sequencing of tumor tissue and blood samples from each patient. Bioinformatic analyses of sequence was carried out to determine mutation in tumors and to evaluate the potential of mutated peptides to bind to MHC HLA class I.

They identified a number of tetra-peptide sequences that occurred in melanoma tumors in patients who responded to anti-CTLA4 therapy. These tetra-peptides were recognized by T cell receptors. They confirmed that the genes that encoded these specific neo-epitopes were widely expressed in the melanoma tumors. Some epitopes resembled antigenic proteins of specific micro-organisms.

Snyder *et al.* emphasized the importance of tumor genetic analyses in determining whether CTLA4 directed therapy would be clinically beneficial. The key point is that tumor neo-epitopes are recognized as non-self in the immune system. Importantly they noted that concepts of driver and passenger mutations should be expanded. Some mutations formerly classified as passengers may be immune determinants.

Topalian *et al.* (2015) reported that agents that block PD1 and its ligand PDL1 have been found to be useful in the treatment of a number of different types of cancer. They report that PD1 is absent from resting naïve and memory T cells and that it is expressed following T cell receptor (TCR) engagement. Its expression requires transcriptional activation. PDL1 is produced by activation of T cells and NK cells. PDL2 is expressed by dendritic cells and macrophages. PDL1 and PDL2 have immune system functions besides being ligands for PD1.

The anti PD1 agent pidilizumab is effective against melanoma and also against kidney and colorectal cancer. There is some evidence that anti CTLA4 and anti-PD1 agents used together have synergistic effects. Topalian *et al.* noted that next generation checkpoint molecules are being investigated. These include lymphocyte activation gene LAG3, a checkpoint molecule that is expressed on activated T cells, B cells, NK cells and dendritic cells. Blocking LAG3 counteracts suppression of T cells by Treg cells.

NK cells can promote apoptosis. However NK functions are impacted through activation of receptors including killer inhibitory receptors (KIR). Topalian *et al.* noted that KIRs can be considered as immune checkpoint molecules and that KIR blocking antibodies are being explored as therapeutic agents to treat tumors. Phase 1 trials are in place to test antibodies against KIRs.

TIM3 (T cell immunoglobulin and mucin 3) is an immune blocking molecule expressed on activated T cells, NK cells and monocytes. It is co-expressed with PD1 and is also being investigated as target for tumor therapy.

7. Lipids, Atherosclerosis, Metabolic Syndromes and Diabetes

7.1 Plasma Lipids

Since lipids are relatively insoluble they are transported in plasma bound to proteins, most frequently apolipoproteins. Lipoproteins are classified based on their size and on their relative content of lipids and proteins (AOCS lipid library). Chylomicrons have the lowest content of protein relative to lipids. Chylomicrons transport exogenous lipids absorbed from the gut. They are initially transported in the lymph. In chylomicrons the proportions of protein to cholesterol, triglycerides and phospholipids are 1:4:90:5. In chylomicrons the proteins include apolipoproteins A and B (APOA and APOB). Apolipoproteins are produced in the intestine and in the liver. In the plasma the chylomicrons acquire apolipoproteins C and E.

Activation of lipoprotein lipase in capillary walls leads to hydrolysis of triglycerides and generation of chylomicron remnants that are taken up in liver. Chylomicron remnants retain APOE and this facilitates their uptake in liver. Triglycerides are taken up by the low-density lipoprotein receptor (LDLR) and related proteins LDLR-related protein 1 (LRP1) and alpha2 microglobulin receptor.

In very low-density lipoproteins (VLDLs) the ratios of protein to cholesterol, triglycerides and phospholipids are 8:25:55:12. VLDLs are excreted from the liver into the plasma. In plasma they bind

APOE, APOB and APOC and undergo cleavage by lipoprotein lipase, leading to the release of triglycerides to form intermediate density lipoproteins and subsequently low density lipoproteins that remain bound to APOB and taken up by LDLRs.

Low-density lipoproteins (LDLs) have ratios of protein to cholesterol, triglycerides and phospholipids of 20:55:5:20. High-density lipoproteins play key roles in the transport of cholesterol from cells and have ratios of protein to cholesterol, triglycerides and phospholipids 50:20:5:25 (APOC lipid library).

7.2 LDLR

Brown and Goldstein (1979) demonstrated that LDLs are taken up into cells through a specific receptor, the LDLR. There is evidence that the LDLR binds the APOB that is located on the outside of the low-density lipoprotein molecule. Following internalization these low-density lipoproteins enter into coated pits and subsequently they enter the lysosomal system. There they are degraded and cholesterol is released into the cytoplasm. Cholesterol released from lipoproteins inhibits the transcription factor SREBP (sterol regulator element binding transcription factor). This then leads to reduction of transcription of the cholesterol synthesizing enzyme HMG CoA reductase and to reduced transcription of the LDLR gene. Cholesterol released from lipoproteins also activates the function of the ACAT (acyl-CoA cholesterol acyltransferase) enzyme that generates cholesterol esters for storage. This system ensures homeostasis so that when cholesterol is released into the cytoplasm from lipoproteins excess cholesterol is not synthesized.

The SREBP protein is anchored in membranes of the endoplasmic reticulum through its interactions with another protein SCAP (SERBP cleavage activating protein). Goldstein and Brown (2015) reported that SCAP is a sensor of the levels of cholesterol in the membrane. When cholesterol levels are high SCAP is bound to another membrane protein INSIG. When cholesterol levels in the membrane fall SCAP changes conformation and no longer interacts with INSIG. The combination of SREBP and SCAP can then be taken up into specific vesicles and moved to the Golgi. SREBP in the

Golgi undergoes two proteolytic cleavages, one that releases it from SCAP and a second releases a soluble fragment from SREBP. It is this soluble fragment that enters the nucleus and acts as a transcription factor.

Goldstein and Brown (2015) reported that the soluble SREBP factor activates transcription of the LDLR genes and transcription of HMG CoA reductase and all genes involved in the cholesterol biosynthetic pathway. Three different genes encode SREBP proteins that all undergo processing in the SREBP pathway.

In familial hypercholesterolemia, due to defects in the LDLR, homeostasis is not maintained since lipoproteins are not efficiently taken up and cholesterol is therefore not released into the cells. In cholesterol depleted cells the SREBP transcription factor molecules are transported to the Golgi where they undergo cleavage and generation of a soluble product. The soluble SREBP then enters the nucleus and activates transcription of genes involved in cholesterol synthesis.

Plasma LDL levels are elevated in individuals who carry deleterious LDLRs in heterozygous forms and cholesterol levels are highly elevated in individuals who are homozygous for deleterious LDLR mutations.

7.3 Cholesterol Efflux

Rothblat and Phillips (2010) described reverse cholesterol transport that involves removal of cholesterol from macrophages in the arterial wall. Rosenson *et al.* (2012) described components of the efflux pathway from macrophages. This pathway is of great importance since most cells cannot directly degrade cholesterol. Cholesterol degradation takes place primarily in the liver. Efflux of cholesterol from macrophages involves the ATP binding cassette transporter ABCA1 and small high density lipoproteins that are apolipoprotein A1 rich.

Cholesterol efflux from macrophages represents one component of the reverse cholesterol transport system. Other components of this system include non-macrophage arterial specific cholesterol efflux, non-arterial wall efflux, and hepatic cholesterol uptake and cholesterol excretion into the intestine.

Cholesterol efflux involves the high density lipoproteins. Rosenson *et al.* noted that the high-density lipoprotein (HDL) cholesterol concentration is inversely correlated to the cardio-vascular disease risk and is positively correlated with the beneficial effects of statins. APOA1 constitutes 65% of the HDL protein mass; between 12 and 15% of the protein mass is APOA2 and in addition there are small quantities of other proteins in HDL. These include proteins that function in complement regulation, protease inhibition and immunity. They note that HDL also contains a variety of different lipids including sphingosine-1-phosphate.

7.4 Atherosclerosis and Atherosclerotic Plaques

Sergin and Razani (2014) reviewed the pathogenesis of atherosclerosis. They described sequential steps in the generation of atherosclerotic plaques. Initial changes occur in the arterial intima, the region between the endothelium and the smooth muscle of the vessel. The initial change is likely lipid accumulation; excess lipids cause endothelial cells to produce chemokines and cytokines that recruit monocytes and macrophages. Macrophages engulf lipids, mostly ApoB containing lipoproteins. Macrophages take up lipoproteins through receptor mediated endocytosis. Macrophages also secrete cytokines leading to additional recruitments of monocytes, macrophages and extracellular matrix. The macrophages progressively ingest lipids and become foam cells. The accumulations of macrophages and extracellular matrix lead to expansion of the impacted intima region and this may eventually lead to rupture.

Sergin and Razani reported that degradation of the macrophage ingested lipoproteins occurs through autophagy and the endosome lysosome system. They noted that autophagy markers such as LC3 (microtubule associated protein 1A/1B light chain) were readily detectable in atherosclerotic plaques. More than 30 proteins are required to promote autophagosome formation and docking of mature endosomes with lysosomes. Particularly important in this process are autophagy proteins ATG5 and ATG7 and Beclin 1. Release of cholesterol from lysosomes following degradation of lipoproteins leads to initiation of the cholesterol efflux mechanism.

Sergin and Razani report that increased ingestion of lipoproteins by macrophages leading to increased burden of cytoplasmic lipids may trigger organelle damage, e.g. mitochondrial damage, leading to the generation of reactive oxygen species (ROS). Furthermore accumulation of aggregates in macrophages triggers an inflammatory response with interleukin IL1B production. ROS may also trigger DNA damage. ROS may also lead to generation of toxic derivatives of cholesterol and formation of cholesterol crystals. The latter may also lead to damage of lysosomal membranes and leakage of hydrolytic enzymes from lysosomes.

Proposed therapeutic interventions to reduce plaques include use of agents that enhance autophagy, e.g. inhibition of TORC1 with rapamycin-related agents. Inhibition of mTOR stimulates activity of the autophagy promoting kinase ULK1.

7.5 Vascular Calcifications

Vascular calcifications involve crystallization of hydroxyapatite in the extra-cellular matrix and in the intima and are associated with atherosclerotic plaques. Lanzer *et al.* (2014) reported that intima calcifications are linked to presence of lipoproteins and cytokines.

In Monckeberg medial sclerosis, vessel calcification is localized to arteries of the extremities and is associated with type 2 diabetes and end stage renal disease. Occurrence of vascular calcification in chronic renal disease led to studies that demonstrated that in these patients hyperphosphatemia occurs and promotes vessel calcification.

Munnur *et al.* (2014) reported that coronary artery computed tomographic angiography (CCTA) is a robust non-invasive technique to assess coronary artery disease.

Sen *et al.* (2014) reported that coronary artery calcification is a definitive morphologic marker of atherosclerosis and that it strongly predicts risk for coronary cardio-vascular emergencies. They carried out DNA and RNA sequencing in patients with severe coronary calcification. They determined that a specific allele in the gene TREML4 in the single nucleotide polymorphism (SNP) rs2803496 led to a 6.4 fold increased risk of coronary artery calcification. Through analysis of necrotic plaques they determined that the TREML4 protein localizes to areas of micro-calcification in coronary artery plaques.

7.6 Dyslipidemias

7.6.1 *LDLR and APOB defects*

High levels of plasma cholesterol present in LDLs characterize the form of familial hypercholesterolemia analyzed by Brown and Goldstein (1979) and found to be due to mutation in LDLR. These patients present with xanthomas of the skin and with early onset coronary heart disease. Patients with similar presentations have been identified and found to not carry LDLR mutations but have mutations in APOB.

It is important to note that specific mutations in APOB lead to hypercholesterolemia while others are associated with hypocholesterolemia.

7.6.2 *Proprotein convertase subtilisin kexin 9 (PCSK9)*

Reports from the Dallas Heart Study (Cohen *et al.*, 2006) revealed that PCSK9 is involved in degradation of LDLR receptors. Gain of function mutations in PCSK9 lead to increased degradation of LDLR receptors, and to high levels of hypercholesterolemia with increased levels of LDL cholesterol. Loss of function mutations in PCSK9 led to higher LDLR numbers and to lower LDL cholesterol levels.

Further studies revealed that PCSK9 increases lysosomal and endosomal degradation of LDLRs.

7.6.3 *Lipoprotein lipase*

Loss of function mutations in lipoprotein lipase lead to rare monogenic hyperchylomicronemia and to impaired catabolism of triglyceride core proteins. Clinical manifestations include xanthomas, pancreatitis and abdominal pain. Both common and rare variants in lipoprotein lipase contribute to hypertriglyceridemia (Kassner *et al.*, 2015).

7.6.4 *Apolipoprotein C3 (APOC3) loss of function mutation and reduced risk of ischemic vascular disease*

Normal APOC3 functions to inhibit the lipolytic activity of lipoprotein lipase. Through this inhibition the plasma levels of atherogenic triglyceride rich lipoproteins are increased. Loss of function mutations in APOC3 are associated with lower triglyceride levels, and the ratio of HDL cholesterol to LDL cholesterol is increased and coronary heart disease risk is reduced. Jørgensen *et al.* (2014) reported that loss of function mutations in APOC3 led to a risk reduction of 41% for ischemic vascular disease.

7.6.5 *Lysosomal membrane proteins required for cholesterol release and Niemann–Pick disease*

Patients with Niemann–Pick disease have very high levels of plasma cholesterol. Patients with Niemann–Pick disease type 1 carry mutations in the NPC1 gene and patients with Niemann–Pick disease type 2 carry mutations in NPC2. Studies on these patients revealed that specific lysosomal membrane proteins encoded by NPC1 and NPC2 are required for correct processing of lipoproteins in lysosomes and for cholesterol release from lysosomes.

Vance and Peake (2011) reported that treatment of NPC deficient mice with cyclodextrin extended their life span and reduced neurodegeneration. Cyclodextrin is a cholesterol binding compound.

7.6.6 *Rare or low frequency variants and dyslipidemias*

Cumulative family studies have revealed that high cholesterol levels can result from mutations in the LDLR, in PCSK9, in ATP binding cassette transporters ABCG5 and ABCG8, and in LDLR autosomal recessive (LDLRAP1) disorder (Lange *et al.*, 2014).

In order to determine if rare or low frequency coding variants impact LDL cholesterol levels these investigators carried out exome sequencing on 2,005 individuals, including 554 individuals with either extremely high (>98th percentile) or extremely low (<2nd percentile) levels of LDL cholesterol.

Lange *et al.* determined a measure they defined as a burden frequency. The burden frequency was the percentage of individuals with at least one copy of a rare or damaging allele in a specific gene and the impact on LDL cholesterol levels. They also separated out loss of function (LoF) variants and non-synonymous (NS) variants. They determined that burden frequencies were statistically significant for four genes. The genes, types of variants and degree of significance were as follows: PCSK9 LoF variants $p = 3 \times 10^{-18}$; PCSK9 NS variants $p = 7 \times 10^{-17}$; LDLR NS variants $p = 3 \times 10^{-13}$; APOB LoF variants $p = 2 \times 10^{-10}$; PNPLA5 NS variants $p = 3 \times 10^{-7}$.

Specific rare coding variants in PCSK9 and in APOB were associated with low LDL cholesterol levels. Rare variants in PNPLA5 and in LDLR were associated with high cholesterol levels. PNPLA5 encodes patatin-like phospholipase that plays a role in autophagy.

7.6.7 *Polygenic dyslipidemia*

Kathiresan *et al.* (2009) reported results of genome-wide association studies (GWASs) in 20,000 individuals with dyslipidemia. They identified 30 distinct loci associated with abnormal lipoprotein concentrations. These included 19 loci previously associated with dyslipidemias and 11 "new" loci. All of the loci had low effect sizes. They concluded that multiple common variants contribute to dyslipidemias.

7.6.8 *Familial combined hyperlipidemia (FCHL)*

Brouwers *et al.* (2012) reported that FCHL is the most common hyperlipidemia in the western world. In FCHL hypercholesterolemia and hypertriglyceridemia occur; however they occur in different degrees even within the same kindred. FCHL is also associated with high body mass index (BMI) and with small dense LDL particles in plasma.

They reported that extensive lists of genes have been associated with FCHL. The identified genes support the concept that FCHL may arise due to adipose tissue dysfunction, hepatic fat accumulation, fat over-production, disturbed metabolism and clearance of APOB containing particles.

Brouwers *et al.* concluded that the FCHL gene pool consists of a mix of very rare genes that account for FCHL in a few pedigrees and common variants that contribute to the FCHL phenotype in the majority of pedigrees.

7.7 Loci Associated with Atherosclerosis

Holdt and Teupser (2015) reviewed aggregate results from 21 GWASs of atherosclerosis. They noted that in total 24% of the loci identified overlapped with lipid loci; 10% of loci overlapped with loci involved in hypertension and 2% overlapped with loci involved in diabetes mellitus. A region of chromosome 9p21.3 contains SNPs that represented the most replicated loci for atherosclerosis. The SNP rs1333049 that maps to 9p21.3 is the SNP most strongly associated with coronary heart disease. This 9p21.3 locus does not overlap with protein coding genes. It does overlap with the 5′ end of a long non-protein coding.

Holdt and Teupser noted that the chromosome 6p24.1 locus associated with atherosclerosis contain the gene PHACTR1. This locus was identified as being associated with coronary heart disease in studies of European, Asian and Middle Eastern populations. It has also been associated with coronary calcification. PHACTR1 encodes a protein phosphatase.

7.8 Exome Sequencing and Risk Loci for Myocardial Infarction

Do *et al.* (2015) carried out exome sequencing in 9,793 patients with early onset myocardial infarction, onset <50 years in males and <60 years in females. In total 2% of cases of early myocardial infarction harbored damaging LDLR mutation. In controls, 1 in 217 had a damaging coding mutation and had LDL cholesterol levels higher than 190mg/dl.

Their studies revealed that carriers of rare non-synonymous variants in the LDLR gene were at 4.2 fold increased risk for myocardial infarction. Carriers for null alleles of LDLR were at 13 fold increased risk for myocardial infarction. Individuals who were carriers of damaging rare variants in the gene APOA5 had a 2.2 fold increased risk for myocardial infarction and damaging rare variants led to an increase in triglyceride levels.

7.9 Adipose Tissue and Metabolism

There is evidence for the occurrence in humans of three types of adipose tissue (Harms and Seale, 2013). White fat is the most common form, brown fat occurs in specific regions (e.g. interscapular regions) and beige fat occurs within white fat.

White fat stores fats, particularly triglycerides. Brown fat burns fat and it generates heat through the mitochondrial uncoupling protein UCP1. Beige fat also has fat burning (thermogenesis) capabilities. Adipocytes within beige fat are sometimes known as inducible brown adipocytes.

Harms and Seale reported that brown and beige adipocytes had similar levels of UCP1. There is evidence that white fat contains precursor cells that be induced to express UCP1. The transcription factor PRMD16 (PR domain zinc finger protein 16) is enriched in brown fat. PRMD16 expression is increased in beige adipocytes. There is evidence from mouse studies that the PPARG (peroxisome proliferator associated receptor gamma) or agonists of PPARG are induced on beta adrenergic stimulation leading to beige adipocyte development. Thiazolidinediones used in treatment of type 2 diabetes are agonists of PPARG and they increase thermogenic activity of beige adipocytes.

Harms and Seale demonstrated that sympathetic nerve activity activates UCP1. Exposure of individuals to cold serves to activate beige adipocyte development. Cold induces PGC1 alpha (PPARG coactivator) activity. A hormone like molecule irisin was shown to be induced by cold in mice and to stimulate browning of white fat. Irisin, also known as FND15 (fibronectin type III domain containing) is an exercise induced myokine also found in humans. FND15 is translated from an unusual translation start site ATT. Jedrychowski *et al.* (2015) used mass spectrometry to analyze plasma from humans and showed that the concentration of FND15 levels increased in plasma following exercise.

UCP1 and other uncoupling proteins (e.g. UCP2 and UCP3) are mitochondrial proton carriers that carry out reverse flow of proteins across the inner mitochondrial membrane. This reversal of proton flow reduces the proton gradient in the intermembrane space and reduces the generation of ATP. UCP1 is expressed in brown and beige fat. UCP2 is widely expressed and UCP3 is primarily expressed in skeletal muscle. The UCP effect of mitochondrial function is sometimes referred to as proton leak.

Lidell *et al.* (2014) reported that in adult humans brown fat is of the beige type, and that beige type adipose tissue is present in most humans. The activity of brown or beige fat is increased in hyperthyroidism; it is decreased in obesity and in type 2 diabetes. Brown or beige fat is increased on exposure to cold. The stimulus for this increased activity is dependent on hypothalamic action and increased norepinephrine production followed by increased stimulation of beta adrenergic receptors on adipocytes. Activation of brown or beige fat involves increased activity of uncoupling protein 1 and increased mitochondrial biogenesis. Increased beta adrenergic activity leads to increased activity of PGC1 alpha that activates increased expression of PPARG, increased expression of UCP1 and increased expression of thyroid hormone expression. In addition there is increased expression of DIO2 (deiodinase iodothyronine) that generates formation of tri-iodothyronine from the less active thyroxin T4. They also reported that increased expression of the transcription factor PRMD16 stimulates browning of white adipocytes and mitochondrial biogenesis. EHMT1 (euchromatic histone lysine methyltransferase) interacts with and stabilizes PRMD16. Increased activity of the

NAD-dependent protein deacetylase sirtuin 1 also induces browning of white adipose tissue.

Stanford *et al.* (2015) reported evidence that physical exercise promotes beiging of white adipose tissue, increased expression of UCP1 and of adipocyte adipokines.

7.10 GWASs in Obesity with Increased BMI

GWASs have revealed that at least 97 loci are significantly associated with increased BMI. However the effect size of these loci is generally low. The fat mass and obesity associated (FTO) locus was found to be significantly associated with BMI in several studies. Pei *et al.* (2014) reported that the significance level of the association of FTO with increased BMI was $p = 1.97 \times 10^{-14}$. However the presence of the risk allele, i.e. the minor allele, only increases the risk of obesity by 1.2%. The genetic variants in the FTO locus that increase obesity risk are located within introns 1 and 2 of the FTO gene.

Claussnitzer *et al.* (2015) carried out detailed studies to investigate the underlying mechanism through which the FTO locus increases obesity risk. They evaluated epigenomic data, allelic activity, DNA sequence motifs, regulatory effects and co-expression patterns. They also further validated their predictions through genome editing studies. In the key FTO polymorphic marker, SNP rs1421085, the non-risk allele is T and the risk allele is C. Their studies on human subjects involved 100 healthy individuals with normal BMI, including 52 participants who were homozygous for the risk allele C and 42 participants who were homozygous for the non-risk T allele.

Their studies revealed firstly that the FTO obesity linked genomic region has enhancer activity. Secondly they determined that the risk allele C disrupts binding of a repressor ARID5B. In the presence of the risk allele C, the repressor fails to bind and there is increased expression of two homeobox genes IRX3 and IRX5. The increased expression of those two genes favors development and differentiation of adipocytes of white adipose tissue. This then shifts fat balance to an energy storage mode rather than to a thermogenesis mode.

7.11 Metabolic Syndrome and other Acquired Metabolic Disorders

Features of metabolic syndrome include central obesity (waist to hip measurements include larger waist than hip circumference), dyslipidemia, hypertension, atherosclerotic heart disease, and increased insulin resistance that may develop into diabetes. Disruptions in a range of different physiological mechanisms have been identified in metabolic syndrome.

7.11.1 *Mitochondrial dysfunction*

Gao *et al.* (2014) reviewed evidence for mitochondrial dysfunction in acquired metabolic disorders. Their review included discussion of energy stress, its impact on mitochondrial function and possibilities for treatment of metabolic diseases with impacted mitochondrial dysfunction. They centered discussions on aspects of metabolic flexibility and maintenance of energy homeostasis.

Gao *et al.* noted that mitochondrial structure and function differ between insulin sensitive and insulin resistant individuals and that deficient mitochondrial function limits the capacity of the individual to oxidize fats. They noted further that accumulation of lipids in non-adipose tissues (including muscle) is associated with insulin resistance and that down regulation of lipid oxidation and mitochondrial metabolism occur in type 2 diabetes mellitus.

These investigators note that for maintenance of energy homeostasis adequate mitochondrial activity is required, and in the longer term, increases in mitochondrial number may be required. Energy homeostasis requires mitochondrial lengthening, fission and fusion processes and disposal of mitochondria when their capacities to generate adequate membrane potential are compromised.

Gao *et al.* report that there is a growing body of evidence on the importance of protein acetylation and deacetylation processes for maintenance of energy homeostasis. The mitochondrial NAD-dependent deacetylase sirtuin 3 (SIRT3) plays a key role in acetylation deacetylation dynamics. Prolonged high fat intake decreases SIRT3 activity and leads to mitochondrial dysfunction.

Fasting activates SIRT3 activity and facilitates fatty acid oxidation and efficient electron transfer. Under conditions of nutrient stress and insufficiency the AMP-activated protein kinase (AMPK) is activated and this activates catabolic processes. Caloric excess and lipid loading decreases efficiency of the mitochondrial fission and fusion processes.

Gao *et al.* note that increased oxidative capacity can be achieved by increasing mitochondrial biogenesis. Expression of PPARG and its coactivator PGC1 alpha increases mitochondrial biogenesis.

7.11.2 *Genetic studies in obesity and metabolic syndrome*

Fall and Ingelsson (2014) reviewed genetic studies in obesity and metabolic syndrome. They noted that in the 1990s loss of function mutations in specific genes was reported in syndromic obesity. These include the pro-opiomelanocortin gene (POMC), the melanocortin 4 receptors (MC4R) and the leptin receptor gene. They noted that MC4R mutations lead to the most common form of monogenic obesity in childhood. However the monogenic forms of obesity are very rare in the general population. Syndromic obesity occurs in the genomic imprinting disorder Prader–Willi syndrome and in the genetically heterogeneous disorder Bardet–Biedl syndrome that can arise due to mutations in any one of 15 genes.

In reviewing results of GWASs Fall and Ingelsson emphasized that association between the phenotype and the genetic variants must reach significance levels of $p = 5 \times 10^{-8}$ to be considered valid. Their review indicated that 51 different genes were significantly associated with increased BMI and that 10 genes were significantly associated with extreme obesity. The genes associated with extreme obesity included FTO locus and variants near the MC4R encoding gene. The MC4R gene is therefore apparently associated with both monogenic forms of obesity and common forms of extreme obesity. The MC4R gene is expressed in the hypothalamus and the encoded protein undergoes post-translational cleavage to produce alpha melanocyte stimulating hormone. This peptide has appetite reducing effects.

For GWASs in metabolic syndrome Fall and Ingelsson (2014) noted that the most significant locus in studies on European

Americans was located near the APOC1 locus and the significance of the association with this locus rs4420638 was $p = 1 \times 10^{-57}$. The significance of association of this locus and atherogenic dyslipidemia was $p = 1 \times 10^{-31}$.

Fall and Ingelsson emphasized that many of the obesity associated SNPs and the metabolic syndrome associated SNPs were in non-protein coding regions of the genome. Furthermore only small risk effects were observed even with significantly associated loci.

7.12 Type 2 Diabetes: Genetic and Life-style Risk Factors

7.12.1 *GWASs in type 2 diabetes*

Rutter (2014) reported that comprehensive GWASs in type 2 diabetes have implicated more than 500 different loci. Rutter noted that the majority of the associated loci fall in the non-protein coding regions of the genome; however there are exceptions. A number of the associated loci were found to affect insulin production rather than insulin action.

The SLC30A8 gene encodes a zinc transporter that is highly enriched in insulin storing granules in pancreatic beta cells. The type 2 diabetes associated polymorphism in the ADCY5 gene lies in an intron of the gene. These polymorphisms impacts expression of ADCY5 in pancreatic islet cells and carriers of the risk allele have elevated fasting glucose levels.

Rutter reported that the diabetes associated gene CDKAL1 (CDK regulatory associated protein like 1) encodes an RNA modifier. The risk allele that affects insulin secretion alters the tRNA for lysine leading to misreading of lysine in the pro-insulin translation product.

A type 2 diabetes associated polymorphism occurs in MTNR1B (melatonin receptor gene). Rutter noted that this finding provided new insights into how perturbation of circadian rhythms impacts type 2 diabetes risk.

Rutter emphasized that many of the other variants found to be significantly associated with type 2 diabetes in GWASs act in different tissues and they might impact insulin levels even though they do not impact pancreatic islet cells.

7.12.2 *Diet lifestyle and genetic risk factors in type 2 diabetes*

Ardisson Korat *et al.* (2014) reviewed results of three large cohort studies carried out in the USA. These cohorts included the Nurses' Health Study which contained data from 121,700 nurses followed from 1976 onwards and a second cohort of 116,000 nurses enrolled since 1989. The third cohort included the Health Professional follow-up study. This study began in 1986 and includes follow-up on 51,529 male health professionals.

Ardisson Korat *et al.* noted that the cohort studies included GWAS analyses. They reported that results from the large cohort studies provided convincing epidemiological evidence that a healthy diet, regular physical activity, maintenance of a healthy weight, moderate alcohol consumption, avoidance of sedentary behaviors and smoking would prevent the majority of type 2 diabetes cases.

7.13 Key Gene Products Involved in Maintenance of Energy Homeostasis

7.13.1 *Peroxisome proliferator activator receptors (PPARs)*

Ahmadian *et al.* (2013) reviewed information on PPARs, nuclear receptors that function as transcription regulators. The superfamily of these ligand inducible transcription factors includes PPAR alpha, PPAR beta/delta and PPARG. PPAR sub-units combine with other sub-units including retinoid receptor X sub-units to control expression of genes involved in adipogenesis, lipid metabolism, metabolic homeostasis and inflammation.

Distinct functional domains in PPAR include the N-terminal transactivation domain, a DNA binding domain and a C-terminal ligand binding domain. Ahmadian *et al.* reported that the three PPARs each have different tissue distribution, respond to different ligands and manifest functional differences.

PPAR alpha is expressed primarily in liver and brown fat and plays a major role in lipid metabolism. PPAR beta/delta activates fatty acid oxidation and is expressed in liver, heart and skeletal muscle. PPARG1 is expressed in many tissues. PPARG2 is expressed primarily in adipose tissue including white and brown fat, and it impacts adipogenesis, lipid metabolism, insulin sensing, glucose homeostasis and glucose transport.

In individuals who have dominant negative mutations in the PPARG gene, insulin resistance and lipodystrophy occurs.

In their review Ahmadian *et al.* focused primarily on PPARG. They noted that oral medication used to treat type 2 diabetes included ligands of PPARG, thiazolidine (TZD) and rosiglitazone; however these compounds have significant deleterious side effects.

TZD increases insulin sensitivity. However side effects include weight gain, fluid retention and heart problems. Ahmadian *et al.* noted that improved understanding of PPARG functions may expedite development of new classes of TZDs with fewer side effects.

7.13.2 *PGC1 family of transcriptional co-activators*

Spiegelman (2007) reported that the PGC1 transcriptional coactivators promote oxidative metabolism and mitochondrial biogenesis. He noted that in addition PGC1 coactivators promote transcription of genes that encode proteins and enzymes that protect against damaging ROS.

Summermatter *et al.* (2013) reported that exercise activates expression of PGC1 alpha and that activation in turn enhances lipid oxidation. They determined that that response was dependent on the PGC1 alpha function as co-activator of nuclear receptor X that promotes expression of fatty acid synthase.

Summermatter *et al.* demonstrated that PGC1 alpha is a key regulator of tissue and systemic lactate homeostasis. Specifically they determined that PGC1 alpha expression leads to remodeling of lactate dehydrogenase enzyme (LDH) so that LDHB subunits predominate and the content of LDHA sub-units is reduced. This shift facilitates use of lactate as oxidative fuel during exercise and that counteracts accumulation of lactic acid and lactate in the blood during exercise.

They also noted that disarranged metabolism of lactate metabolism arises in skeletal muscle in obesity and in type 2 diabetes. They proposed that exercise and increased expression of PGC1 alpha might improve the metabolic state in these cases.

7.13.3 *Sirtuins*

Many studies revealed that calorie restriction can delay onset of aging and aging related disorders including diabetes. Hall *et al.* (2013) reported that there is some evidence that sirtuins may play roles in the beneficial effects of calorie reduction. They noted that there are seven sirtuins in mammals. The different sirtuins share a conserved NAD binding domain but differ with respect to other domains. They also differ with respect to the sub-cellular domains in which they are expressed. Different sirtuins also differ to some extent in their biological functions. However all sirtuins have protein deacetylase activity and deacetylate lysine residues in proteins.

SIRT3 is expressed in mitochondria and is perhaps the best characterized sirtuin. Hall *et al.* reported that SIRT3 impacts enzymes that play roles in oxidation of long chain fatty acids. In addition SIRT3 deacetylates protein components of the mitochondrial electron transport complexes and this expedites oxidative phosphorylation.

McDonnell *et al.* (2015) reported that SIRT3 knockout in mice led to impaired fatty acid oxidation. They also reported that mice fed a high fat diet have reduced SIRT3 activity and they develop impaired glucose tolerance and insulin resistance. They noted however that the molecular mechanisms that alter SIRT3 activity are not fully understood.

7.13.4 *Adenosine monophosphate activated kinase (AMPK)*

AMPK activity stimulates glucose transport, mitochondrial function and fatty acid oxidation and it inhibits mTORC1 activity. AMPK activity can enhance autophagy and thereby impact endoplasmic reticulum stress. Ruderman *et al.* (2013) reported that there is a close relationship between AMPK dysregulation and insulin resistance. Sustained decreases in AMPK activity lead to insulin resistance. AMPK activation also leads to increased insulin sensitivity.

AMPK is composed of three sub-units: The catalytic alpha sub-unit, and regulatory beta and gamma sub-units. In addition each subunit gives rise to multiple isoforms.

7.13.5 *Reactive oxygen species (ROS)*

ROS are derived in part from electron transport processes in mitochondria. In these processes superoxide radicals arise particularly from activities of complex I and complex III and are generated both on the matrix side and in the inner mitochondrial space (Holmström and Finkel, 2014). They noted that ROS are also generated in mitochondria through the activities of glycerol-3-phosphate dehydrogenase (GPDH), flavoprotein ubiquinone oxidoreductase (FQR), pyruvate dehydrogenase and 2-oxoglutarate dehydrogenase.

ROS are also generated through activities of NADPH oxidases NOX1, 2, 3 and 4. These enzymes transfer electrons from the cellular interior across the cell membrane. Transferred electrons are coupled to molecular oxygen to produce superoxide radicals. This process plays important roles in the innate immune systems in destruction of micro-organisms. Atomic oxygen has two unpaired electrons that occupy separate orbits in the outer shell. Reduction through the addition of electrons leads to superoxide radicals. These include superoxide where only one unpaired electron is present in the outer shell and peroxide where there are no unpaired electrons present.

It is important to note that ROS do play important roles in physiological processes; however excessive quantities of ROS have damaging effects on lipids, proteins, DNA and RNA. Zuo *et al.* (2015) reported that although all amino acids are vulnerable to oxidative damage, sulfur containing amino acids (particularly cysteine) and aromatic amino acids are most susceptible. Oxidative damage can lead to abnormal crosslinking, denaturation, and aggregation of proteins. Kermanizadeh *et al.* (2015) reported that cell membranes are particularly vulnerable to oxidation due to the presence of high concentration of poly-unsaturated fats with double bonds. ROS can also damage DNA and RNA, and guanine is particularly sensitive to ROS damage and oxidized triphosphates are generated (e.g. 8-oxo-dGTP or 8-oxguanine). These derivatives may be removed through hydrolysis or in DNA through the action of repair mechanisms. Damaged RNA is removed through action of ribonuclease.

Damaging effects of ROS produced by NOX enzymes has also been reported. Gray *et al.* (2013) reported that in hyperglycemic stress there is increased expression of NOX enzymes in epithelial cells. They reported that mouse models with hyperglycemia and APOE deficiency develop profound atherosclerosis. The atherosclerosis is mitigated if NOX1 is knocked out.

Nunomura *et al.* (2012) emphasized that the brain is particularly susceptible to oxidative damage because of the high content of fatty acids and high oxygen consumption.

Kermanizadeh *et al.* noted that a number of physiological systems are in place to reduce ROS; gamma-glutamyl-L-cysteinyl-glycine is an abundant thiol that in its reduced form is an anti-oxidant. Two molecules of reduced form glutathione (GSH) are oxidized to form glutathione disulfide (GSSG). They note that GSH occurs in the nucleus, mitochondria and endoplasmic reticulum. GSH can act directly or indirectly with ROS.

Superoxide dismutase enzymes (SOD1, 2 and 3) are important in defense against ROS. SOD1 occurs in the cytoplasm, and SOD2 and SOD3 are located in the mitochondria and in the extra-cellular matrix. The enzyme catalase is important in neutralizing the hydrogen peroxide.

8. Early-Onset Neurocognitive and Neurobehavioral Disorders

Intellectual impairment combined with congenital malformations in some cases result from monogenic defects. However early-onset intellectual impairment may be multifactorial in origin. The complexity of multifactorial disorders derives from the fact that many different genetic factors, and in some instances, interacting environmental factors, play important roles in their etiology. Furthermore, despite heterogeneity in causative mechanisms the phenotypes may be indistinguishable. Early-onset cognitive disorders serve as examples of this complexity.

8.1 Structural Brain Abnormalities, and Insights into Genes and Pathways Involved in Brain Development

Insights into genes and pathways involved in brain development have been obtained in part through studies in patients with structural brain abnormalities. Earlier classification defined three major groups of malformations of cortical development. These include malformations of cell proliferation, defects in neuronal migration and post-migration defects in cortical organization and connectivity.

Guerrini and Dobyns (2014) noted however that these boundaries might be artificial since a broad range of brain malformations occur with mutation in a specific gene, e.g. with mutations in WDR62

(WD repeat protein 62), DYNC1H1 (dynein cytoplasmic heavy chain) or TUBG1 (Tubulin gamma 1).

8.1.1 *Clinical presentations of malformations and biological pathways of cortical development*

Guerrini and Dobyns (2014) reported that malformations of cortical development (MCD) have variable presentations and variable degrees of disability. Infants may present with feeding difficulties, seizures, developmental delay or with unusual head size or head growth. Abnormal movements may be present. Defects in autonomic functions sometimes occur and additional congenital anomalies may be present.

They noted that more than 100 different gene defects have been documented in cases with malformations of cortical development. These include metabolic defects such as mitochondrial and pyruvate metabolic defects, glycosylation defects, non-ketotic hyperglycinemia and peroxisomal biogenesis defects.

Biological pathways implicated in malformations of cortical development include cell cycle regulation, cell growth and proliferation, and cytoskeletal functions. In addition defects at a number of different sites in the phosphatidyl inositol-AKT-MTOR signaling pathway occur in these conditions. It is important to note that germline defects and post-mitotic mutations in genes in that signaling pathway lead to structural brain defects.

Specific gene defects in signaling pathways documented in cases of MCD include mutations listed in Table 8.1(a). Normal functioning in the Reelin pathway is essential to normal brain development. Reelin pathway genes shown to be mutated in MCD are listed in Table 8.1(b). Additional gene defects leading to structural brain malformations and microcephaly are listed in Table 8.1(c).

8.2 Segmental Chromosomal Copy Number Variants (CNVs) in Developmental Delay

There is abundant evidence for increased burdens of CNVs, including deletions and duplications in cases of neurodevelopmental delay. Cooper *et al.* (2011) analyzed 15,767 DNA samples from individuals

Table 8.1. Specific genes and pathways in which mutations occur leading to cortical malformations.

(a) **Signaling pathways**

FGFR3 fibroblast growth receptor 3
PIK3R2 phospho-inositide-3-kinase regulatory sub-unit 2
PIK3CA phosphoinositol-4-5 biphosphate- 3 kinase catalytic sub-unit alpha
PTEN phosphatase and tensin homolog
AKT1 and AKT3 serine threonine kinases
EZH2 polycomb repressor complex secondarily impacts PI3K MTOR pathway
TSC1 and TSC2 tuberous sclerosis genes
MTOR mechanistic target of rapamycin

(b) **Reelin pathway defects in structural brain defects**

RELN reelin extra-cellular matrix protein
VLDLR very low density lipoprotein receptor
LIS1 (PAFAH1B1) platelet activated acetyl hydrolase 1B
NDE1 nudE neurodevelopmental protein
DYNC1H1 dynein cytoplasmic heavy chain
KIF5C kinesin family member 5C
KIF2A kinesin heavy chain 2A
DCX doublecortin
WDR62 WD repeat domain 62
TUBA1A tubulin alpha 1A
TUBA8 tubulin alpha 8
TUBB2B tubulin beta 2B class IIb
TUBB3 tubulin beta class II
TUBG1 tubulin gamma 1

(c) **Other genes associated with structural brain defects including microcephaly**

RAB3GAP1, RAB3 GTPase
RAB3GAP2, RAB GTPase hydrolysis RABGTP to RABGDP
RAB18 RAS family gene
EOMES (TBR2) eomesoderm T box brain protein (mesoderm protein)
PAX6 paired box 6

---

with a diagnosis of intellectual disability and/or developmental delay. They noted that within this cohort there were also patients with congenital malformations, hypotonia, feeding difficulties, growth retardation and epilepsy. Furthermore 7.3% of cases were defined as intellectual disability and autism. They compared CNV content of cases and controls. Controls included 83,229 DNA samples from adults.

The studies reported by Cooper *et al.* revealed a frequency of CNVs larger than 400kb in 25.7% of cases with developmental delay and in 11% of controls. Furthermore disease risk increased as the number of CNVs in the case increased.

Coe *et al.* (2012) reported that cases with congenital malformations manifested the greatest increase in CNVs. The second largest category of patients with increased burdens of CNVs included cases of intellectual disability or developmental delay where the diagnostic yield of pathogenic CNV is 14.2%; the diagnostic yield of pathogenic CNVs in cases of autism is 6.8%. They reported that CNVs larger than 500 kb and *de novo* CNVs in probands but not present in parents could most convincingly be classified as pathologic. Smaller variants were more likely to be inherited. Coe *et al.* noted however that in some cases small CNVs, and in particular, deletions that impact specific sites in the genome were likely pathogenic. Coe *et al.* noted that there was also evidence that combinations of CNVs in a specific patient were likely pathogenic.

Coe *et al.* emphasized that the same primary genetic lesion can result in different outcomes in different individuals. They proposed that these differences in outcomes might result from differences in genetic background. There is evidence that identical large CNVs may result in different disease outcomes. Another important finding reported was that a CNV in a specific position in the genome may be associated with different disease phenotypes. One example involves CNVs that lead to deletions in chromosome 15q13.3. These occur in cases of autism, epilepsy and in some patients with schizophrenia.

Unbalanced translocations also occur in some cases of intellectual disability. Figure 8.1 illustrates an unbalanced translocation resulting in increased dosage of chromosome 2pter genes (labeled green). Figure 8.2 shows results of microarray analysis revealing increased dosage of the 15q12-q13 genes.

8.3 X-linked Intellectual Disability (XLID)

In 2009, Gécz *et al.* reviewed the genetic landscape of XLID. They reported that this disorder affected between 1 in 600 and 1 in 1,000 males. This disorder may be part of a syndrome or it may be

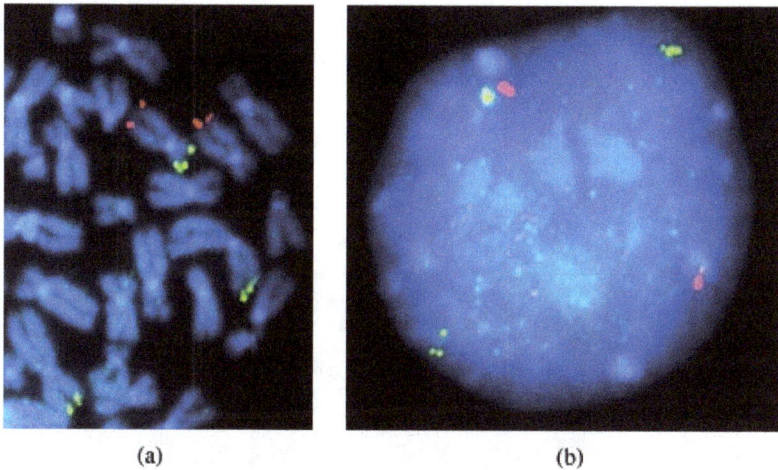

(a)   (b)

Figure 8.1. Partial metaphase (a) and interphase nucleus (b) from patient with intellectual disability. Gene on chromosome 2p terminal region labeled with green dye. Green on chromosome 12q terminal region labeled with red dye. Fish studies reveal presence of two copies of chromosome 2. In addition a segment of chromosome 2p has translocated to chromosome 12p region. Dosage imbalance occurs with increased copies of the chromosome 2pter region. Original case photo by Moyra Smith.

non-syndromic. They documented that defects in any one of 90 genes on the X chromosome (11% of the total number of X chromosome genes) could lead to intellectual disability. In 25% of cases protein truncating mutations were found. They noted that CNVs or missense variants likely cause disease. However, the functional relevance of these mutations needed to be verified. In addition, in 50% of the families they studied no clear cut mutations were found.

Gécz *et al.* noted that three major hotspots of structural variants were associated with X-linked cognitive impairments. One occurred at the MECP2 locus on chromosome Xq28. The second occurred in the hydroxysteroid dehydrogenase locus at Xp11.22. The third was located in the Xq26.2-q27 region and it involved the SOX3 locus and was associated with hypopituitarism.

Duplications at the MECP2 locus lead to mild to moderate intellectual disability. Deletions in MECP2 lead to Rett syndrome in females and this syndrome is associated with loss of cognitive

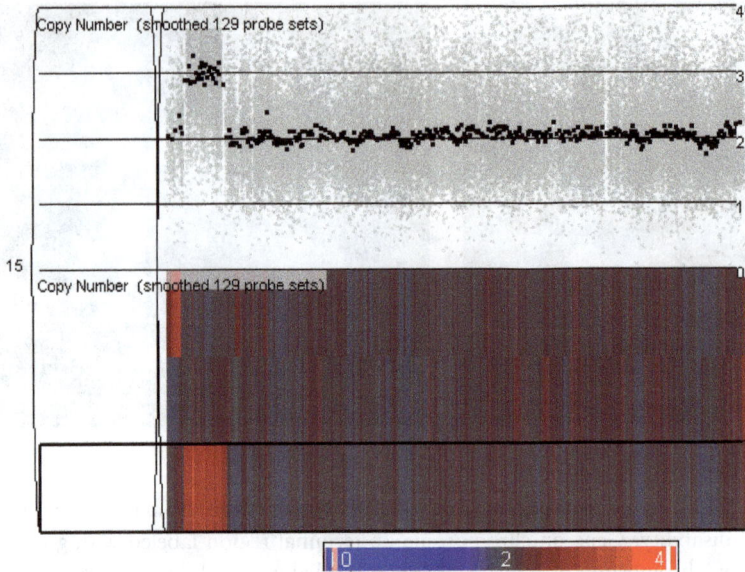

Figure 8.2.   Image of micro-array analysis of chromosome 15 of autism subject and parents. There is evidence of increased red signal due to increased dosage of genes in the 15 q12-q13 region in the autistic proband (bottom lane). Dosage changes in the 15q12-q13 region represent one of the most common CNVs in cases of autism. One parent (top lane) has increased signal in the 15q11 region. The region is poly-morphic and increased signal in 15q11.1 region is not pathogenic.

abilities, autism and neurological symptoms including impaired hand use and movement. Deletions of MECP2 are usually lethal in male fetuses.

Pelizaeus–Merzbacher disease is a demyelinating disorder due to duplication within the PLP1 gene (encodes proteolipid protein 1) on Xq21.22, This disease is associated with neurological impairments and mild intellectual impairment. Gécz *et al.* tabulated functional categories of genes involved in non-syndromic intellectual disability.

The functional categories were: Transcriptional regulation 9 genes; signal transduction 10 genes; kinase activity and post-translational modification 3 genes; regulation of actin cytoskeleton 2 genes; cell adhesion 2 genes; ubiquitination 2 genes; ion transport 1 gene; other functions 3 genes.

8.3.1 *Family studies in XLID*

Hu *et al.* (2015) reported that results of their prior studies on 208 families with XLID utilized linkage studies, Sanger sequencing and CNV analysis. Those studies led to identification of pathogenic genetic or genomic variants in about half of the families studied. In 2015 these investigators reported the results of high throughput sequencing of the X chromosome in the families without prior genetic or genomic diagnoses. All cases had normal karyotype analyses and fragile X FMR testing results were normal.

Hu *et al.* carried out exome sequencing on 745 X chromosome genes. They then carried out analyses to determine if potentially clinical relevant sequence changes co-segregated with intellectual disability in specific families. In 19 families they detected likely causative sequence variants. Results revealed likely causative variants in established XLID disease in some cases. In other cases novel deleterious changes were found in X-linked genes. One of these genes, CLCN4, encodes a chloride channel that was found to carry deleterious mutations in five unrelated families. Studies on a model organism (mouse) revealed that defects in homologs of the human CLCN4 gene caused neuronal defect. In knock-out mice depleted of Clcn4, neuronal differentiation was impaired. In studies in mice Hu *et al.* reported defects in other confirmed XLID genes Cnksr2 (Pdz10) that encodes a scaffold protein for mitogen-activated protein kinase (MAPK) genes and Fmrpd4 (Pdzk10) that encodes a regulator of dendritic spine morphogenesis. These products of these genes interact with post-synaptic density proteins.

Hu *et al.* used the Ingenuity program and literature searches to examine protein-protein connections and networks of XLID gene products. They identified three major cellular networks/pathways in which XLID gene products were located. The networks they defined also included gene products that were not encoded by X-linked genes. The three networks were:

(i) PSD95 (post-synaptic density 95), Rho/Rac signaling
(ii) Transcription translation network that includes RNA polymerase II, ribosomal sub-units and eIF (eukaryotic initiation factor)-type translation initiation factors EIF1B, EIF3A, EIF2S1,

EIF2B1, EIF6, mediator sub-units MED12, MED13, and histone methyltransferases

(iii) Ubiquitination network that includes KLHL15 (Kelch-like,) CUL4B CUL2 and USP27X (E3 ubiquitin ligase)

Hu *et al.* considered possible explanations for the fact that in one third of families with XLID, genetic defects were not detected on exome sequencing. Possible explanations included importance of variants in introns and variants in non-protein coding regions of the genome. Regulatory variants might play roles. They also noted that in some cases the intellectual disability might be multigenic in origin and could involve defects in autosomal genes despite the fact that the inheritance pattern suggested X-linked inheritance.

8.4 Autosomal Recessive Disorders Manifesting Primarily with Developmental Delay and/or Cognitive Impairment

Najmabadi *et al.* (2011) studied causative genetic factors in autosomal recessive cognitive impairment. They carried out homozygosity mapping followed by exon enrichment and next generation sequencing in consanguineous families with cases of intellectual disability. Most of the 90 families in their study were from Iran; 10% of the families had Turkish or Arabic backgrounds. They noted that most cases they studied had moderate or severe intellectual disability and in some cases patients were also diagnosed with autism spectrum disorders. These authors emphasized that homozygosity mapping is the strategy of choice for mapping recessively inherited diseases in consanguineous families. They utilized single nucleotide arrays to identify homozygous genomic intervals. They noted that in many families homozygous intervals were large. They also carried out linkage studies if both affected and unaffected offspring were available.

Najmabadi *et al.* reported that in 115 of the 136 affected cases plausible causal defects were found. In 78 of the 115 cases specific disease causing mutations were found that impacted single genes. These 78 cases included 26 cases with rare metabolic defects or defined syndromic genetic disorders. The metabolic defects included folate receptor deficiency, mild Tay–Sachs disease, hydroxyglutaric aciduria, Sanfilippo mucopolysaccharidoses, pyruvate dehydrogenase deficiency, peroxisome biogenesis disorder, congenital disorder of glycosylation, glucose transporter (SLC2A1) defects, and Leigh syndrome. Monogenic syndromic disorders identified included Joubert syndrome (AHI deficiency) Bardet–Biedl syndrome due to BBS7 deficiency, spinocerebellar ataxia (PRKCG protein kinase C gamma) deficiency, and ponto-cerebellar hypoplasia (VRK1 protein kinase deficiency). Taking neurological symptoms into account is clearly important in establishing diagnoses in cases with intellectual disability. Najmabadi *et al.* noted that in highly inbred families two different recessive defects sometimes co-occur.

Other defects encountered in this cohort of patients include defects in chromatin modifiers and in chromatin regulatory genes. These defects include defects in mediator sub-unit MED13L and defects in LARP, a regulator of RNA polymerase II transcription. In addition homozygous defects in POLR3B occurred in some patients. POLR3B encodes a core component of the enzyme involved in synthesis of transfer RNA (tRNA). Pathogenic mutations were identified in TRMT1 RNA methyltransferase that acts on tRNA. Homozygous mutation occurred in EEF1B2 that is involved in translation and specifically in transport of amino acyl tRNA to ribosomes. Mutations in the histone demethylases KDM5 and KDM6 were found. In three families pathogenic mutations occurred in histone encoding gene HISTH1, HISTH4B and HIST3H3.

Homozygous mutations in CCNA2 (cyclin A2) impacted cell cycle control, mutation in LAMA1 impacted cell migration and protein degradation was impacted by mutations in UBR7. Mutations in the calcium ion channel component CACNA1G occurred. Defects also occurred in signaling pathway components in the RAS-RHO and RALGDS pathways.

8.4.1 *Types of mutations*

Putative disease causing mutations found in this study included 28 protein truncating mutations, frame shift, splice site and nonsense mutations, in frame deletions, and exon deletions. Najmabadi *et al.* noted that the mutations they found frequently occurred in genes that are involved in fundamental cellular processes such as transcription and translation, cell-cycle control, energy metabolism and fatty-acid synthesis. They emphasized the extreme causative heterogeneity in cases of autosomal recessive intellectual disability.

8.5 Whole Genome Sequencing in Severe Intellectual Disability

Gilissen *et al.* (2014) carried out whole genome sequencing in 50 patients with severe intellectual disability (IQ <50). They also sequenced parental DNA. Sequencing was carried out with 80 fold coverage. They focused their analyses on single nucleotide variants (SNVs) and on CNVs. They then investigated whether variants they found had been reported in at least five other cases with intellectual disability, or overlapping phenotypes including autism, schizophrenia or epilepsy. The gene list they checked against included 528 known intellectual disability genes.

Gilissen *et al.* determined that 84 of the *de novo* mutations they found in their whole genome-sequenced cases occurred in previously described intellectual disability genes. These mutations include insertion deletion mutations, nonsense mutation and missense mutations in highly conserved nucleotides. Three of the *de novo* SNVs seemed likely to be present in mosaic form.

They also identified structural variants, including CNVs detected on sequence analysis although they had not been detected in prior microarray analyses. Three of the CNVs were smaller than 10kb and four of the CNVs impacted known intellectual disability genes and in six cases the CNVs encompassed exons. Alignment of sequence provided accurate breakpoint information.

Gilissen *et al.* noted that conclusive diagnoses were made in 21 of the 50 patients they studied.

8.6 Autism Spectrum Disorders

The core features of autism include impaired social interactions, impaired communication, repetitive stereotypic behaviors and restricted interests. Many investigators have noted that there is heterogeneity in clinical features between patients and even in affected members in the same family. In addition symptoms may extend beyond the core manifestations and include one or more of the following: Attention deficit hyperactivity, intellectual disabilities, mood disorders, motor disorders including hypotonia and apraxia, sensory abnormalities and eating disorders (Brandler and Sebat, 2015). In addition autism may occur in syndromic disorders (e.g. tuberous sclerosis and fragile X syndrome). In those cases patients have manifestations of the specific syndrome and autism manifestations.

8.6.1 *Autism heritability*

A number of studies designed to determine heritability in autism have focused on twins. Colvert *et al.* (2015) analyzed data from the United Kingdom Twins Early Development Study. Twins underwent a number of tests including the childhood autism spectrum test (CAST, 6,423 pairs), autism diagnostic observation schedule (ADOS, 203 pairs) and autism diagnostic interview (ADI, 207 pairs). These investigators reported that on all autism spectrum disorder measures, diagnostic correlation for monozygotic twin pairs were much higher (0.77–0.99) than correlates for dizygotic twin pairs (0.22–0.65).

Sandin *et al.* (2014) examined data from a Swedish population cohort of 2,049,973 children born between 1982 and 2006. Autism disorder was diagnosed according to ICD (International Classification of Diseases) codes. In this cohort 14,516 (0.7%) were diagnosed with autism spectrum disorder and 5,689 (0.2%) had autistic disorder. In this study the relative recurrence rate increased with genetic relatedness. Sandin *et al.* determined that the relative recurrence rate for full siblings was 10.3, for maternal half-siblings it was 3.3, and for paternal half-siblings it was 2.9. The relative recurrence rate for cousins was 2.0. These studies revealed that genetics accounted for 50% of autism liability.

8.6.2 *Different rare autism predisposing variants in siblings*

Yuen *et al.* (2015) carried out whole genome sequencing in 85 quartet families, each with two siblings with autism. They examined results in 36 families to search for mutations in genes defined in prior studies as autism risk genes or as putative autism risk genes. Their studies revealed that in 25 of the 36 families, the two affected siblings did not share the same rare penetrant risk variant.

It is tempting to conclude from this study that rare variants found were not the only risk factors that contributed to autism in these cases.

8.6.2.1 *Genetic variants that play roles in autism*

A number of studies that utilized exome sequencing have led to the discovery of rare *de novo* damaging gene variants in autism probands. A few studies have also identified inherited damaging variants in autism probands. De Rubeis *et al.* (2014) identified a set of 33 transmitted and *de novo* damaging variants in autism probands. These genes function primarily in synapses, in transcription regulation and in chromatin remodeling. These genes were grouped into three categories based on levels of their false discovery rate (FDR) predictions between <0.01, 0.05 and <0.1. The FDR is an estimation of type I errors in null hypothesis testing when conducting multiple comparisons.

It is important to note that 20 of the 33 genes also appear in the Simons Foundation Autism Research Initiative (SFARI) list of autism risk genes (see Table 8.2).

8.6.2.1.1 Estimates of contribution of different types of variants to autism

De Rubeis and Buxbaum (2015) reported that microscopic chromosome abnormalities occur in approximately 5% of cases of autism. The most frequent of these abnormalities are 15q11-q13 duplication, 22q11.1 deletions (DiGeorge velocardiofacial syndrome), and 22q13 deletion (Phelan-McDermid syndrome). Submicroscopic structural chromosome abnormalities (CNVs) occur more frequently in autism cases than in controls. However the impact of these depends on their size and on their gene content.

Table 8.2. SFARI list of autism genes based on SFARI scoring system. Gene functions are included.

(a) High confidence genes

ADNP	Activity-dependent neuro-protector homeobox
ANK2	Ankyrin 2, neuronal
ARID1B	component of the SWI/SNF chromatin remodeling complex
ASXL3	transcriptional regulator 3
CHD8	chromodomain helicase DNA binding protein 8
DYRK1A	dual-specificity tyrosine phosphorylation-regulated kinase
GRIN2B	N-methyl-D-aspartate (NMDA) receptors are a class of ionotropic glutamate receptors
POGZ	transposable element s protein was found to interact with the transcription factor SP1
PTEN	phosphatase and tensin, phosphatidylinositol-3,4,5-trisphosphate 3-phosphatase
SCN2A	sodium channel, voltage gated, type II alpha subunit
SETD5	SET domain containing 5 (also associated with intellectual disability)
SHANK3	SH3 and Ankyrin repeat domain, associated with post-synaptic density
SUV420H1	suppressor of variegation 4-20 homolog 1, has protein-protein interaction domains
SYNGAP	synaptic Ras GTPase activating protein 1, component of post-synaptic density
TBR1 T	box brain 1, transcription factor involved in the regulation of developmental processes

(b) Strong candidate genes

ANKRD11	ankyrin repeat domain 11, inhibits ligand-dependent activation of transcription
BCL11A	B lymphoid cell protein but also plays a role in neurodevelopment
CACNA1H	calcium channel, voltage-dependent, T type, alpha 1H sub-unit (epilepsy-associated)
CACNA2D3	alpha-2/delta subunit, a protein in the voltage-dependent calcium channel complex
CHD2	chromodomain helicase DNA binding protein 2
CNTN4	contactin 4 cell adhesion molecules that function in neuronal network formation
CNTNAP2	contactin associated protein-like 2, nervous system adhesion molecule
CTNND2	catenin (cadherin-associated protein), delta 2, implicated in development

---

(*Continued*)

Table 8.2. (*Continued*)

CUL3	cullin 3, role in the polyubiquitination and degradation of specific protein substrates
DEAF1	regulator of transcription
DSCAM	Down syndrome adhesion molecule nervous system development
FOXP1	forkhead box P1, regulation of tissue- and cell type-specific gene transcription
GABRB3	gamma-aminobutyric acid (GABA) A receptor, beta 3
GRIP1	glutamate receptor interacting protein 1, membrane organization and trafficking
KATNAL2	katanin p60 subunit A-like 2 highly expressed in central nervous system
KDM5B	lysine (K)-specific demethylase 5B (chromatin modification)
KMT2A	lysine 4 (H3K4) methyltransferase activity which mediates chromatin modifications
KMT2C	(MLL) lysine (K)-specific methyltransferase 2C
MAGEL2	paternally imprinted gene in Prader–Willi region
MED13L	sub-unit of Mediator complex transcriptional coactivator
MET	proto-oncogene, receptor tyrosine kinase
MSNP1AS	moesin A pseudogene 1, antisense
MYT1L	myelin transcription factor 1-like
NRXN1	neurexin 1 nervous system as cell adhesion molecule and receptor
PTCHD1	patched-domain containing 1, receptor for the morphogen sonic hedgehog
RELN	reelin, critical for cell positioning and neuronal migration during brain development
SHANK2	SH and ankyrin domain 2 protein, functions in the postsynaptic density

CNVs greater than 100kb, and in particular, deletion CNVs, contribute to approximately 15% of cases of autism. However many of the CNVs identified in autism cases also occur in intellectual disability, in epilepsy and in schizophrenia and may occasionally be present in individuals without symptoms. There is recent evidence that smaller deletions and duplications are more frequent in autism cases than in controls.

Many sequencing studies have concentrated on loss of function variants in autism probands that include stop-codon (nonsense), frame-shift and splice-site mutations. De Rubeis and Buxbaum noted that it is also important to consider damaging missense mutations that impact gene function. They estimated that the contribution of rare damaging variants to autism was 3% and that this is half of the variation contributed by common variants. They also raised the possibility that rare inherited variants can act additively.

8.6.2.1.2 Impact of common and additive genetic variants on autism

Klei *et al.* (2012) carried out analyses in 2,106 fully genotyped families of European ancestry with autism probands. Each family consisted of unaffected parents with at least one autism affected child and at least one unaffected sibling. For their analyses Klei *et al.* examined 5,156 single nucleotide polymorphisms (SNPs) with minor allele frequencies greater than 0.05 and the SNPs were at least 0.5 megabases apart. The latter measure was undertaken to decrease artifacts introduced through linkage disequilibrium between the alleles.

Klei *et al.* hypothesized that if additive genetic variation predisposes to autism liability they would be spread across the genome and the number of contributing variants on a specific chromosome would be proportional to the length of that chromosome. Results achieved through their analyses followed this prediction except in the case of the X chromosome where fewer common variants predisposing to autism were found than expected based on the length of that chromosome.

These studies by Klei *et al.* revealed that common genetic polymorphisms exert substantial additive effects on autism liability. In simplex families with one affected autism individual, 40% of liability traced to additive effects. In multiplex families with more than one autism affected individual, 60% of the liability traced to additive genetic variants. Heritability due to additive effects of genetic variants is referred to as narrow sense heritability.

De Rubeis and Buxbaum (2015) noted that autism risk shaped by common variants may be significantly altered by occurrence of a *de novo* damaging variant.

Wang *et al.* (2015) identified seven SNPs near the CYFIP1 gene. This gene encodes a protein that interacts with fragile X mental retardation protein (FMRP) and modulates its function. Wang *et al.* established that the CYFIP1 variants they identified modulate mRNA expression of CYFIP1 mRNA. They also noted that variants that increase CYFIP1 production increase risk for autism. The CYFIP1 gene maps with the chromosome at the 15q11.2 region that is duplicated in a number of cases with autism. Duplication in this region is illustrated in Figure 8.2.

8.6.2.1.3 Autism spectrum disorder severity and contribution of *de novo* and familial influences

Robinson *et al.* (2014) carried out a study to determine whether differences in cognitive impairments and symptom severity in autism reflect *de novo* genetic changes or familial history of broadly defined psychiatric disorders. They used data from more than 2,000 simplex cases of autism (single autism case per family). They determined that proband IQ was positively associated with family history of psychiatric illness. Proband IQ was negatively associated with a number of loss of function genomic changes. In low-IQ cases there was an increased risk of loss of function genomic changes.

8.6.2.1.4 Update on genetic variants in autism

Geschwind and State (2015) reported that the contribution of common genetic variants to autism is estimated to be 40–60%. They noted that to be able to identify common risk loci through association studies they would require very large cohorts that together included between 10,000 and 50,000 individuals.

Geschwind and Flint (2015) reported that large rare *de novo* CNVs (>750kb) occurred in 5–7% of cases of autism and therefore more frequently than in controls. They reported that whole exome sequencing on autistic probands and their parents revealed that *de novo* mutations likely contributed to disease. In studies on 5,000 autistic probands they noted that 71 genes were found to be pathogenic.

Sanders *et al.* (2015) published results of genomic and genetic studies in 2,591 cases of autism. Their analyses revealed that in some patients in this cohort *de novo* CNVs were present and these occurred primarily at six risk loci: 1q21.1, 3q29, 7q11.23, 16p11.2, 15q11-q13 and 22q11. In the autism cohort they identified 65 genes that carried pathogenic variants. These risk genes could be placed into two major networks.

One major network included neuronal elements related to synapse, neuronal projections, SRC homology signaling, long-term potentiation, post-synaptic density and cytoskeleton.

The second network included elements related to chromatin organization, SWI/SNF nucleosome remodeling complexes, bromodomain histone modification complexes, transcription related elements, SMAD domain (signaling) elements.

The SFARI database (https://gene.sfari.org/) lists genes (Table 8.2) relevant to autism spectrum disorders based on rigorous statistical comparison between cases and controls, yielding genome-wide statistical significance, with independent replication to be the strongest possible evidence for a gene.

Chen *et al.* (2015) reviewed genetics and pathogenesis of autism spectrum disorders. Comprehensive genetic studies have revealed that structural genomic variants, including CNVs, rare nucleotide variants including *de novo* variants and inherited variants play roles in autism etiology. These investigators emphasized that large numbers of common nucleotide variants, each likely of small effect, also impact the risk of autism. It also seems likely that these variants lead to individuality of manifestations even in patients where rare variants or structural variants are thought to play key roles in autism pathogenesis.

8.6.2.2 *Pathology and transcription studies in autism*

Studies on brain pathology by Chow *et al.* (2012) revealed age dependent changes in autism cases. They reported changes in cortical patterning and cell number in brains of young children with autism. Pathologic changes in limbic structures and in cerebellar Purkinje cell number were reported by several investigators (Kemper and Bauman, 1993; and Bailey *et al.*, 1998).

Macrocephaly occurs in 15–20% of autism cases and these cases manifest increased brain volume on MRI. The increased brain size is apparently due to increased white matter volume particularly in the frontal and temporal lobes, amygdala and hippocampus (Chen *et al.*, 2015).

8.6.2.2.1 Transcription studies of the brain

Studies on gene expression of genes in brain and on brain pathology have shed light on autism pathogenesis. Conclusions from these studies relate of course to post-mortem case studies and represent therefore a restricted number of cases. Voineagu *et al.* (2011) identified gene modules that were down regulated in autism cases and others that were up regulated. The down regulated modules included genes that encoded products that are expressed in synapses and in parvalbumin positive interneurons. These investigators noted that the down regulated genes had also been identified as autism risk genes in association studies. They include SHANK3, NRXN1, NLGN3, NLGN1 and CNTNAP1.

The genes that Voineagu *et al.* identified as up regulated in autism brain specimens had not been identified as autism risk genes in association studies. The up regulated genes were largely expressed in microglia and astrocytes and some of these genes had immune related functions. These findings on up regulated genes led Voineagu *et al.* to propose that they represented secondary findings.

It is important to note that there are studies that indicate immune dysregulation in autism. In brain transcriptome analysis Gupta *et al.* (2014) reported a gene expression pattern that indicated a dysregulated microglial response that co-occurred with altered expression patterns of specific neuronal activity genes. A number of studies have revealed altered blood cytokine levels in children with autism, including altered levels of interleukins and granulocyte colony stimulating factor (Estes and McAllister, 2015).

The Blencowe laboratory (Xiong *et al.*, 2015) carried out studies on the mRNA transcripts in the brain. Their studies revealed that brain exome splicing of many genes differs from that in other tissues. In addition they reported differences between autism cases and controls. Irimia *et al.* (2014) reported that highly conserved program of neuronal microexons presence is misregulated in autistic brains. There is also evidence for differential expression of the neuron specific splicing factor RBFOX1 between autism cases and controls (Fogel *et al.*, 2012).

8.6.2.3 *Pathway analyses in autism spectrum disorders*

Many mouse models have been developed that have changes in genes implicated in autism in attempts to understand the pathophysiology of autism. It is however important to note that defects in 200–1,000 genes have been implicated in autism (Chen *et al.*, 2015). Clearly it is of great importance to determine whether genes implicated in autism converge on specific cellular pathways, molecular processes or on specific developmental processes.

Ben-David and Shifman (2013) carried out analysis of *de novo* single nucleotide changes in 965 autism probands sequenced in four different studies. They examined data on 121 genes with *de novo* nucleotide changes in the autism cases that led to nonsense, frame shift or splice site mutations. They utilized databases with information on gene ontology (GO) and cellular processes (DAVID, UniProtKB). Their analyses revealed significant enrichments in autism cases versus control for changes that impacted chromatin regulators. Furthermore their analyses revealed that genes that were most frequently impacted in autism were highly expressed in the brain during prenatal development. Further functional categorization of genes altered in autism revealed that these were characterized as nuclear expressed and transcription regulators, chromatin regulators and chromatin modifiers. The second smaller module of gene impacted in autism included genes that were characterized as having increased expression after birth. These genes did not fit into any specific functional category.

Ben-David and Shifman therefore concluded that many of the genes that show damaging *de novo* mutations in patients with autism are involved in transcription regulation and chromatin regulation during development. They noted further that chromatin regulatory genes have also been shown to be impaired in intellectual disability.

Parikshak *et al.* (2013) undertook studies to map genes implicated in autism to specific biological pathways and to specific developmental time period and to specific brain areas. They concluded that genes impacted in autism spectrum disorders were expressed primarily during the 10th to the 20th week of gestation. Their studies revealed that autism implicated genes were primarily expressed during two specific phases of fetal development. One set of genes was expressed early in post-conceptual development and these genes primarily impacted transcription and chromatin regulation of gene expression. Autism implicated genes in the second module expressed later in prenatal development primarily impacted synaptic development. The autism implicated genes were primarily expressed in superficial cortical layers and in glutamatergic projection neurons. Furthermore many of the products encoded by these genes interact with FMRP1, the protein implicated in fragile X syndrome, and converged functionally on activity dependent gene regulation and synaptic plasticity pathways.

Parikshak *et al.* noted that their analyses of autism implicated genes concentrated primarily on *de novo* damaging variants in protein coding genes. However they conceded that different classes of variants including inherited damaging variants and damaging variants in non-protein coding genes could also contribute to pathogenicity in autism.

8.6.2.3.1 Molecular pathways frequently implicated in autism

Gene studies have led to the identification of molecular pathways implicated in autism. These include pathways involved in chromatin modification, transcription regulation, synaptic functions and receptor signaling. A connection has also been established between genes that regulate synthesis of the FMRP or proteins that interact with FMRP mRNA or protein.

8.6.2.3.2 FMRP target transcripts and autism

A number of investigators, including Parikshak *et al.* (2013), reported that genes that encode RNA transcripts that are targets of

FMRP polyribosome associated RNA binding protein were often designated as autism risk genes.

Darnell and Klann (2013) reported that FMRP, the protein product of the fragile X locus, is primarily associated with polyribosomes and that its main function is to repress translation of specific target mRNAs. Darnell and Klann identified 842 mRNA targets in the mouse brain. They determined that these mRNAs were expressed in pre-synapse and post-synapses neuronal regions. The post-synaptic targets included mRNAs for approximately 30% of the post-synaptic density proteome and included NMDA receptor sub-units and the mGLUR5 receptor. Post synaptic density (PSD) mRNAs included SHANK1–3, Homer, SynGap1, PSD95, neuroligins 1–3. In contrast their studies of mouse brains revealed that FMRP did not impact mRNAs for AMPA receptor subunits, Shisa 9 or CACNG. Presynaptic targets of FMRP included synapsins, synaptic vesicle glycoproteins, synaptotagmins, syntaxin, SNAP25 and many voltage gated calcium channels.

Darnell and Klann proposed that the fragile X syndrome that decreases FMRP levels should be considered a disease of dysregulated translation.

8.6.2.3.3 Multiple-hit etiology of autism

It is important to note that some investigators have proposed a multiple-hit etiology for autism. Steinberg and Webber (2013) developed the Trend test to detect multiple-hit etiologies. Using this test they obtained evidence in support of contribution from mutations in multiple FMRP related genes to the etiology of autism in subsets of patients.

8.6.2.3.4 Autism neuronal activity calcium signaling, activity dependent transcription and translation

Matamales (2012) reviewed models designed to determine how synaptic activity can trigger gene expression and specifically how stimulation of dendritic synapses can be coupled to the transcription of genes in the distant nucleus of the neuron. She concluded that the most plausible mechanism for relaying synaptic activity to nucleus involved action potentials. Excitatory or inhibitory signals are initiated through influx of calcium from the extra-cellular space through ligand gated receptors. The rise in intra-cellular calcium leads to activation of signaling, such as activated ERK kinase and NFkB, to pass to the nucleus. Furthermore the intra-cellular calcium and other second messenger systems, for example inositol phosphate (IP3), also stimulate receptors in the endoplasmic reticulum lumen leading to further release of calcium from the endoplasmic reticulum. Matamales proposed that inhibitory signals can impact the effects of stimulatory signals. Integration of excitatory and inhibitory signals impact opening of voltage gated calcium channels.

8.7 Epilepsy, Genetics and Genomics

Williams and Battaglia (2013) reviewed the molecular biology of epilepsy genetics. They emphasized that the common epilepsies are predominantly polygenic and multifactorial in etiology. They also noted that there are a number of monogenic forms of epilepsy. In addition structural changes and particular CNVs have been shown to play roles in some forms of epilepsy. They reported that the overall population frequency of epilepsy is 0.5 to 1%. Different classifications exist for epilepsy. Common epilepsies are distinguished from epilepsies that are associated with structural brain abnormalities or with specific additional clinical abnormalities. Another classification distinguishes between genetic disorders, structural abnormalities, metabolic disorders, and epilepsy of unknown causation.

It is important to note however that even in specific syndromes that are frequently characterized by epilepsy, the penetrance of the disease causing mutation even in a specific family is variable.

Williams and Battaglia noted that although the biological causes of epilepsy are heterogeneous all apparently disturb the balance between excitatory and inhibitory neuronal activity. Some classifications of epilepsy are based on the type of abnormal activity seen in electroencephalograms, on the age of onset and on the timing of seizures. Examples of categories include neonatal onset seizures, myoclonic seizures, cortical seizures, temporal seizures, absence seizures and febrile seizures.

8.7.1 Monogenic forms of epilepsy

Studies in families helped in identification of monogenic forms of epilepsy. A number of these manifested as autosomal dominant conditions and these involved a number of different specific ion channels including sodium ion channels SCN1A, SCN2A amd SCN1B; potassium ion channels KCNQ2 and KCNQ3; chloride channel CLCN2; calcium ion channel CACNB4 and calcium homeostasis factor EFHC1. Other autosomal dominant monogenic causes of epilepsy include mutations in neurotransmitter receptors GABRG2 and GABRD2. Additional monogenic epilepsy genes include STXBP1 which encodes a product involved in synaptic vesicle release and signaling-related factor LGI1 which regulates activity of voltage gated ion channels.

8.7.2 Polygenic effects in epilepsy

Williams and Battaglia reported that a polygenic and multifactorial threshold model has most frequently been applied to epilepsy. This model indicates that multiple genes have variants that can act as risk factors for epilepsy and those additional factors, e.g. environmental factors (exogenous or endogenous), interact with the gene variants to lead to the epilepsy phenotype.

8.7.3 Epilepsy associated with XLID

Shoubridge *et al.* (2010) reported that the ARX1 gene is one of the most frequently mutated genes in XLID and in many of these cases epilepsy occurs. In some cases with ARX1 mutation brain malformations occur and these may include lissencephaly. ARX1 is a homeobox gene. Studies in animal models revealed that it plays a role in specification of Gabaergic interneurons. However ARX1 mutations are also associated with additional abnormalities including ambiguous genitalia.

The ARX gene has four intra-genic poly alanine tracts. Shoubridge *et al.* noted that more than half of the ARX1 mutations involved expansion of the first two polyalanine tracts. Shoubridge *et al.* reported that in addition to polyalanine tract related defects other pathogenic mutations that occurred in ARX1 included splice site mutations, insertion deletions and a number of different missense mutations and that these sometimes lead to brain malformations. Mutations predominantly led to clinical manifestations in males though in some severe mutations led to manifestations in females.

Another gene involved in X-linked epilepsy is PCDH19 which maps on Xp22.1. Defects in this gene lead to female restricted intellectual disability. PCDH19 encodes a calcium dependent cell adhesion protein that is primarily expressed in the brain.

8.7.4 *Genomic CNVs in epilepsy*

Mullen *et al.* (2013) reported that three genomic microdeletions constitute established risk factors for genetic generalized epilepsy (GGE). These include microdeletions on 15q13.3, 15q11.2 and 16p13.11. They reported that each of these microdeletions occur in 0.5% to 1% of cases of GGE but are rare in controls. It is however important to note that these microdeletions are not always associated with epilepsy and their presence in control populations indicates that they are risk factors and that additional factors are likely required to precipitate epilepsy.

Mullen *et al.* carried out studies in two cohorts of patients; one cohort comprised 359 probands with GGE and the second cohort included patients with intellectual disability and GGE. They determined that the three recurrent microdeletions described above occurred more frequently in the intellectual disability GGE cohort. In 18% of cases in that cohort they also detected other rare CNVs and these non-recurrent CNVs involved several different chromosomes. These CNVs were primarily deletions that overlapped with one or more genes; however some of the CNVs did not overlap with protein coding genes.

8.7.5 *Targeting specific epilepsy phenotypes for genetic analyses*

Berg and Dobyns (2015) noted that epilepsy studies should take into account age of onset, seizure type and electrographic features. In their studies they treated different forms of epilepsy as discrete diseases. They suggested that this approach facilitated identification of epilepsy genes.

Mercimek-Mahmutoglu *et al.* (2015) reviewed the diagnostic yield of genetic testing in childhood epilepsy. They reviewed data from 110 patients and noted that genetic causes were identified in 28% of patients; 7% of patients were found to have inherited metabolic disorders. These included pyridoxine deficient epilepsy caused by mutations in the ALDH7A1 gene, Menkes disease, pyridoxamine-5-phosphate oxidase deficiency, cobalamin G deficiency, MTHFR (methylene tetrahydrofolate reductase) deficiency, glucose transporter deficiency, glycine encephalopathy, and pyruvate dehydrogenase complex deficiency.

They identified putatively pathogenic CNVs in three cases. These included a 1 MB (megabase) *de novo* deletion in 17p13.3, a 5 MB *de novo* deletion that encompasses the region 10p15.3-10p15.1, and a 1 MB *de novo* deletion on chromosome 6p25.2.

Two patients received clinical diagnoses of Simpson–Golabi–Behmel syndrome. CHARGE syndrome with *de novo* CHD7 mutations was diagnosed in one patient. Rett syndrome with a MECP2 mutation occurred in one patient. One patient had lissencephaly and a TUB1A mutation. Damaging mutations in the sodium channel encoding gene SCN1A occurred in four patients. Damaging mutations in SCN2A occurred in three patients and in two of these patients the mutations were *de novo*. Parents of the third case were not tested. *De novo* deletion mutation occurred in SCN8A in one patient. Three patients had damaging deleterious mutations in the potassium channel encoding gene KCNQ2. Three patients had *de novo* damaging mutations in the STXBP1 gene that encodes a synaptic vesicle release factor. One patient had *de novo* damaging mutations in protocadherin encoding gene PCDH19. One patient had a damaging *de novo* mutation in the solute carrier encoded by the SLC9A6 gene.

The authors noted that approximately 50% of the genetically diagnosed patients had dominant *de novo* mutations.

It is important to note that a number of laboratories offer screening for epilepsy gene mutations through the use of panels. The Greenwood Genetic Center for example has a panel that screens for mutations in 145 different genes.

It is also important to note that "new" epilepsy causing gene mutations are constantly being found. The EPICURE consortium (2012) reported epilepsy causing mutations in five additional genes, VRK2 (vaccinia related effector of signal kinase), ZEB2 (zinc finger E homeobox binding repressor), PNPO (pyridoxamine phosphate oxidase, involved in vitamin B6 synthesis), CHRM3 (muscarinic cholinergic receptor) and SCN1A (sodium channel 1A).

8.7.6 *Epilepsy genetics and genomics*

"This is the bronze age of epilepsy neurogenetics when single genes are no longer interrogated one at a time but caste with dozens of interrogators and reassembled to define neural network behavior."

J. Noebels, 2015

Noebels (2015) reported that at least 500 different genes are associated with epilepsy. He noted that epilepsy is at the cross roads of neuronal synchronization disorders. Key factors implicated in the expansion of information concerning genes involved in epilepsy include exome sequencing and targeted mutagenesis in model organisms. Gene studies contribute in model organism and also contribute to pinpointing the circuitry involved in epilepsy pathogenesis. Possible mechanisms through which genes lead to epilepsy include:

(i)   Altered membrane excitability
(ii) Altered neurotransmitter release and synaptic functions
(iii) Altered neural wiring properties

Noebels emphasized that the discovery of gene defects in epilepsy patients should be followed by determination of effects of mutations on functions, analysis of the impact of neurogenetics pathways and application of next generation experimental opportunities. These include the use of induced pluripotent stem cells and the application of mutagenesis in cells and model organisms to validate the importance of gene targets.

He also emphasized the importance of studies designed to investigate the pathogenicity of gene mutations and their possible generation of epileptic encephalopathy.

8.7.7 *History of discovery of genes related to epilepsy*

Noebels noted that epilepsy associated with specific inborn errors of metabolism provided initial clues to genes involved. Myoclonic epilepsy associated with ragged red fibers provided evidence of the importance of abnormal mitochondrial function in epilepsy causation. The description of generalized epilepsy syndrome in patients and families with neuronal nicotinic acetyl choline receptor alpha 4 sub-unit defects provided evidence of the importance of defects in voltage gated ion channels in epilepsy generation.

The classification of subtypes of epilepsy has evolved. Noebels noted however that mutations in a specific gene lead to different types of epilepsy. One example is KCNQ2 ion channel protein mutations. Mutations in the sodium channel beta 1 sub-unit SCN1B occur in febrile seizure syndrome, and SCN1B and SCN1A mutations lead to Dravet syndrome epileptic encephalopathy. The spectrum of different epileptic phenotypes that occur in patients with mutations in GLUT1 and in patients with KCNT1 mutations is broad. One question that arises is whether the specific type of mutation influences the epilepsy phenotype.

Noebels noted further that epilepsies also occur in association with other neurological syndromes, cognitive disorders, autism and specific dementias. He postulated that as specific brain circuits fail in these disorders, inhibitory defects follow. He proposed that epilepsy represents a synchronization disorder rather than a manifestation of static hyper-excitability.

234 Unravelling Complexities in Genetics and Genomics

8.7.8 *Sudden unexpected death in epilepsy*

Mutations in SCN5A sodium channel and mutations in the KCNQ1 potassium channel gene lead to long QT syndrome in humans and are associated with sudden death. Noebels reported that in the mouse models of these mutations abnormalities occur in the brain. There is also evidence that SENP2 (SUMO protease protein 2) gene mutations impact potassium ion channels in the heart and brain.

Noebels emphasized that no specific gene is commonly mutated in epilepsy and that it is important to consider gene convergence and the impact of gene mutations on neuronal networks. Another aspect that needs to be considered is whether somatic mutations may play roles in epilepsy and in particular, focal cerebral dysplasias. Examples of genes and syndromes in which somatic mutations in specific genes lead to focal cerebral dysplasias including DCX, PAFAH1 (LIS1) genes where specific mutations lead to sub-cortical band heterotopias and polymicrogyria. Noebels proposed that somatic mutations lead to seizures that arise in specific foci. He noted further that even in the absence of imaging abnormalities, analysis of resected epileptic foci for somatic mutations is important for gene discovery.

8.7.9 *Driver mutations in epilepsy*

Noebels proposed that hundreds of driver genes in epilepsy impact cellular pathways involved in membrane excitability, neurotransmission and neuronal migration. Mutations in genes that impact inhibitory processes (including ion channel function and phasic inhibition of the thalamocortical circuit) are important in absence epilepsy. Important monogenic drivers involve gene products that impact synaptic inhibitory GABA (gamma-amino-butyric acid) neurotransmission. Noebels emphasized that the overall position of many functionally assigned gene products in the inhibitory pathway need to be determined.

Important aspects of the neuroinhibitory pathway include neurotransmitter synthesis, vesicular packaging, neurotransmitter release and reuptake and post-synaptic receptor activation. Intra-cellular chloride gradient contributes to synaptic plasticity.

Noebels noted that inflammatory pro-epileptic antibodies target GABA receptors, NMDA receptors, potassium channels and calcium channels.

8.7.10 *Analysis of cell specific effects and effects in model organisms*

Selective destruction in mice of Cacna1a ion channels in parvalbumin positive GABA interneurons led to seizure pattern generation. Noebels also reported that elimination of Scn1 from inhibitory neurons leads to seizures in mouse models of Dravet syndrome. Seizures do not occur when Scn1 is eliminated in excitatory neurons.

Mecp2 removal from inhibitory interneurons leads to seizures in mouse models of Rett syndrome. However, Noebels noted that seizures occur when Mecp2 is removed from excitatory neurons possibly through impact on other elements involved in neurotransmission. He noted that mutations in a single gene might impact multiple proteins. Other examples include the Nova2 gene that encodes an mRNA binding protein that impacts splicing of mRNA of a number of synaptic genes.

8.7.11 *Epistatic complexity in epilepsy*

Examples of epistatic effects include interactions between ion channel proteins. Noebels noted that over 400 ion channel proteins are known. There is evidence that different ion channel proteins can act as co-modifiers. He presented evidence of combinatorial effects of multiple ion channel variants and evidence of channel type differences (i.e. multiple non-synonymous SNVs) in cells of individuals with epilepsy. The pattern of variants impacted firing patterns.

8.7.12 *Suppressor and modifier genes*

Noebels presented examples of epilepsy modifiers. Studies in mouse model organism revealed that deletion of Tau (MAPT) that colocalizes with ion channels Kcna1 and Scn1A suppressed epilepsy in mice where these genes were mutated.

8.8 FOXG1 Defects Brain Malformations and Epilepsy

Duplications or deletions in a specific gene can both lead to abnormal phenotypes. It is also important to note that the disruption of specific protein coding elements in a gene and disruption of regulatory elements of the gene can lead to phenotypic abnormalities.

Deletion of the gene region on chromosome 14q12 surrounding the FOXG1 gene or intragenic mutations in FOXG1 lead to postnatal microcephaly, frontal lobe simplified gyral pattern and abnormalities of the corpus callosum, including hypogenesis or agenesis. Associated clinical abnormalities include intellectual disability, absent language, abnormal movements, including choreiform movements (Seltzer *et al.*, 2014). Duplications in 14q12 that also involve FOXG1 were reported to be associated with normal head size, normal corpus callosum but with clinically determined autistic behaviors, absent language and intellectual disabilities.

Seltzer *et al.* noted that epilepsy had previously been described in cases with FOXG1 defects. They analyzed epileptic features and abnormal development in 23 subjects with FOXG1 deletions or intragenic mutations and they studied 7 subjects with FOXG1 duplications. In total epilepsy was diagnosed in 87% of patients with FOXG1 abnormalities. The subjects with FOXG1 deletions or mutations had a variety of different types of epilepsy including complex partial seizures, myoclonic seizures and generalized tonic seizures. They identified siblings with the same FOXG1 protein deletion mutation, p. Gly 172_Met 192 del and determined that only one member of the sib pair had seizures.

Seltzer *et al.* documented a specific epilepsy type, infantile spasms, in patients with duplications that involved FOXG1. In these patients treatment with ACTH led to cessation of seizures. They reported that overall subjects with FOXG1 disorders had severe neurodevelopmental disabilities by three years of age.

8.8.1 *Functions of FOXG1 and regulatory elements*

Brancaccio *et al.* (2010) reported that the product of the FOXG1 gene promotes neurogenesis and inhibits gliogenesis. Over-expression of FOXG1 in forebrain led to thickening of the neuroepithelium likely due to decreased neuroepithelial apoptosis.

Allou *et al.* (2012) identified three cases that had manifestations similar to those in patients with FOXG1 deletions with 14q12 micro-deletions but these patients did not include the protein coding FOXG1 gene. They searched for regulatory elements of FOXG1 and identified a long range silencer of FOXG1 expression. This silencer element was located 0.6 MB away from the FOXG1 protein coding gene.

8.9 Glycine Receptor Abnormalities in Hyperekplexia, Intellectual Disability with Impaired Speech

Bode and Lynch (2014) reviewed information on inhibitory glycine receptors and abnormalities of function of these receptors. Mutations that impair function of inhibitory glycine receptors may lead to exaggerated startle response (hyperekplexia), intellectual disability and impaired speech development. They noted that hyperekplexia is characterized in early infancy by episodic and generalized stiffness that subsequently decreases in severity. However additional manifestations subsequently develop including episodes of stiffness, startle response and apnea without loss of consciousness.

Inhibitory glycine receptors are composed of alpha and beta sub-units. Shiang *et al.* (1993) first revealed that mutations in the gene that encoded the alpha1 sub-unit of the glycine receptor lead to hyperekplexia. Bode and Lynch noted that hyperekplexia is induced by mutations in the GLRB gene that encodes the beta subunit of the inhibitory glycine receptors. Other gene defects may also lead to hyperekplexia. These include mutations in the SLC6 gene that encodes the glycine transporter and mutations in the synaptic clustering proteins gephyrin and colistin. In addition there are cases of hyperekplexia where the underlying gene defect has not been discovered.

There is now evidence for three genes that encode alpha sub-units of glycine receptors and for one alpha glycine receptor pseudogene GLRA4. The spinal cord and brain stem motor neurons are the major sites of glycine receptors. Glycine receptors are also present in pain sensitivity neurons in the dorsal horn of the spine and in the cerebral cortex and hippocampus.

Bode and Lynch reported that hyperekplexia causing alpha1 glycine receptor mutations are commonly missense mutations that lead to codon changes or to premature stop codons. The mutations are commonly homozygous or compound heterozygous mutations and hyperekplexia is inherited as a recessive trait. They noted that recessive mutations frequently reduce the glycine receptor current amplitude. Autosomal dominant alpha glycine receptor mutations also occur and lead to increased channel activity in some cases.

Specific GLRB mutations including both autosomal dominant and autosomal recessive mutations also lead to hyperekplexia. Bode and Lynch reported that specific GLRB1 mutations cause delays in gross motor development and delayed speech development.

9. Integrating Advances in Neurobiology and Genetics into Psychiatry

"The collective actions of individual nerve cells linked by a dense web of intricate connectivity guide behavior, shape thoughts, form and retrieve memories and create consciousness"

O. Sporns, 2010

9.1 Rethinking Psychiatric Diagnoses

Cuthbert and Insel (2013) reported that a "tectonic shift" has taken place in psychiatry. They promoted the incorporation of modern concepts of cognitive, affective, social and molecular neuroscience into psychiatric diagnoses. They also stressed that disorders traditionally defined as "mental disorders" should be considered as "brain disorders", as syndromes of disrupted neural, cognitive and behavioral systems.

Cuthbert and Insel emphasized that this tectonic shift was necessary given growing evidence that, when older traditional psychiatric classifications were used to distinguish different psychiatric disorders, there are often cross disorder features in patients and frequently a variety of different cognitive and behavioral functions are disrupted in a particular patient. In addition current therapeutic interventions for psychiatric disorders are often inadequate and treatments are non-specific.

They proposed that increased insight into mechanistic processes impaired in a specific patient should lead to matched or targeted therapy. The National Institutes of Mental Health (NIMH) in the USA proposed a new classification system based on information on underlying brain systems that sub-serve perception, motivation, cognition, emotion and social behavior. This new system was designated as Research Domain Criteria (RDoC). The specific domains addressed in RDoC include: Negative/positive valence, cognitive system, social process systems, arousal and modulatory systems.

9.1.1 *NIMH RDoC for psychiatric diagnosis*

The NIMH team noted the changes in focus in medicine during the last few decades. These changes include incorporation of information on genomes, cell biology, pathophysiology and discovery of identifiable subtypes within clinical phenotypes. They noted further that identification of subtypes improves possibilities for more effective treatment and management.

The NIMH team considered it imperative that new information on brain circuits and on genomic factors be taken into account in diagnosis and management of mental illness. They noted that current and long established criteria for diagnosis of mental illness (e.g. the Diagnostic and Statistical Manual of Mental Disorders, 5th Edition (DSM-5) and the International Statistical Classification of Diseases and Related Health Problems, 10th Edition (ICD-10)) antedate contemporary neuroscience, molecular and cell biology and genetics. DSM-5 and ICD-10 criteria are based on phenomenological symptoms and observations. They proposed the RDoC criteria initially as a research tool.

Key principles of the RDoC criteria are firstly that they should be based on knowledge of brain circuits and behaviors. Secondly they should be agnostic of current disorder categories. Thirdly they should incorporate information on imaging, physiological activity and behavior. The RDoC assessments are based on specified dimensions of function and domains of function relating to motivation, cognition and social behavior. Motivation includes negative valence systems (aversive) and positive valance systems.

In addition to cognitive systems and social process systems an arousal regulatory system is included. Constructs were selected based on evidence that a particular brain circuit or region could be associated with that dimension. Additional information included physiological measurements such as heart rate, startle reflex, cortisol measurements, and analysis of genes and genomes.

The negative valence matrix includes assessment of acute threat, fear, potential threat, sustained threat, anxiety, loss, frustration and non-reward.

The positive valence matrix includes reward valuation, effort valuation, willingness to work, responsiveness to reward, prediction of reward and learning habit.

The cognitive system matrix includes perception (visual, auditory, olfactory, somatosensory), declarative memory, language, goal selection and updating, representation, response selection, inhibition suppression, performance monitoring, working memory, flexible updating, capacity and interference control.

The social process domain includes attachment formation and maintenance, reception and production of facial and non-facial communication, perception, understanding mental states and self-knowledge.

Arousal and regulatory systems include circadian rhythms, sleep and wakefulness.

9.1.2 *Brain regions and circuits associated with RDoC domains*

The NIMH RDoC document identifies the following associated regions and circuits. The negative valence system (or its opposite pole, fearlessness) is associated with the amygdala and hippocampus that interacts with the ventromedial prefrontal cortex, the bed nucleus of the striatum, the hypothalamic-pituitary axis, cortisol releasing factor (CRF) and cortisol.

The positive valence system (and opposite pole anhedonia) are associated with the mesolimbic dopamine pathway. Habit-based behavior (and obsession compulsion spectrum) is associated with the orbitofrontal cortex, thalamus and dorsal striatum.

Cognitive systems include working memory and involve dorso-lateral pre-frontal cortex and other prefrontal cortex areas; cognitive control (opposite pole impulsivity, disinhibition, externalization), involve anterior cingulate gyrus, medial and lateral prefrontal cortex.

Social processes involve distributed cortical activity, mesolimbic and dopamine systems. Facial expression and recognition involve the ventral visual system and fusiform gyrus. Self-representational circuits include dorsal and posterior anterior cingulate gyrus and insula.

Arousal regulatory processes and stress regulation are dependent on the raphe nuclear circuits. Facilitated stimulus processing is dependent on the locus coeruleus circuits and norepinephrine. Readiness for stimulus processing and responding is dependent on the brain resting network.

In addition the NIMH team emphasized that neural circuits must be considered in dimensions of function, their molecular and cellular building blocks, and their specific interactions with the environment.

9.2 Neurotransmitters and Receptors

9.2.1 *Catecholamine (monoamine) system*

The initial definition of the monoamine catecholamine neuronal systems took place during the 1960s and 1970s. More recent studies have led to delineation of the dopamine neuronal cell groups and their distinctive properties and new information on key enzymes involved in catecholamine synthesis (Björklund and Dunnett, 2007).

New information on key enzymes and cofactors involved in monoamine synthesis has been discovered more recently in part through studies on patients with inborn errors of metabolism (Kurian *et al.*, 2011). Key cofactors for synthesis of monoamine neurotransmitters are tetrahydrobioterin (BH4) and pyridoxal phosphate (vitamin B6). The key precursors for serotonin synthesis are the amino acid tryptophan and the enzyme tryptophan hydroxylase. For synthesis of dopamine and its derivatives norepinephrine and epinephrine, the key precursor amino acid is tyrosine and the key enzyme is tyrosine hydroxylase. Serotonin synthesis and synthesis of dopamine require the activity of aromatic acid decarboxylase

(AADC). Monoamine oxidase is essential for degradation of serotonin. Dopamine, norepinephrine and epinephrine require catecholamine methyltransferase (COMT) and monoamine oxidase (MAO) for degradation.

Chandler *et al.* (2013) reported that the midbrain ventral tegmental area is a major source of dopamine and the locus coeruleus is a major source of epinephrine. They reviewed evidence that prefrontal cortex cognitive functions, including working memory and attention, are strongly influenced by dopamine and norepinephrine release. They noted further that reciprocal connection exist between the locus coeruleus and the ventral tegmental area part of the midbrain close to the substantia nigra.

9.2.2 *Dopamine pathways*

Björklund and Dunnett (2007) noted that dopaminergic systems play key roles in regulating motor control, motivation, emotion, cognition and neuro-endocrine functions. Through impact on sympathetic regulation, dopamine can impact cardiovascular, renal and gastrointestinal function.

Four major dopamine pathways have been defined. The mesolimbic pathway transmits dopamine to the limbic system via the nucleus accumbens. The mesocortical system connects the ventral tegmental area (VTA) to the frontal cortex. The nigrostriatal pathway extends from the substantia nigra pars compacta and the dorsal striatum to the caudate nucleus and putamen and impacts motor control. The tubero-infundibulum pathway extends from the hypothalamus to the pituitary.

Dopamine functions through five dopamine receptors DRD1-DRD5. In addition there is evidence that it can function through other receptors, through ion channels or through tyrosine kinase activation.

9.2.3 *Downstream effects of dopamine signaling*

Signaling through dopamine receptors can modulate cyclic AMP synthesis and protein kinase A, phospholipase C, and protein kinase

C synthesis. Activation of protein kinase A leads to mTOR activity. There is also evidence that dopamine receptor activation can trans-activate the brain-derived neurotrophic factor (BDNF) receptors.

Beaulieu *et al.* (2015) reported that although dopamine receptors were initially described as monomeric, recent studies revealed that heterodimers occur. These heteromeric receptors contain subunits derived from the different dopamine receptor encoding genes and heteromeric receptors may vary in physiological roles. They noted that dopamine receptors are modulated by pharmacological agents used in the treatments of schizophrenia, bipolar disorder, depression, Parkinson's disease and sympathetic nervous system related disorders. They noted that essentially all approved antipsychotic drugs target the DRD2 receptor. Newer atypical antipsychotic drugs target DRD2 and other receptors particularly the serotonin (5HT2A) receptors. However these antipsychotics also have metabolic effects. They noted that emerging antipsychotics have higher affinity for DRD3 receptors than for DRD2 receptors. Other antipsychotic drugs in development target DRD2, 5HT2A serotonin receptors and adrenoreceptors.

Dopamine released into the synaptic cleft can be taken up again into the presynapse through the activity of dopamine transporter. Specific drugs that inhibit dopamine reuptake are used in the treatment of disorders including attention deficit hyperactivity disorder (ADHD).

9.2.4 *Noradrenergic system*

Chandler *et al.* (2013) noted that the noradrenergic neurons in the locus coeruleus, located in the pons, receives input from numerous neurotransmitters, including glutamate, GABA, glycine, dopamine, serotonin and acetylcholine, and from neuropeptides including corticotrophin releasing hormone and encephalin. They noted further that fibers that leave the locus coeruleus innervate many brain regions.

Noradrenaline released from presynaptic vesicles into the synaptic cleft signals through alpha or beta adrenoreceptors. Noradrenaline in the synaptic cleft may be taken up again into presynapses through norepinephrine transporters or it may degrade through monoamine oxidase.

Chandler *et al.* reported that noradrenaline signaling is involved in stress responses, attention, memory, sleep-wake cycle, decision-making and sympathetic states. They noted further that the noradrenergic system and corticotrophin releasing hormone CRI are main systems involved in the stress response. Memory is enhanced by adrenergic agents (e.g. clonidine). Alpha 2 adrenergic antagonists suppress working memory. The anti-hypertensive drug reserpine impacts monoamine oxidase and it negatively impacts prefrontal cortex activity and can induce depression.

9.2.5 *Serotonin and serotonin receptors*

Saulin *et al.* (2012) reviewed the serotoninergic system and noted that it is an important modulatory neurotransmitter system. Serotonin (5-hydroxy-tryptamine, 5HT) is synthesized in the raphe nuclei located in the midline of the brainstem. 5HT1A receptors are inhibitory receptors and there is evidence that they play important roles in determining synaptic plasticity. The 5HT2 receptors are excitatory. Serotonin released into the synaptic cleft is taken up again into presynapses by the serotonin transporter SERT. SERT inhibitors are used in the treatment of psychiatric disorders.

9.2.6 *Acetylcholine signaling and cholinergic receptors*

Acetyl choline regulates excitability through nicotinic receptors (nACHR) or through muscarinic receptors. Both types of receptors occur in the corpus striatum, cerebral cortex, hippocampus, thalamus, hypothalamus and cerebellum. Nicotinic receptors occur in the substantia nigra and locus coeruleus. Nicotinic acetylcholine receptors are involved in cognition, motor activity and reward systems. Muscarinic receptors occur in the midbrain, medulla and pons.

There is evidence that cholinergic receptors are reduced in Alzheimer's disease (Buckingham *et al.*, 2009).

9.3 Genetic and Genomic Studies in Psychiatric Disorders

9.3.1 *Genome-wide association studies (GWASs) in schizophrenia*

The Schizophrenia Working Group of the Psychiatric Genomics Consortium (2014) reported results of a multi-stage GWAS in up to 36,989 cases of schizophrenia and 113,075 controls. The study also included family trio studies. They identified 128 independent associations in 108 loci. 75% of the loci included protein coding genes and 8% of the associated variants were within 20kb of genes. They reported notable associations of schizophrenia with gene encoding products involved in neurotransmission and synaptic plasticity including dopaminergic receptor DRD2, and genes that that encode glutamatergic neurotransmitter receptor sub-units, GRM3 (metabotropic glutamate receptor), GRN2A (glutamate receptor ionotropic NMDA), GRIA1 (ionotropic glutamate receptor AMPA) and SRR serine racemase. Serine racemase catalyzes the synthesis of D-serine from L-serine and D-serine is an agonist with glutamate of NMDA receptors. They also identified association with schizophrenia and genes that encode voltage coded calcium channel sub-units CACNA1C, CACNB2 and CACNA1I. They also reported that associations were enriched in genes that encode products involved in immunity. The greatest degree of association was found between schizophrenia and the extended MHC region on chromosome 6. The authors also carried out polygenic risk score profiling. They concluded that studies of common and rare variant studies in schizophrenia are complementary. Glutamatergic and calcium ion channel gene loci have also been implicated as causative in schizophrenia through studies on rare variants.

Other significantly associated loci in these GWASs that are tied to mechanism include the activity-regulated cytoskeleton-associated protein (ARC) complex loci and loci that encode binding targets of fragile X mental retardation protein (FMRP).

It is important to note that the MHC histocompatibility locus achieved the highest level of association significance in GWASs in schizophrenia, indicating that the immune system may play a role in this disorder.

9.3.2 *Exome sequencing and de novo mutations in schizophrenia associated genes*

Fromer *et al.* (2014) carried out exome sequencing on trios (affected probands and their unaffected parents) and analyzed data to determine whether or not *de novo* mutations occurred at higher rates than expected in genes that had been implicated in schizophrenia based on other genomic approaches. They also reviewed reported data on gene mutations in autism and intellectual disability.

They determined that there was a statistically significant increase in the rate of non-synonymous rare mutations in schizophrenia cases versus controls (significance $p = 0.0003$) and an increase in the number of loss of function mutations in cases versus controls ($p = 0.0097$). There was a statistically significant enrichment of mutations in the ARC complex and NMDAR glutamate receptor gene in cases versus controls; significance levels $p = 0.0008$ for non-synonymous mutations and $p = 0.006$ for loss of function mutations. There was also statistically significant overlap between schizophrenia and autism in associated genes with loss of function mutations ($p = 0.0007$) and overlap in schizophrenia and autism in genes with non-synonymous mutation ($p = 0.02$).

Enrichment for *de novo* mutations in genes that impact the ARC complex was particularly significant in schizophrenia, non-synonymous mutations ($p = 0.00048$) and loss of function mutations ($p = 0.0015$). Mutations in NMDAR complex and post-synaptic density (PSD) genes occurred more frequently in autism and intellectual disability cases than in schizophrenia. However mutations in NMDAR complex were significant in schizophrenia with significance at the 0.025 level.

9.3.3 *Copy number variants (CNVs) in schizophrenia*

Szatkiewicz *et al.* (2014) undertook a genome wide survey for CNVs in 4,719 cases of schizophrenia and 5,927 controls. They confirmed that the overall CNV burden was higher in schizophrenia patients than in controls. However there was evidence of non-penetrance since some of the CNVs present in patients were also observed in controls. However CNVs larger than 500kb were relatively infrequent in controls.

The most common CNVs that occurred in schizophrenia patients included 15q13.3 deletion, 1q21.1 deletions, 3q29 deletion, 22q11.2 deletion and 16p11.2 duplication. Szatkiewicz *et al.* reported that in addition the CNVs in schizophrenia patients were enriched for genes active at the post-synaptic junction, in products involved in calcium signaling and in mitochondrial function.

Kirov (2015) reviewed CNV studies in more than 10,000 cases of schizophrenia. He reported that 11 CNVs were of particular importance as causative factors. He noted these 11 CNVs occurred with a cumulative frequency of 2.4% in patients with schizophrenia and in 0.5% of controls. The 11 CNVs included 1q21.1 deletion; 1q21.1 duplication, deletion in the NRXN1 gene 2p16.3, 3q29 deletion, 7q11.2 duplication, 15q11.2 deletion, 15q11.2-q11.3 duplication, 15q13 deletion, 16p13.11 duplication, 16p11.2 duplication and 22q11.2 deletion. The 22q11.2 deletion was not detected in controls.

Pocklington *et al.* (2015) reported results of functionally informed CNV studies in 5,745 schizophrenia patients and 10,675 controls. They searched for CNVs greater than 100kb in length that occurred at frequencies lower than 1% in the general population. Their analyses revealed disruption of Gabaergic signaling modalities and disruptions of glutamatergic signaling modalities. They also found CNVs that disrupted post synaptic density complexes (PSD). NMDAR complexes were enriched in duplications. PSD95 and ARC complexes were impacted in deletions. ARC is required for synaptic plasticity and memory. The ARC complex localizes to active synapses. ARC mediates trafficking of AMPA type glutamate receptors (Day and Shepherd, 2015).

In the studies reported by Pocklington *et al.*, serotoninergic and nicotinic alpha7 receptor complexes were not found to be significantly involved.

9.3.4 *Overlap in risk loci in different psychiatric disorders*

Kavanagh *et al.* (2015) concluded that there are more than 100 common risk loci for schizophrenia and at least 11 rare risk loci. They emphasized further that genetic risk does not map neatly to psychiatric diagnoses and that there is shared risk between bipolar disorder, autism spectrum disorders, intellectual disability and ADHD.

9.3.5 *Genetic risk for schizophrenia, exome sequencing and plasticity pathways*

In reviewing genetic risk for schizophrenia, Hall *et al.* (2015) noted that a number of studies have revealed that multiple risk variants each contributed a small component of accounted for risk of schizophrenia. In addition rare CNVs also increased risk. They noted further that trio sequencing studies (two unaffected parents with one affected offspring) revealed that rare *de novo* mutations in offspring also contributed to risk.

In addition, they noted that exome sequencing in affected cases and controls have revealed important findings regarding pathways involved in schizophrenia pathogenesis. Hall *et al.* noted that although rare variants only account for a small proportion of overall risk, findings of different studies converge. Genes implicated in schizophrenia, including neurotransmitter receptor complex and calcium ion channel encoding genes (CACNA1C and CACNB2), are also implicated in neuronal plasticity. They noted further that the ARC gene product that is implicated in schizophrenia plays key roles at dendrites. ARC expression and F-actin dynamics are associated with synaptic plasticity. They emphasized that the ARC pathway emerges as a key regulator of structural and molecular plasticity of dendrites.

Hall *et al.* noted further that the degree to which the synaptic plasticity associated pathway involvement crosses diagnostic boundaries in psychiatric disease also needs to be fully established. In addition the interaction of synaptic plasticity with other defects in schizophrenia such as impaired dopamine signaling in psychosis needs to be explored.

9.3.6 *CNVs in bipolar disorder*

Green *et al.* (2015) analyzed CNVs on genotyping arrays in 2,591 cases of bipolar disorder and 9,942 controls. They reported that the overall frequency of CNVs in bipolar patients was low. Specific CNVs that were found more frequently in cases than in controls included 1q21.1 duplication, 3q29 deletion and 16p11.2 duplication.

9.3.7 *Bipolar disorder: Genome sequencing studies*

Ament *et al.* (2015) carried out genome sequencing in 200 individuals from 41 families with multiple members affected with bipolar disorder. In their analyses of sequence data they initially focused on 3,087 candidate genes known to have synaptic functions. They identified increased burdens of rare variants in GABA neurotransmitter function related genes GABRA4, GABRA1, GABRA5 and in calcium ion channel sub-unit encoding genes CACNB1, CACND CACN1G and CACNA1C and also in GPHN (gephyrin) and SRC. GPHN protein anchors neurotransmitter receptors to the post-synaptic cytoskeleton; the SRC gene encodes a non-receptor tyrosine kinase.

9.3.8 *Analysis of GWAS data across psychiatric disorders to identify biological pathways*

The Schizophrenia Working Group of the Psychiatric Genomic Consortium (2014) examined genome wide single nucleotide polymorphism (SNP) data that had been gathered across five disorders: Schizophrenia, major depressive disorder, bipolar disorder, autism spectrum disorder and ADHD. They classified significant genes found in association with these disorders using five different functional databases: GO, KEGG, Panther, Reactome and TargetScan. The top integrative functional pathway for genes associated with bipolar disorder, major depressive disorder and schizophrenia was the histone lysine 4 (H3K4) methylation pathway. Statistically significant associations for these three adult onset disorders were also found with immune, neuronal signaling and PSD pathways.

In the cross-disorder analysis of SNPs in 33,332 cases of schizophrenia, bipolar disorder, major depressive disorder, autism spectrum disorder and ADHD and in 27,888 controls, the Psychiatric Genomics Consortium determined that four loci surpassed cutoff for genome-wide significance. These included SNPs in two chromosomal regions (3p21 and 10q24) and SNPs in two genes that encode L-type voltage gated calcium channels. These included CACNA1C that maps to chromosome 12p (significance $p = 1.87 \times 10^{-8}$), and CACNB2 that maps to chromosome 10 (position 10:18,641,934).

For the five disorders considered jointly the highest LOD score was achieved on chromosome 3p21 with the SNP rs2535629 located in a genomic region close to many genes, including ITIH3 gene that encodes a protein that stabilizes extra-cellular matrix. The chromosome 10 locus at position 10:104,649,994 maps close to many protein coding gene loci including AS3MT that encodes a methyl transferase.

The investigators in the Cross-Disorder Group of the Psychiatric Genomics Consortium (2013) proposed that genetic variation in specific basic systems, e.g. neuronal calcium handling, might increase susceptibility to psychiatric disorder and that additional factors, including environmental factors, might then play roles in promoting the development of specific psychiatric disorders. They noted that in addition compensatory mechanisms, including adaptive factors, might counteract the deleterious effects of the underlying core factors.

9.4 Translating Genetic and Genomic Findings in Psychiatric Disorders to Development of Therapies

Schubert *et al.* (2014) described key steps in translating human genetic findings into treatment targets. These included, firstly, identification of genetic loci that are robustly shown to contribute to disease risk or to protect against risk. Secondly, causative genes within associated loci must be identified. Thirdly, analyses must be undertaken to understand the biological or pathological mechanisms through which putative causal genes act and actionable hypotheses for treatment must be developed. Fourth, there must be opportunity to recruit appropriate patients to test therapies.

Schubert *et al.* noted that calcium ion channels are potentially druggable targets. They noted that there are concerns about the effects of such drugs on the cardiovascular system. However they emphasized that further analyses should be undertaken of cell and tissue distributions and specificities of actions of different calcium channel sub-units. They noted that the glutamatergic system has been targeted for therapies in schizophrenia with mixed results.

Schubert *et al.* noted that the schizophrenia associated gene SRR on 17p13 encodes serine racemase. This enzyme catalyzes conversion of L-serine to D-serine. D-serine is a co-agonist and activator of NMDA glutamate receptors. They noted that D-serine depletion induced schizophrenia like symptoms. Another possible therapeutic target discussed was CLCN3, the voltage sensitive chloride channel that impacts neurotransmitter loading of GABA-ergic synaptic vesicles and modulates synaptic plasticity.

Schubert *et al.* emphasized the importance of studies designed to assess the impact of risk genotypes on brain function as assessed through various techniques including imaging. They noted that there are few cohorts available comparing individuals who have had genetic studies and deep phenotyping, including the use of neuroimaging and neurophysiology.

9.5 The BRAIN Project and the Allen Brain Atlas

BRAIN is an acronym for Brain Research through Advancing Innovative Neurotechniques. The goals of the BRAIN project are described as determining how molecules, cells, circuits and systems enable behavior, thought and emotion (Jorgenson *et al.*, 2015). Investigators plan to apply technologies to define brain function and to use chemical and molecular approaches to analyze structural biology. High-resolution studies will be undertaken to examine genes, molecules, synapses and neurons. In addition whole brain imaging will be undertaken.

Jorgenson *et al.* noted that a consensus definition of the taxonomy of brain cells awaits development. Questions that need to be answered include determining the relationship of specific cell types to circuit function, and determining how information is coded and processed in different brain circuits. In addition macro and micro connectivity maps need to be constructed. Information needs to be gathered on how learning changes circuits and how memory is accessed. In addition neural activity patterns under different emotional and perceptual conditions need to be analyzed.

Cross boundary and interdisciplinary collaborations will be essential.

9.5.1 *The Allen Brain Atlas and transcription studies*

The developers of the Allen Brain Atlas seek to determine the connection between brain architecture and brain function in part through development of a comprehensive atlas of the brain transcriptome.

In 2012, Hawrylycz *et al.* published information on the transcription of brain regions and structures defined through 3-dimensional magnetic resonance imaging (MRI), tissue macro and micro dissection and histology. RNA was extracted from tissue sections of well-defined regions and structures and this RNA was analyzed on microarrays. They defined gene expression patterns related to specific cell types and structures.

Studies on neuronal PSDs revealed that 740 genes were expressed there and 31% of the genes showed differential expression in PSDs from different regions. The PSD gene set they defined included many genes that are defined in the Gene Ontology database as "synapse associated". As an example of regional differences in PSD located genes the authors presented results of studies of the precentral gyrus, the site of origin of neurons with the longest projections. They noted that these neurons were enriched in expression of RNA for the neurofilament proteins NEFL, NEFM and NEFH (light, medium and heavy chain neurofilament proteins).

Hawrylycz *et al.* defined different modules where each module contained genes that were co-expressed in different locations. For example modules M1 and M2 included genes expressed in neurons in neocortex regions. Another module included genes expressed in the sub-cortical regions, e.g. in the dentate gyrus. One module, M5, included genes expressed in the ventricular ependymal lining and choroid plexus. Another region specific module included genes expressed in the striatum, e.g. dopamine receptors. The globus paliidus had a unique pattern of gene expression. Analysis of the hippocampus revealed that RNA for the calcium binding protein CALB1 was highly expressed.

Of particular importance was the discovery that a large percentage of the differentially expressed transcripts did not map to defined GENCODE genes though they did map to specific genomic contigs. These transcripts likely represent non-protein coding RNA transcripts.

Miller *et al.* (2014) used the Allen Brain Atlas resources including magnetic resonance imaging, laser microdissection, histological analysis and microarray analysis to study human brain from fetuses at 15, 16 and 21 weeks post-conception. They examined approximately 300 sections and utilized these for RNA isolation and microarray analysis.

In studies of the neocortex they analyzed nine layers; from the deepest to the surface; these included ventricular zone, inner and outer sub-ventricular zones, intermediate zone, inner and outer cortical plates, marginal zone and sub-pial granular zone.

Miller *et al.* reported that 96% of genes expressed in the neocortex were Reference Sequence (RefSeq) genes. They reported that different layers showed unique molecular signatures. The cell marker of Cajal Retzius cells, calbindin calcium binding protein (CALB2) was enriched in the marginal zone and in the sub-pial granular zone. The cortical progenitor cell markers TBR2 (T Box transcription factor) and PAX6 homeobox were enriched in the germinal layers, i.e. the ventricular and sub-ventricular zones. These zones also showed abundant expression of the transcription regulators TGIF1 and SIX3 homeobox transcription factor. In the pial layer that overlies the cortex the zinc finger transcription factor ZIC1 showed abundant expression.

Expression patterns in the post-mitotic layers reflected developmental maturation. Miller *et al.* documented expression of neuronal markers of synaptic transmission in cells in the sub-plate layer that contained the earliest generated neurons. In the inner cortical plate genes involved in encoding proteins responsible for connections formation were highly expressed. In the inner cortical plate the youngest neurons occurred and showed enrichment for metabolic functions.

Miller *et al.* identified a distinct module of expressed genes that was characteristic of interneurons. This module included DLX1 and DLX2 homeobox gene transcripts that are known to be important in interneuron migration. This specific gene expression pattern was found where radial glial cells were present.

Miller *et al.* concluded that cortical patterning is likely the result of intrinsic signaling controlled through graded transcription factor expression. They also indicated that differential expression of human lineage conserved non-protein coding sequences (huCNS sequences) also plays roles in the regulation of developmental patterns of gene expression in the neocortex. These huCNS sequences were more abundantly expressed in the rostral (frontal) cortex than in the caudal cortex consistent with expansion of the frontal cortex in primates.

9.6 Transcription Analyses of the Human Brain

9.6.1 *Examples of human-specific genes expressed in the neocortex*

In primates, especially in humans, expansion of the neocortex occurs due to increased proliferation of neural stem cells and progenitor neurons. This is primarily due to amplification of basal progenitors in the fetal sub-ventricular zone. Florio *et al.* (2015) noted that cortical expanse in humans has been attributed to increased generation of basal progenitor cells. They noted that neural progenitor cells include apical progenitors at the ventricular side and basal progenitors in the sub-ventricular zone and sub-apical cells that have contact with the sub-ventricular and ventricular zones. The basal progenitor cells primarily undergo expanded proliferation in primates.

Florio *et al.* carried out transcriptome analyses of neural progenitor cells and of the ventricular zones and sub-ventricular zone in mice and primates including humans. They determined that 56 different genes were primarily expressed in human apical and basal ventricular regions and in the derived radial glial cells. They identified ARHGAP11B as the human gene that leads to increased amplification of basal progenitors. The ARHGAP11B gene was duplicated from the ARHGAP11A gene in the human lineage after branching from the chimpanzee lineage. ARHGAP11B encodes a RHO guanosine triphosphatase activating protein.

Florio *et al.* expressed human ARHGAP11B in mouse neocortex after fetal life. In addition they injected ARHGAP11B into organotypic slice cultures of mouse neocortex. These procedures induced increased generation of radial glial cells, increased the cortical plate area and induced gyrification of the cortex.

9.6.2 *Developmental regulation of transcription in the prefrontal cortex*

Jaffe *et al.* (2015) reported results of RNA sequencing and analyses on 72 dorsal prefrontal cortex samples obtained across six life stages. The analyses were designed to detect differentially expressed sequences. They then carried out analyses on 36 additional samples.

They determined that 41.5% of the differentially expressed regions (DERs) corresponded to non-protein coding genomic segments including introns. The highest number of differentially expressed genes occurred in fetal brains; fetal brains also manifested the largest expressed genome fraction.

Jaffe *et al.* noted that the highly expressed mRNA in fetal brains include the SOX11 transcription factor and DCX that is involved in the migration of neuroblasts. SLC6A1 (GAT1 GABA transporter) was also highly expressed in fetal brains. NRGN (neuregulin) and CAMK2A (calmodulin kinase 2A) became highly expressed in infant and teenage brains. RGS4 (regulator of G-protein signaling 4) had highest expression during adolescence and CNTNAP1 (contactin associated protein) reached highest expression in adulthood.

9.7 The Connectome

The NIMH connectome project is designed to acquire and share data about structural and functional connectivity of the brain. In 2015 the project would have completed studies of 1,200 healthy adults and this study includes 300 twin pairs. In neurobiology there is currently increased emphasis on networks and connectivity and recognition that networks are critically important in cognition and neuropsychology.

Sporns (2010) emphasized that understanding networks requires knowledge of system components, knowledge of how components interact and emergent properties of interactions and appreciation that patterns of activity result from selective coupling between elements.

Sporns noted that historically Santiago Ramón y Cajal determined that neurons are polarized cells, that dendrites receive signals and relay them to a transmitting structure, the axons. Later neurons were shown to communicate through electrical activity and this was observed through electro- encephalography. Later other techniques were designed to register brain activity, and these included metabolic studies with labeled glucose and hemodynamic studies coupled with magnetic resonance brain imaging.

He noted that multiple scales of connection exist. On a microscale there are cells and synapses; cell groups are located in regions and regions are connected into systems. Histologic connectivity can be defined. Macro connectivity includes use of structural MRI studies, and diffusion tensor MRI.

Sporns emphasized that functional connectivity is time dependent and that function is dependent on the biophysical properties of neurons. This includes their morphology, molecular components and chemical modifications.

9.7.1 *The human brain connectome*

Approximately one hundred billion (10^{-11}) neurons are present in the human brain. Buckholtz and Meyer-Lindenberg (2012) noted that the number of connections between neurons dwarfs the number of neurons. They emphasized that the creation and adaptation of networks underlies cognitive, affective, motivational and social processes. Furthermore aberrant connectivity patterns occur in mental disorders.

In their review of the human brain connectome, Buckholtz and Meyer-Lindenberg evaluated approaches to measure neural connectivity and to analyze connectivity in psychiatric disease. They emphasized that transdiagnostic patterns of dysconnectivity occur and could explain co-morbidity in diagnostic categories. These investigators also presented evidence that specific genetic variants impacted connectivity circuits. COMT and CNTNAP2 have variants that are associated with slower activity or abnormal activity within the default mode network (DMN). This network is active under waking restful conditions when individuals are not focused.

They emphasized functional segregation and functional integration. Functional segregation refers to specialized processing. They noted that MRI blood oxygen level analysis could be used to measure resting state connectivity and task based functional activity; analysis of connectivity needs to take into account that there is an inherent constant adaptive reconfiguration of functional integration.

Importantly, network mechanisms have been identified that underlie cognitive, affective motivation and social function domains and these domains are included in the RDoC diagnostic criteria described above.

Activity of the lateral prefrontal cortex, dorsal anterior cingulate and dorsal parietal cortex participate in core executive functions. This network is utilized for working memory and goal-directed attention. Buckholtz and Meyer-Lindenberg reported that defects in executive function occur in schizophrenia, ADHD, in major depression and in substance abuse.

The lateral pre-frontal cortex, ventromedial pre-frontal cortex, orbito-frontal cortex and amygdala and the cingulate cortex comprise a corticolimbic circuit involved in regulating vigilance, arousal, affect, anxiety and anger; this network is impacted in mood, anxiety and personality disorders.

The prefrontal cortex and striatum are important in determining responses to reward, value based learning and these brain regions are impaired in ADHD, personality disorder and substance abuse.

The ventromedial pre-frontal cortex, posterior cingulate cortex, the temporal-parietal junction and intra-parietal sulcus are active in the social domains of cognition.

The temporal-parietal junction, posterior cingulate and ventro-medial prefrontal cortex are sometimes considered together as the DMN. This may be involved in thought, beliefs and emotions.

Buckholtz and Meyer-Lindenberg reported that heritable differences in brain connectivity likely contribute to mental illness. They concluded also that polygenic inheritance likely contributes to mental illness and that there are many risk alleles, each of small effect. In addition rare highly penetrant alleles may have impact. CNVs represent one form of rare variant that may have an impact. However CNVs have variable penetrance and pleiotropic effects. They proposed that across the population the burden of deleterious alleles was a continuous quantitative variable.

They emphasized that environmental factors are likely critical in determining susceptibility to psychopathology. They noted that childhood maltreatment had been shown to impact structure and function of the cingulate and amygdala regions.

A few examples of network-related studies are reported below.

Ebisch *et al.* 2011 reported that the insula cortex is involved in emotional awareness of self and others. They presented evidence that functional connectivity properties of the insula are altered in individuals with high functioning forms of autism.

Tan *et al.* (2012) defined a working memory network that engages pre-frontal, parietal and sub-cortical brain regions. They determined that specific functional polymorphisms in COMT (catechol methyl transferase), DRD2 (dopamine receptor) and AKT1 kinase impact working memory. The COMT Val158Met polymorphism impacted prefrontal control, parietal, and striatal processing of working memory. Tan *et al.* determined that DRD2 and AKT1 polymorphisms influenced the pre-frontal and striatal parts of the network.

Hahamy *et al.* (2015) studied adult subjects with high functioning autism. They examined inter-hemispheric and intra-hemispheric functional connectivity using functional MRI (fMRI) data. They reported that the topographic pattern of resting state connectivity in the autism subjects was distorted relative to the more consistent functional connectivity patterns observed in controls. They noted further that the pattern of connectivity differed between individual autism subjects and that the patterns were idiosyncratic. However the degrees to which connectivity patterns were distorted correlated with the level of autism severity as measured by the Autism Diagnostic Observation Schedule (ADOS).

Narr and Leaver (2015) concluded that variations in clinical, behavioral and cognitive characteristics occurred in patients and that these resulted from disruptions in particular connectivity networks. They noted further that in particular patients with schizophrenia symptoms often overlapped with clinical features of other neuropsychiatric disorders.

9.7.2 *Methods in determining structural and functional connectivity*

Diffusion weighted imaging (DWI) is based on diffusion of H2O that occurs along axons. Newer methodologies enable assessment of direction of diffusion. Whole tractography extracts information on all tracts at once and specific tracts can be followed individually. Dennis and Thompson (2013) noted that resting MRI and functional MRI are used to assess activity in the absence of or in the presence of a specific task. Coupling between brain regions during a specific task can be assessed through blood-oxygen-level dependent (BOLD) imaging.

9.8 Synaptic Pruning, Synaptic Plasticity and Circuits

There is abundant evidence that synaptic pruning during development is essential for appropriate brain function. Formation of mature neuronal circuits requires pruning. Koyama and Ikegaya (2015) reported that this process is critical for establishing appropriate excitation/inhibition balance in the brain. They noted further that microglial dysfunction can impair pruning and that this plays roles in neurodevelopmental disorders.

There is evidence that members of the complement cascade can activate microglia, the innate immune cells of the central nervous system. In addition activation of microglia by complement was reported to play roles in synaptic pruning (Schafer and Stevens, 2013).

9.8.1 *Synaptic plasticity*

In studies on the molecular biology of memory, Kandel (2001) reported that experience-dependent changes in synaptic structure played key roles. Leal *et al.* (2014) noted that the most studied form of synaptic plasticity is in the hippocampus and that three steps in this process include short term potentiation, early and late long term potentiation. Early steps in potentiation involve local protein synthesis in dendritic spines that are remote from the nucleus and this involves translation of transcripts in polyribosomal complexes (Steward and Levy, 1982).

FMRP and Staufen proteins 1 and 2, double stranded RNA binding proteins, are involved in binding mRNA in polyribosomal complexes and targeting these to dendrites.

There is evidence that BDNF is important in inducing long term potentiation. Leal *et al.* reported that the timing and release of BDNF to promote long term potentiation were still under study. BDNF upregulates neurotransmitter release from synaptic vesicles and this release also requires a small GTP binding protein RAB3A. B binds to and activates TRKB receptors on the post-synaptic membrane. TRKB receptors then activate the tyrosine kinase FYN that in turn phosphorylates NMDA receptor sub-units. Leal *et al.* reported that TRKB activates phospholipase gamma (PGCgamma) and it promotes conversion of phosphatidyl inositol-4-5-biphosphate to diacylglycerol (DAG) and inositol triphosphate (IP3). IP3 then activates receptors in the endoplasmic reticulum that release stored calcium (Ca2). Diacylglycerol then activates a specific transient receptor channel (TRPC3) to further increase intracellular calcium.

Calcium then activates calcium calmodulin dependent protein kinases (CAMKK1, CAMKK2). These then phosphorylate adenylate kinase (AKT). Leal *et al.* noted that AKT is also activated through PI3K signaling. AKT signaling upregulates MTOR that in turn activates the translational machinery.

There is evidence that BDNF regulates sub-units that compose the AMPA receptors, and calmodulin kinase facilitates insertion of AMPA receptors into the post-synaptic membrane in part through alteration of actin polymerization.

BDNF activates translation of transcripts of ARC phosphorylates cofilin and this facilitates F actin polymerization. ARC plays a major role in cytoskeleton remodeling.

The impact of BDNF on protein synthesis depends in part on its phosphorylation of EIF4E binding protein and initiation of formation of the complex of EIF4E (eukaryotic translation initiation complex) and S6 protein.

In addition BDNF impacts long term potentiation through its effects on transcription of genes in the nucleus. This effect is dependent on signaling from BDNF-activated TRK receptors and through activation of mitogen activated protein kinase (MAPK) and ERK kinase 1/2.

9.8.2 *ERK MAPKs and neuronal gene transcription*

Wiegert and Bading (2011) reviewed the roles of the MAPK cascade as the signaling system that transduces information from calcium signaling generated at the post-synaptic membrane to the nucleus. They noted that activity dependent gene expression is necessary for the formation and maturation of neural networks and for protein modification of the mature network.

Calcium influx through ionotropic NMDA receptors and through voltage gated calcium channels impacts gene transcription that is necessary for long term potentiation and for long term depression and for remodeling synaptic connections.

Wiegert and Bading noted that calcium dependent activation of gene transcription requires signal translocation across the nuclear membrane. Small molecules can enter the nucleus through nuclear pores, but larger molecules cannot. Calcium activates protein kinases, calmodulin kinase and MAPKs and these are involved in indirect transmission of calcium signaling to the nucleus. One of the key results of signaling in the indirectly activated cascades is phosphorylation of the cyclic AMP response element (CREB) and CREB binding protein (CREBBP).

Wiegert and Bading reported that the ERK MAP cascade is the major signaling pathway following synaptic entry of calcium. ERK

MAP signaling regulates transcription factors that enhance expression of specific genes including BDNF, ARC CFOS and CJUN. In addition there is evidence that ERK MAP signaling plays roles in chromatin remodeling. ERK1/S activity promotes acetylation of lysine 14 on histone H3 (H3K14).

9.9 Epigenetics and Memory Consolidation

Jarome *et al.* (2014) reviewed recent studies aimed at understanding the molecular, cellular and genetic basis for memory formation, storage and retrieval. There is evidence that different processes are involved in memory acquisition, transfer to labile short term memory and later development of long term memory defined as consolidation. They noted further that following recall a consolidated memory enters a labile period and it then requires reconsolidation. During reconsolidation the memory can potentially be erased.

Jarome *et al.* noted that memory storage requires alterations in gene transcription and translation in neurons. These processes lead to changes in synaptic plasticity. They emphasized that epigenetic mechanisms likely play important roles in the regulation of gene transcription during long term memory formation and storage. Epigenetic mechanisms are known to be critical regulators of learning-dependent synaptic plasticity. Epigenetic mechanisms regulate chromatin structure through histone modifications and also impact DNA methylation. Histone phosphorylation and histone acetylation have been found to be important in memory consolidation; histone 3 lysine 9 (H3K9) demethylation enhanced memory. DNA methylation also plays roles in memory formation.

Different forms of cytosine methylation occur. These include methyl CpG where the guanine nucleotide follows the methylated cytosine. Hydroxymethyl cytosine may also occur, referred to as hmCG.

There is evidence that in the brain methylation occurs not only at CpG (cytosine guanine) sites but also at CpH sites where H adenine, cytosine or thymine (A, C or T). These configurations are referred to as mCH. Lister *et al.* (2013) reported that extensive methylome configuration occurs during development; mCH methylation is absent from fetal brain. However it accumulates during post-natal development and this increase coincides with the period of neuronal synaptogenesis. Accumulation of mCH also coincides with the period of increased expression of the methyl transferase DNMT3A. Lister *et al.* noted that levels of hMC accumulate in adult brain particularly in the prefrontal cortex. A special form of bisulfite sequencing, TET-assisted bisulfite sequencing, is required to determine the presence of hMC.

Jarome *et al.* emphasized that much remains to be understood about epigenetics and memory and about connections of epigenetics to upstream signaling processes.

9.9.1 *Long genes expressed in neurons and important in synaptic function*

There is evidence that very long genes are particularly expressed in neurons and are involved in synaptic function. King *et al.* (2013) reported that transcription of very long genes is facilitated by topoisomerase that resolves DNA supercoiling. They demonstrated that topoisomerase inhibitor topotecan particularly impaired expression of very long genes.

10. Neurodegenerative Diseases

10.1 Clinical Manifestations in Late Onset Neurodegenerative Disease

The most common clinical diagnoses assigned to neurodegenerative diseases include Parkinson's disease, Alzheimer's disease, frontotemporal dementia (FTD) and amyotrophic lateral sclerosis (ALS). Parkinson's disease symptoms include tremors, increasing muscle rigidity, balance problems, slowness of voluntary movements and shuffling gait. Later, confusion and difficulties with swallowing may develop. Disorders with clinical features of Parkinson's disease are sometimes referred to as Parkinsonism. These include tauopathies and synucleinopathies.

Clinical manifestations of Alzheimer's disease include short term memory difficulties and cognitive impairment.

Other age-related neurodegenerative diseases include dementia with Lewy bodies (DLB) and in that disorder there are also pathological features of Alzheimer's disease and clinical features of Parkinson's disease.

FTD is associated with behavioral problems, disinhibition, apathy, stereotyped behavior, cognitive impairment, difficulties with executive functions; later psychosis may develop. Other manifestations include language defects or loss of language. FTD may arise as a result of mutations in tau; however it also results from mutations in other genes. Clinical features of FTD and ALS may arise as a result of expansion of the repeat sequence G6C2 in C9ORF72. ALS

265

is associated primarily with symptoms resulting from damage to motor neurons.

Paralysis of upward gaze, referred to as progressive supra-orbital paralysis, may occur FTDs or in tauopathies.

10.2 Primary Neuropathologies in Alzheimer's and Parkinson's Diseases

Serrano-Pozo *et al.* (2011) reviewed neuropathological hallmarks of Alzheimer's disease. They defined positive lesions and these included amyloid plaques, cerebral amyloid angiopathy, neurofibrillary tangles and astrogliosis. They also defined negative lesions and included neuronal and synaptic loss in that category. They provided evidence that amyloid plaque build-up occurs prior to the onset of cognitive symptoms and noted that cognitive decline worsened as neurofibrillary tangles accumulated and neuronal loss occurred. In addition they noted that congophilic angiopathy was frequently present.

Serrano-Pozo *et al.* defined two types of amyloid plaques, the extra-cellular deposits of beta amyloid, and the 40 and 42 amino acid peptides derived from the amyloid precursor protein (APP). They included dense core plaques that were stained with Congo Red and Thioflavin S, and diffuse plaques, deposits with ill-defined contours, that failed to stain with Congo Red or Thioflavin S. Importantly they noted that such diffuse plaques commonly occurred in the brains of normal individuals.

In addition they defined two types of deposits of the microtubule associated protein tau. These include neurofibrillary tangles, intra-neuronal aggregates of hyperphosphorylated and misfolded tau and neuropil threads in axonal and dendritic segments. The neurofibrillary tangles can be demonstrated with silver staining and with Thioflavin S.

The neurodegenerative processes involved damage to synapses and retrograde degeneration of axons and later atrophy of dendrites. Serrano-Pozo *et al.* referred to the work of DeKosky *et al.* (1996) and the postulation that loss of synapses in the neocortex and the limbic system is strongly correlated with cognitive loss. They noted that glial response with reactive astrocytes and microglial cells occurred throughout the course of the disease. They noted that correlation of clinical and pathological findings indicated that some individuals are more resilient than others to the toxic effects of amyloid plaques and neurofibrillary tangles.

The occurrence of manifestations of Alzheimer's disease and Lewy body formation is sometimes referred to as DLB. Serrano-Pozo *et al.* reported that in DLB cases there were extensive deposits of amyloid beta and synuclein in the striatum and hippocampus. They noted that 25% of patients with Alzheimer's disease develop clinical symptoms of Parkinson's disease and that some studies have revealed that 70% of cases of sporadic Alzheimer's disease develop alpha synuclein positive Lewy body structures in the amygdala and in the limbic system. They noted further that cases with familial Alzheimer's disease, cases with Down's syndrome and Alzheimer's disease also have Lewy body pathology in their brain, and clinically they have features of Parkinsonism.

Goedert (2015) reviewed historical aspect of Alzheimer's and Parkinson's disease pathologies. He noted that original descriptions defined Alzheimer's disease as a condition associated with amyloid plaques and neurofibrillary tangles. Parkinson's disease was defined as a disorder associated with Lewy bodies and Lewy neurites.

In the 1980s, amyloid was determined to be the cleavage product of APP encoded by the APP gene. Neurofibrillary tangles were found to be comprised of tau protein, the product of the microtubule associated protein tau (MAPT) gene.

Studies on rare inherited forms of these diseases facilitated the analysis of their molecular etiologies. In some cases, familial Alzheimer's disease was shown to be due to the over-production of APP while in other familial cases it was due to defects in sub-units of the amyloid cleavage enzyme gamma secretase. Familial cases of Parkinson's disease were found to be due to synuclein mutations. Goedert noted that 1,998 mutations in the tau protein were identified in patients with Parkinson's disease and in patients with FTD.

Goedert emphasized that these proteins, beta-amyloid, tau and synuclein, occur within aggregates associated with disease and that the proteins adopt alternative conformations. Amyloidogenic peptides that accumulate in Alzheimer's disease adopt beta sheet conformations.

Filamentous aggregates of tau accumulate in chronic traumatic encephalopathies, and also in tauopathies (atypical Parkinsonism), progressive supranuclear palsy, corticobasal degeneration, FTD (Pick's disease) and in a condition referred to as argyrophilic grain disease.

Goedert reported that 95% of cases of classical Parkinson's disease have Lewy body pathology. Lewy bodies are composed of alpha-synuclein deposits. Alpha-synuclein deposits particularly impact dopaminergic neurons in the pars compacta of the striatum. The biological spectrum of Parkinson-related disorders includes Parkinson's disease with dementia, DLB, and multiple system atrophy, which features Parkinsonism, autonomic nervous system dysfunction and ataxia. In multiple systems atrophy synuclein deposits also accumulate in glial cells.

Goedert emphasized that the mechanisms of the origin of Alzheimer's disease and Parkinson's disease were previously thought to be cell autonomous with abnormal aggregates forming at multiple sites. He emphasized that more recent studies suggest that inclusions arise in a small number of cells and then spread to other brain regions. He postulated that these findings imply a prion concept for neurodegenerative diseases that are characterized by abnormal protein assemblies and aggregates.

The prion concept first evolved through studies of animal diseases such as scrapie in sheep and bovine spongiform encephalopathy, and through human studies on Creutzfeld–Jakob disease (CJD). Goedert emphasized that less than 1% of cases of CJD are due to infection with contaminated food or through administration of contaminated forms of cadaveric human growth hormone. Inherited forms of CJD are due to mutations in the PRNP gene. PRNP encodes a sialoglycoprotein that gives rise to a beta sheet. Goedert emphasized that amyloid beta and tau give rise to similar conformations. Inherited PRNP mutations are rare causes of CJD. However in Jews of Libyan ancestry there is a founder mutation E200K in PRNP protein. Carriers of this mutation are asymptomatic until middle age when they become progressively symptomatic (Cohen *et al.*, 2015).

Further discussions on prions and neurodegenerative diseases are provided at the end of this chapter.

10.3 Mixed Neuropathologies and more than One form in Patients with Dementia

10.3.1 *Mixed neuropathologies*

There is a growing body of evidence that mixed neuropathologies are often found in patients who succumbed to dementia. In a post-mortem study of brains from 3,303 individuals with a clinical diagnosis of dementia, Kovacs *et al.* (2013) reported that 53% of cases had mixed pathologies, including evidence of Alzheimer's disease, vascular pathology, synucleinopathies including Lewy bodies, phosphorylated tau and argyrophilic (silver staining) inclusions that impact hippocampus and basal ganglia.

Schneider *et al.* (2007) reported results of a community-based longitudinal study of older patients who agreed to clinical evaluation and brain donation. Post-mortem brain studies revealed that in patients diagnosed clinically with dementia prior to death, 38% had evidence of Alzheimer's disease and vascular defects leading to infarcts. Pathological evidence of Alzheimer's disease only, was found in 30% of patients with dementia and in 12% of dementia patients there was only evidence of vascular disease. In 12% of the dementia patients there was evidence of Alzheimer's disease and Parkinson's disease with Lewy bodies.

10.3.2 *More than one form of protein aggregates in neurodegenerative disease*

In Alzheimer's disease beta amyloid and tau aggregates occur. In Parkinson's disease synuclein and tau aggregates occur. Moussaud *et al.* (2014) noted cross talk between proteinopathies. They reported that tau aggregates are characteristically found in patients with dementia. They proposed that unfolded tau and synuclein impact the solubility of each and promote fibrillization of each other. Furthermore they suggested that specific common triggers promote aggregation and that the triggers may differ in differ in different cell types.

In Parkinson's disease aggregated phosphorylated alpha synuclein accumulates in Lewy bodies in sub-cortical brain regions. In Alzheimer's hyper-phosphorylated aggregated tau accumulates in neurofibrillary tangles. Familial forms of Parkinson's disease result in some cases from mutations or duplications in synuclein (SNCA), or mutations in the MAPT gene that encodes tau. Common genetic variants in at least 24 different genes, including SNCA and MAPT, increase the risk for Parkinson's disease. Two major haplotypes H1 and H2 occur in the MAPT region on chromosome 17. The different haplotypes derive from an ancient inversion in the genomic region that surrounds MAPT. The H2 inversion haplotype correlates with lower tau expression and a lower risk of neurodegeneration (Wade-Martins, 2012).

Alpha-synuclein aggregates promote tau phosphorylation and the kinase GSK3 also plays an important role in tau phosphorylation.

There is growing evidence for prion-like cell-to-cell propagation of aggregates of tau, alpha synuclein and beta amyloid aggregates.

Soluble forms of amyloid beta and tau that form the building blocks of insoluble plaques and tangles are currently thought to be the key elements that drive neural cell death (Bloom, 2014). There is also evidence to support the contention that tau is required for amyloid beta neuronal toxicity. Furthermore misfolded soluble amyloid beta and tau can act as prions and spread from one region to another.

10.3.3 *Amyloid and tau accumulation, impaired peri-vascular drainage and angiopathy*

Impaired removal of amyloid may result from defective function of the perivascular drainage system. This system normally moves interstitial fluid and solutes within the brain. More pronounced age-related impairment of this system occurs in individuals who carry the apolipoprotein E4 (APOE4) allele. Hawkes *et al.* (2011) reported that amyloid beta removal is further impeded by changes in the basement membranes in blood vessels. They noted that changes in drainage are often associated with formation of fibrillar amyloid in blood vessel walls in cerebral amyloid angiopathy.

10.4 Types of Protein Aggregates

10.4.1 *Tau protein and tauopathies*

Filamentous inclusions of hyperphosphorylated tau occur in a number of different human neurodegenerative disorders. Clavaguera *et al.* (2013) reported that the list of these diseases includes Alzheimer's disease, tau tangle only disease, Pick's disease (a form of FTD), argyrophilic granule disease, progressive supra-nuclear palsy, and corticobasal degeneration. These authors injected brain extracts from patients who died of tauopathies into mouse brain. They reported that the injections induced tau aggregates that could then be propagated between mouse brains. They concluded that tau aggregates could be self-propagating in a prion like manner.

Tau protein aggregates also occur in FTD and Parkinson's disease, particularly forms of that disease that map to chromosome 17 to the locus tau locus MAPT.

Goedert and Jakes (2005) reported that the H1 haplotype of the tau gene region on chromosome 17 was associated with increased risk for supra-nuclear palsy and corticobasal degeneration. They reported that tau mutations might in some cases predispose to tau aggregation.

Tau is encoding by the MAPT gene that has 16 exons that undergo alternate splicing and where six isoforms occur. Specific transcripts may be missing exon 2 or exon 3. In addition exons toward the 3' end have repeats and the specific numbers of repeats in transcripts vary, between 4 and 6.

In tau aggregates in Alzheimer's all six repeats are present. However in progressive supra-nuclear palsy the tau aggregate that accumulates has four repeats, while in Pick's disease and FTD three repeat isoforms predominate.

Spillantini and Goedert (2013) reported that in all diseases where tau filaments accumulate tau is hyperphosphorylated. It is important to note that although mutant tau accumulates in some forms of neurodegenerative disease, in other neurodegenerative disorders where tau accumulates it is not mutated.

It is also important to note that FTD may also arise due to mutations in other genes (e.g. granulin).

Spillantini and Goedert listed 25 different forms of neurodegeneration characterized by tau inclusions:

1. Alzheimer's disease
2. ALS and Parkinsonism dementia complex
3. Argyrophilic grain disease
4. Chronic traumatic encephalopathy
5. Down syndrome
6. Familial British dementia
7. Familial Danish dementia
8. FTD and Parkinsonism linked to MAPT
9. FTD caused by C9orf72 mutations
10. Gerstmann–Straussler–Schenker disease
11. Guadeloupian Parkinsonism
12. Myotonic dystrophy
13. Neurodegeneration with brain iron accumulation
14. Niemann–Pick disease type C
15. Non-Guamanian motor neuron disease with neurofibrillary tangles
16. Pick's disease
17. Postencephalytic Parkinsonism
18. Prion protein cerebral amyloid angiopathy
19. Progressive sub-cortical gliosis
20. Progressive supranuclear palsy
21. SLC9 related mental retardation
22. Subacute sclerosing panencephalitis
23. Tangle only dementia
24. White matter tauopathy with globular glial inclusions
25. Diffuse neurofibrillary tangles with calcification

Therapeutic strategies are being designed to utilize small molecules to inhibit tau accumulation. These include polyphenols, methylene blue. Kinase inhibitors that inhibit tau phosphorylation may also be useful as therapeutic agents. However tau specific kinases and tau specific phosphatases have not yet been identified. Methods to decrease tau formation include activation of ubiquitination and proteasomal function, which may be useful. Aggregates of tau may also be susceptible to degradation by antibodies.

10.4.1.1 *Tau aggregates and tauopathies*

Tauopathies include FTD, progressive supranuclear palsy, corticobasilar degeneration, and Pick's disease. Contact sports associated with head trauma lead to increased accumulation of tau and neurofibrillary tangles in brain. Prusiner (2013) also presented evidence for increased neurofibrillary tangle accumulation as a result of blast exposure, e.g. during warfare.

There is evidence that anesthetic agent administration leads to tau hyperphosphorylation and development of neurofibrillary pathology (Whittington *et al.*, 2013).

Familial tauopathies arise due to mutations in the MAPT gene. Other genes that have mutations that lead to familial tauopathies include TDP43 (TAR binding protein, transcriptional repressor), FUS (component of ribonucleoprotein complex), progranulin and C9orf72.

10.4.2 *Synuclein and synucleinopathies*

10.4.2.1 *Defining a broad category of alpha synucleinopathies*

Brück *et al.* (2015) reviewed diseases they classified as alpha synucleinopathies that included Parkinson's disease, DLB, and multiple system atrophy. In Parkinson's disease alpha synuclein aggregates occur in neurons and are referred to as Lewy bodies. Alpha synuclein aggregates occur in the cytoplasm of glia, including microglia and astroglia, in multiple system atrophy and are sometimes referred to as Papp–Lantos bodies.

Aggregation of alpha-synuclein impairs autophagy. There is also evidence that alpha synuclein can undergo prion-like propagation.

Alpha synuclein occurs primarily in the hippocampus, striatum, thalamus, cerebellum and neocortex. Parkinson's disease patients show marked neuronal loss in the substantia nigra.

Multiple system atrophy is characterized by features of Parkinsonism, balance disorders and ataxia. In addition patients manifest features of autonomic dysfunction including postural hypotension and blood pressure abnormalities.

10.4.3 *Stress granules and RNA binding proteins*

Stress granules are aggregates that are commonly found in ALS and in FTD. They may also occur in Alzheimer's disease and in dementia associated with hexanucleotide repeat expansions and C9ORF2 linked dementia. Stress granules contain RNA binding proteins. Ash *et al.* (2014) reported that stress granules initially play physiological roles in neurons to control translation of mRNAs at particular time points and to move RNAs to specific locations. However they also form pathological detergent insoluble granules under certain condition.

TAR binding protein TDP43 binds RNA and this protein is defective in familial ALS and in tau negative FTD. Mutated forms of RNA binding proteins also occur in familial motor neuron disease. These proteins include TDP43, FUS (RNA binding protein), SMN1 (survival motor neuron protein), ataxin 2, optineurin (OPTN), and angiogenin (ANG).

Ash *et al.* reported that stress granules also occur in sporadic forms of neurodegenerative diseases including sporadic ALS. Possible environmental cues that lead to stress granule formation include enhanced oxidative stress and the unfolded proteins response.

Abnormal RNA forms occur in stress granules. These may be derived from transcription of microsatellite repeat expansions. Such repeat expansions occur in C9ORF72 associated dementia. It is interesting to note that bidirectional transcription of repeat expansions occur where the repeats occur in non-protein coding regions of the genome. Repeat expansion occurs in the 5′ non-coding regions in the fragile X (FMR) gene and in the 3′ non-coding region of the myotonic dystrophy gene. The abnormal transcripts may also serve to sequester RNA binding proteins (Mohan *et al.*, 2014).

The autophagic pathway is important in removing stress granules. This pathway involves activity of valosin containing protein (VCP).

10.5 Alzheimer's Disease Genes and Risk Factors

10.5.1 *Alzheimer's disease: Risk factors*

Dominantly inherited forms of Alzheimer's disease have been determined to be due to defects in APP, or due to mutations in presenilin 1 or 2, the components of the gamma secretase complex that proteolytically cleaves APP. APP cleavage is illustrated in Figure 10.1. Approximately 5–10% of Alzheimer's disease cases are due to highly penetrant autosomal dominant mutations in the APP or in presenilin 1 and 2. The autosomal dominant forms of Alzheimer's disease are characterized by an earlier age of onset, between 25 and 65 years of age (Van Cauwenberghe *et al.*, 2015).

They noted that late onset Alzheimer's disease is multifactorial with complex genetic contributions. They reviewed results of genome-

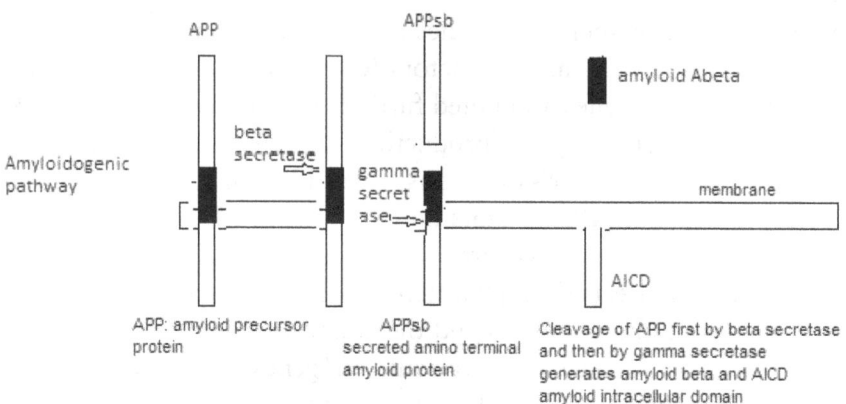

Figure 10.1. Cleavage of APP by beta secretase and gamma secretase and generation of amyloid beta.

wide association studies (GWASs) that identified 20 different loci significantly associated with Alzheimer's disease. However these loci alter risk of Alzheimer's disease with effects ranging from 1.06 to 1.3. The 20 low risk loci do not include the apolipoprotein E (APOE) locus. The impact of the associated allele at the APOE locus is discussed below.

Van Cauwenberghe *et al.* emphasized that identification of the associated loci provided insight into Alzheimer's disease pathogenesis. The risk loci clustered in three pathways, cholesterol and lipid metabolism, immune systems and inflammatory response, and endosomal vesicle signaling.

Reitz and Mayeux (2014) reported that the 21 GWASs detected risk loci for Alzheimer's disease encode products that cluster in specific functional pathways. These pathways include immune response, APP processing, lipid metabolism, endocytosis, cell surface receptors and cell adhesion. Most of these variants increase risk only 1.02–1.06 fold. However an ABCA7 locus variant increases risk in the African American population to almost the same degree as APOE4. ABCA7 (ATP binding cassette transporter) is involved in lipid metabolism. Pahnke *et al.* (2014) reported that ABCA7 plays roles in the perivascular drainage system and in clearance of amyloid deposits.

Karch and Goate (2015) reported that both early onset and late onset Alzheimer's disease have strong genetic components. They noted that large association studies have revealed that both rare and common variants act as risk factors for Alzheimer's disease. They identified 30 risk genes and noted further that most Alzheimer's risk genes impact beta amyloid production and clearance. Four genes harbored rare Alzheimer's disease associated variants. These included presenilin 1 and 2 APP variants and variants in ADAM10 metallopeptidase, a protease. Less rare and common Alzheimer's risk variants exert their effects through different pathways such as cholesterol pathways, immune pathways and the endocytosis pathway.

Alzheimer's disease risk variants in genes in the cholesterol related pathway include APOE4, and variants in ABCA7 (ATP binding cassette 7), transporter, SLC24A4 (sodium potassium calcium exchanger), CLU clusterin (chaperone) and PDL3 (phospholipase). APOE4 remains the strongest risk factor for Alzheimer's disease.

Alzheimer's risk variants in genes that function in the immune pathway occurred in CR1 (complement receptor 1), HLADRB5-1 (histocompatibility locus), CD33 (leucocyte expressed), clusterin (chaperone protein), MS4A (membrane spanning protein), EPHA1 (ephrin receptor) and INPP5D (inositol phosphatase).

Alzheimer's risk variants in genes that function in the endocytosis pathway include variants in RIN3 (RAS RAB interactor), PICALM (phosphoinositol binding protein), PTK2B (protein kinase 2B), BIN1 (binding integrator in vesicle endocytosis) and MEF2C (transcription enhancer).

Mapstone *et al.* (2014) reported studies on individuals who phenoconverted to amnesia, mild cognitive impairment or Alzheimer's disease over a 2–3 year period. They determined that individuals who phenoconverted had decreased plasma levels of phosphatidylcholine and decreased levels of acylcarnitine relative to controls. In addition these individuals also manifested decreased levels of amino acids proline, lysine, serotonin, taurine and phenylalanine. Mapstone *et al.* concluded that the altered lipid profile reflected the altered membrane integrity.

Whiley *et al.* (2014) carried out a plasma lipid profile screening study on 35 patients and a follow-up validation study on 141 patients with Alzheimer's disease and mild cognitive impairment and compared results with those from controls. Their study revealed decreased levels of specific phosphatidylcholines (PC) complexes associated with Alzheimer's disease and mild cognitive impairment. Most significant decreases were in PC (16:0/22:5) and in PC (16:0/22:6), and there were also decreased levels of PC (18:0/22:6).

APP in the endoplasmic reticulum undergoes post-translational modification by glycosylation, phosphorylation and modification with sulfide containing residues. Plácido *et al.* (2014) reported that the majority of synthesized APP remains in the Golgi network. Specific vesicles transport APP to the cell surface. They noted that APP in the trans-Golgi network and early endosomes undergoes amyloidogenic processing and cleavage with secretase enzymes presenilin 1 and 2. Studies by Joshi *et al.* (2014) reported that accumulation of amyloid beta peptides led to fragmentation of Golgi in nerve cells. They noted further that the protein SORL1 9 (sortilin-related receptor) regulates APP trafficking. These studies indicate that the endoplasmic reticulum and Golgi are potential targets for treatment of Alzheimer's disease.

10.5.2 *APOE and Alzheimer's disease*

Individuals who are heterozygous for the APOE4 allele have a two to three times increased risk of Alzheimer's disease and homozygosity for the APOE4 allele increases risk five times above the population risk. The mean age of onset of Alzheimer's symptoms differs and is dependent on the APOE alleles present in individuals. In APOE4 non-carriers the average age of onset of symptoms is 84; in APOE3/4 heterozygotes it is 76 and in APOE4 homozygotes it is 68. APOE2 alleles apparently reduce risk for Alzheimer's disease (Strittmatter *et al.*, 1993).

Huang and Mahley (2014) reviewed the structure and function of APOE and its role in Alzheimer's disease. Synthesis of APOE occurs in a number of different cells and in different tissues, particularly liver and brain. APOE is also present in peripheral nerves. APOE binds avidly to lipids and it transports lipids. In addition it mediates uptake of the bound lipids by cell surface receptors including low density lipoprotein (LDL) receptors, very low density lipoprotein (VLDL) receptors, and APOE receptor (LRP8) and to the heparan sulfate proteoglycan (HSPG) receptor GP330. Huang and Mahley noted that interactive binding with HSPG facilitates subsequent LDL receptor binding. APOE bound lipids can also be directly transferred into cells through HSPG receptors.

Huang and Mahley determined that APOE has two structural domains linked by a hinge. The N-terminal domain and two-thirds of the APOE protein form the receptor binding domain. The C-terminal domain of APOE is the lipid binding domain.

Critical residues that distinguish the different APOE isoforms APOE2, APOE3 and APOE4 are in the N-terminal domain. APOE3 is considered the normal form and has cysteine at position 112 and arginine at position 158. APOE2 has cysteine at position 112 and cysteine at position 158. This change compromises the binding of

APOE2 lipoproteins to receptors and leads to hyperlipoproteinemia. APOE4 has arginine at 112 and arginine at position 158. This amino acid substitution significantly alters the three-dimensional structure of the APOE4 protein. As a result of the altered structure in APOE4, the arginine at position 61 can interact with glutamate in position 255 in the lipid binding domain. Huang and Mahley demonstrated that this alteration causes APOE4 to preferentially bind to VLDLs. The altered three-dimensional structure likely influences several characteristics of APOE function.

There is evidence that APOE binds to amyloid-beta (Abeta). APOE4 binds more avidly to Abeta and it apparently decreases Abeta clearance. Huang and Mahley concluded that APOE also has Abeta independent neuronal and behavioral effects. They emphasized that APOE fragments likely have neurotoxic effects. APOE4 is more susceptible to proteolytic cleavage than APOE3. They noted further that APOE4 fragments accumulate in amyloid plaques and in neurofibrillary tangles. There is also evidence that in the presence of APOE4, tau phosphorylation is increased and mitochondrial dysfunction is more pronounced. Additional studies indicate that APOE4 compromises adult neurogenesis and blood brain barrier integrity.

Huang and Mahley emphasized that multiple pathways are impacted in Alzheimer's disease and that combinations of therapies will likely be required. APOE4 represents one therapeutic target.

High plasma levels of high density lipoprotein (HDL) and particularly of its component APOA1 protect against cardiovascular disease. There is some evidence that high levels of HDL and APOA1 also protect central nervous system function and those therapies that increase their concentrations are worthy of consideration (Hottman *et al.*, 2014).

10.6 Parkinson's Disease

10.6.1 *Parkinson's disease and synuclein*

Familial Parkinson's disease is in some cases associated with mutations in tau proteins, synuclein or LRRK2 (leucine rich repeat kinase 2). In rare cases duplications in synuclein gene play roles in the etiology of Parkinson's disease.

There is also evidence that specific conditions can impact the conformational states of synuclein. Aggregates of synuclein are common features of Parkinson's disease. There is also evidence that synuclein behaves as a prion. Olanow and Brundin (2013) carried out autopsy studies on Parkinson's disease patients who had received fetal neuron transplant a number of years before death. Their studies revealed that the transplanted neurons had accumulated toxic synuclein oligomers. These studies revealed that synuclein aggregates undergo prion-like propagation.

Sporadic Parkinson's disease and familial-inherited Parkinson's disease are both associated with progressive destruction of dopaminergic neurons in the pars compacta of the substantia nigra, with Lewy bodies and Lewy neurites rich in aggregated synuclein and with neuron loss in other brain regions (Kumaran and Cookson, 2015). These investigators noted that monogenic (inherited) forms of Parkinson's disease account for only 5–10% of cases of this disease.

Kumaran and Cookson reported that 19 different loci segregate with familial Parkinson's disease. Seven of these loci encode products that impact mitochondrial function; four loci encode products involved in lysosomal autophagy.

GWASs have revealed that there is some overlap with sporadic and familial Parkinson's disease in the genes that are involved. Mutations in SNCA, MAPT, and LRRK2 genes occur in specific cases with familial Parkinson's disease. There are alleles in the genes SNCA, MAPT and LRRK2 that are linked to sporadic Parkinson's disease. The sites of these alleles within the specific genes are different from the sites involved in mutations in familial cases.

Twenty four different loci were identified in GWASs that each slightly increase the risk for Parkinson's disease. Of these 24 loci five are involved in endosome lysosome function. Significant association of glucocerebrosidase (GBA) variants with Parkinson's disease supports the conclusion that the endosome lysosome system is involved in pathogenesis of that disease. It is interesting to note that a variant in the gene SCARB2 is associated with Parkinson's disease (significance $p = 5.85 \times 10^{-11}$). SCARB2 encodes a protein that transports compounds including GBA across the lysosomal membrane.

Kumaran and Cookson also emphasized the roles of neuroinflammation, the innate immune system, and microglial activation in Parkinson's disease pathogenesis. Association studies have revealed an association of the HLA locus with Parkinson's disease.

Another interesting pathway relevant to Parkinson's disease symptomatology is the dopamine synthesis pathway. Variants in the gene that encodes glycine cyclohydrolase (GCH1) increase risk (significance $p = 5.85 \times 10^{-11}$).

The proteins PINK1 (PTEN-induced putative kinase 1) and Parkin were discovered in gene mapping studies that determined loci for genes that when mutated led to autosomal recessive Parkinson's disease. There is evidence that the PINK1 and Parkin play roles in mitochondrial function and that mitochondrial function is disrupted in Parkinsonism (Exner *et al.*, 2012). PINK1 loss of function leads to decreases in mitochondrial membrane potential and that the PINK1–Parkin pathway is involved in mitochondrial autophagy (Amo *et al.*, 2014). PINK1 deficiency leads to impaired mitochondrial respiratory chain function. Morais *et al.* (2014) reported that PINK1 is required for mitochondrial complex 1 activity and for phosphorylation of mitochondrial complex 1 component NDUFA10 (NADH ubiquinone complex alpha 10). The mitochondrial electron transfer complex is illustrated in Figure 10.2.

A number of investigators have proposed that mitochondrial dysfunction in Parkinson's disease is associated with increased production of reactive oxygen species.

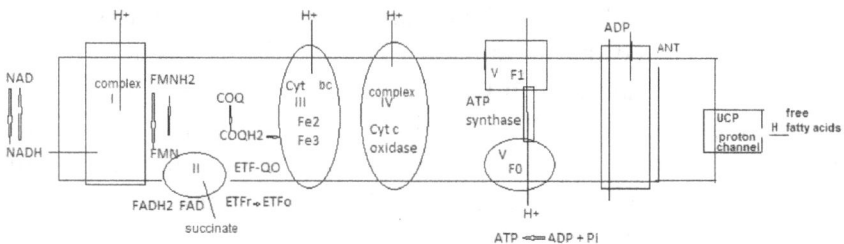

Figure 10.2. Mitochondrial function and electron transport complexes.

Compound heterozygous mutations in PINK1 lead to Parkinson's disease. In 80% of cases with PINK1 mutations there are deletions and in other cases mutations occur. PINK1 is expressed in all tissues and is most abundant in neurons and glia (Zuo and Motherwell, 2013).

Parkin mutations have been found in autosomal recessive inherited Parkinson's disease and in some cases of adult onset sporadic Parkinson's disease. Parkin plays key roles in autophagy of damaged mitochondria.

10.7 ALS, FTD and C9ORF72

Morris *et al.* (2012) reported that there is evidence that clinical and pathological overlap exists between ALS and FTD. Patients with mutations in any one of the following proteins may have manifestations of ALS and/or FTD: Progranulin (GRN) VCP, TAR binding protein (TARDBP1), ubiquitin-like protein 2 (UBQLN2), or C9ORF72. Features of Parkinson's disease may also occur in these disorders.

Renton *et al.* (2014) reviewed genes implicated in ALS. They reported that eight major genes were involved. It is however important to note that in many cases of ALS no specific genes have been shown to be involved. In familial ALS, Renton *et al.* reported that 40% of cases had defects in C9ORF72, 12% of cases had defects in SOD1 (superoxide dismutase 1), and 4% of cases had defects in TARDBP1. For each of the five remaining genes defects occurred in 1% or fewer cases of ALS.

In sporadic ALS, Renton *et al.* reported that 7% of cases had defects in C9ORF72 and defects in each of the remaining genes occurred in less than 1% of cases.

Ng *et al.* (2015) reviewed FTD and genes defective in that disorder. They reported that in familial FTD C9ORF72 was defective in 25% of cases. Mutations in GRN were reported in 5–11% of cases in different studies. MAPT mutations were reported in 2–11% of cases. Mutation in SQSTM1 (sequestosome 1 multifunctional and ubiquitin related protein) occurred in 2–4.4% of cases. Mutations in

VCP and CHMP2B (charged multi-vesicular protein) occurred in less than 1% of cases.

In sporadic FTD cases, Ng *et al.* reported that C9ORF72 mutations occurred in 30% of cases. Mutations in SQSTM1 occurred in 2–2.7% of cases. Defects in each of the genes that encode GRN, MAPT, VCP, CHMPB2 and UBQLN2 rarely occurred.

10.8 C9ORF72

There is evidence that C9ORF72 repeat expansions account for a significant proportion of cases with ALS and FTD cases (Rohrer *et al.*, 2015). The C9ORF72 locus gives rise to a number of different transcripts and depending on the transcript generated, the hexanucleotide repeat is located either in the promoter region or the first intron. Most healthy individuals in the European population have two to ten copies of the GGGGCC (G4C2) repeat. In patients several hundred or even thousands of copies of the repeat may be present. The exact cutoff point at which disease symptoms manifest is not clearly defined; however this is likely between 30 and 50 copies.

Rohrer *et al.* reported that the most common clinical phenotype in patients with expanded C9ORF72 repeats is that of FTD with predominant behavior problems and cognitive deficits. Patients with C9ORF72 repeat expansions may also present with manifestations of ALS and prominent motor neuron deficit. Patients with C9ORF72 expansions may also have features of Parkinsonism.

There have been different theories about how the C9ORF72 repeat expansions cause pathology. A number of investigators have proposed that abnormal mRNA transcripts derived from C9ORF72 form protein binding aggregates. Other investigators noted that abnormal dipeptide proteins are derived from the repeat expansions and translation.

Haeusler *et al.* (2014) determined that transcription of C9ORF72 is aberrant and that in the presence of the hexanucleotide repeat expansion abnormal RNA-DNA hybrids occur (R-loops) and prematurely terminated transcripts occur. They noted further that these aberrant transcripts bind other riboproteins and proteins, including nucleolin.

Other investigators have drawn attention to the fact that aberrant transcription of the C9ORF72 locus due to the hexanucleotide repeat expansion leads to loss of expression of the C9ORF72 open reading frame. Waite *et al.* (2014) demonstrated decreased C9ORF72 protein in the frontal cortex of patients.

Van Blitterswijk *et al.* (2015) reported that decreased expression of C9ORF72 open reading frame occurred in specific brain regions in patients with FTD. They proposed that higher levels of expression of the C9ORF72 open reading frame may have beneficial effects.

In 2015 research from three different groups of investigators, each using different study methods, determined that pathology due to C9ORF72 repeat expansion is likely due to the impact of the expansion on nuclear pore function. Nuclear pores and their functions are described below.

10.8.1 *Nuclear pores*

Cautain *et al.* (2015) reviewed components of the nuclear pore and transport processes. They emphasized that the compartmentalizing in eukaryotic cells and separation of the nucleus with mRNA generation capacity from the cytoplasm with protein synthesis capacity, necessitates the availability of a system to transport macromolecules. This transport includes export of RNA and ribonucleoproteins from the nucleus to the cytoplasm and the import of proteins from cytoplasm to the nucleus. The RAN guanosine triphosphate system (RANGTP) provides energy for this transport. For many of the macromolecules specific transport receptors interact with components of the nuclear pore complex.

The nucleus has a double membrane. The outer membrane faces the cytoplasm and interacts with the endoplasmic reticulum. The inner nuclear membrane interacts with lamin fibrils that in turn interact with chromosomes. The nuclear pore complex (NPC) forms a cylindrical structure and there are between 2,000–5,000 nuclear pores per cell. Cautain *et al.* noted that the NPC is composed of a three-ringed structure that surrounds a central core. The cytoplasmic ring has fibrils that extend to the cytoplasm. The nuclear ring is connected to a basket-like structure in the nucleus. At least 30 different proteins constitute the NPC. Membrane nuclear porins anchor the NPC to the nuclear membrane. Scaffold proteins form the skeleton of the core. Specific barrier proteins line the innermost cylindrical layer. Asymmetric proteins form the filament of the basket and the cytoplasmic filaments. Cautain *et al.* reported that macromolecular proteins have specific signal sequences that are recognized by karyopherins and their adapters.

RANGTP undergoes hydrolysis to RANGDP in a reaction facilitated by activation through RAN GAP, RAN GTPase activating protein. In this process energy is released. RANGEF (RAN guanine nucleotide exchange factor) facilitates regeneration of RANGTP from RANGDP.

Export of RNA and ribonucleoproteins from the nucleus requires special systems. The mRNA for transport requires prior splicing, a polyadenylation signal at the 3' end and addition of guanosine to the 5' end of the transcript and this guanosine is often methylated at the 7th position (5'm7G). Katahira (2012) reported that a specific complex TREX (transcription export complex) travels with RNA polymerase II as the mRNA transcript is generated. The mRNA is coupled to the TREX complex. TREX contains several subunits including THO-C (transcription factor nuclear complex) sub-units and it also contains a helicase and adapter proteins. The TREX complex interacts with the nuclear basket of the NPC.

The TREX complex mRNA and ribonucleoproteins may be transported across the pores through interaction with a specific receptor NXF1. Cautain *et al.* reported that serine arginine repeat proteins also mediate export of mRNA complexes.

10.8.2 *C9ORF72 and nuclear pores*

Three recent studies have indicated that impaired function of nuclear pores likely play key roles in the pathogenesis of disorders such as FTD and ALS that result from abnormal expansion of the G4C2 repeats in C9ORF72. Fox and Tibbets (2015) noted that prior studies revealed that nuclear pore activity is compromised in aging.

Zhang *et al.* (2015) investigated effects of the C9ORF72 30 G4C2 repeats in a Drosophila system. They were readily able to detect repeat damage in the Drosophila eye in their system. They then carried out a candidate screen for modifying mutation. Through their screening they determined that the Drosophila RanGAP, a homolog of the human RANGAP1, was a modifier. RanGAP enhanced expression could suppress the abnormal eye defect induced by the G4C2 repeat. Enhanced RanGAP levels were also found to lead to reduction in accumulation of G4C2 repeat aggregates in pluripotent stems cells derived from ALS patients. They concluded that proteins such as RNAGAP bind abnormally to G4C2 repeats.

Zhang *et al.* also demonstrated that small molecule and antisense oligonucleotides targeted to the repeat quadriduplexes reduced neurotoxicity.

Freibaum *et al.* (2015) generated transgenic Drosophila strains that expressed between 28 and 58 copies of the C9ORF72 G4C2 repeat. The transgene they utilized had a green fluorescent protein tagged open reading frame that gave rise to tagged dipeptide repeat proteins. They demonstrated that neurodegeneration occurred in transfected Drosophila and that the degree of neurodegeneration was dosage dependent on repeat length. They then carried out transfections in a series of Drosophila strains with different mutations in order to identify modifiers that influenced neurodegeneration induced by C9ORF72 repeats. This screen led to identification of 18 genetic modifiers and these encoded components of the NPC.

Jovičić *et al.* (2015) inserted C9ORF72 repeats into yeast. These produced dipeptide repeat proteins under the influence of a powerful promoter and produced toxicity. They then introduced the C9ORF72 repeats constructs into mutant yeast strains to identify genes that would modify the toxic effects of the (G4C2) dipeptide proteins. They identified a suppressor of toxicity MTR10 that mediates transport of serine-arginine containing proteins and bound ribonucleoprotein complexes. Jovičić *et al.* suggested that the blockade of nuclear pore transport by G4C2 repeat expansion was overcome through increased expression of the serine-arginine receptor. They also identified a modifier of toxicity that encoded a karyopherin transportin 1.

These results are consistent with disruption of nucleocytoplasmic transport through accumulation of dipeptide repeat proteins derived from abnormally long G4C2 repeats in C9ORF72.

10.9 Evidence for Activation of Endogenous Retrovirus in ALS

Studies on post-mortem brain specimens from cases of sporadic ALS revealed aberrant expression of an envelope protein encoded by human endogenous retrovirus HERV K (Li *et al.*, 2015). These investigators demonstrated that this protein impacted neurons in culture and led to retraction of neuronal dendrites. They also demonstrated that expression of the HERV K envelope protein in transgenic animals led to motor dysfunction, anterior horn cell damage and muscular atrophy. In transgenic animals the HERV K envelope protein induced double stranded DNA breaks. Of particular interest was the observation that activated expression of the HERV K envelope protein was found to be regulated by TDP43 that is over-expressed in ALS and forms cytoplasmic aggregates in the cytoplasm.

10.10 Organelles and Systems

10.10.1 *Mitochondria and neurodegenerative diseases*

Mitochondrial DNA is present in nucleoids and each nucleoid frequently contains multiple mitochondrial genomes. Bogenhagen (2012) reported that the mtDNAs are dispersed throughout the mitochondrial network as histone-free nucleoids containing single copies or small clusters of genomes. Youle and van der Bliek (2012) emphasized that deletions and mutations commonly occur in mitochondrial DNA so that each cell has a heteroplasmic mixture of wild type and defective mitochondria. Mitochondrial DNA damage also occurs as a result of exposure to reactive oxygen species generated through electron transfer activity in mitochondria. There is some evidence that damaged mitochondrial DNA can undergo direct repair.

There is evidence that asymmetric sorting of damaged material occurs so that following fission the more damaged mitochondrion can be disposed of through autophagy. Johansen and Lamark (2014) reported that aggregation of damaged misfolded mitochondrial protein is facilitated through activity of sequestosome protein (SQSTM1 and NBR1, autophagy receptor protein).

The proteins PINK1 and PARKIN were discovered in gene mapping studies that determined loci for genes that when mutated led to autosomal recessive Parkinson's disease. Function of these proteins and their relevance to Parkinson's disease were discussed above.

10.10.1.1 *Maintenance of functional mitochondria*

Outer and inner membranes surround mitochondria. The inner membrane folds to give rise to cristae to which the electron transfer complexes in oxidative phosphorylation are bound. Mitochondrial fission and fusion play key roles in maintenance of functional mitochondria. Youle and van der Bliek reviewed mechanisms and proteins involved in dynamic of fission and fusion. They reported that mitochondrial morphologies vary in different cell types and even within one cell type. Dynamic morphologies reflect different metabolic conditions.

The fission process gives rise to new mitochondria. It is dependent upon cytoplasmic dynamin protein (DRP1) that forms spirals around the mitochondria and severs outer and inner membranes.

Hoppins and Nunnari (2012) reported that endoplasmic reticulum-associated regions of the outer and inner mitochondrial membranes (ERMD) act as particular sites for dynamin aggregation and mitochondrial constriction. Fission is essential in proliferating cells but it is also important in non-proliferating neuronal cells. Mitochondrial fission defects in neuronal cells occur in Charcot–Marie–Tooth type 2A disease.

Youle and van der Bliek reported that fusion between the outer membranes of two adjacent mitochondria is facilitated by mitofusin proteins MFN1 and MFN2 in mammals, and fusion of the inner mitochondrial membranes of two mitochondria requires the OPA1 protein (defective in optic atrophy type 1). Fusion plays an important role in promoting homeostasis since a damaged mitochondrion can fuse with a normal mitochondrion and function of the damaged organelle is thereby restored. Fusion and fission rates are impacted by cellular metabolic conditions. However several mechanisms are in place to prevent severely damaged mitochondria from fusing with normal mitochondria. In severely damaged mitochondria the OPA1 protein on the inner mitochondrial membrane undergoes proteolysis through activity of the OMA1 (metallopeptidase) protein. OMA1 expression is apparently activated by low potential in the intermembrane space.

There is also evidence that PINK1 and PARKIN play important roles in prevention of fusion of damaged mitochondria with normal mitochondria. Activation of the PINK1 PARKIN pathway leads to ubiquitination and proteolysis of the outer membrane fusion proteins MFN1 and MFN2 and of MIRO (RAS homolog). In normal cells, levels of PINK1 are kept low through its degradation by PARL (presenilin associated protein). In damaged cells PINK1 is sequestered on the outer membrane where it recruits PARKIN and E3 ubiquitin ligase. There is also evidence that PARKIN accumulation facilitates autophagy of damaged mitochondria.

Hoppins and Nunnari (2012) reported that in damaged mitochondria dynamin (DRP1) is recruited to the outer membranes and it assembles into foci. In addition the BAX protein is accumulated to these foci and together these proteins promote permeabilization of the outer membrane. They noted further that these aggregates occur particularly at positions where the endoplasmic domains are attached to mitochondria. The ER mitochondrial related domains therefore constitute key regulatory hubs. BAX recruits BAK to the aggregates, BAX and BAK facilitate opening of the mitochondrial membrane channel. The permeabilization of the outer mitochondrial membrane promotes cytochrome release and triggers the apoptosis pathway.

Zampese *et al.* (2011) reported that PRESENILIN2 mutations impact mitochondrial calcium homeostasis and impact calcium ER shuttling. Presenilin mutations are rare causes of Alzheimer's disease.

10.10.1.2 *Mitophagy*

In mitophagy, damaged mitochondria are taken up into double membrane autophagosomes. Autophagosomes subsequently fuse with lysosomes leading to degradation of damaged mitochondria. PINK1 and PARKIN play key roles in this process. PINK1 localizes to the outer mitochondrial membrane and recruits PARKIN and E3 ubiquitin ligase. Ubiquitination subsequently leads to autophagosome formation. Wong and Holzbaur (2014) reported that PARKIN bound to outer mitochondrial membranes recruits optineurin, DFCP1 (membrane trafficking protein) and LC3 (microtubule associated protein) and that these proteins facilitate autophagosome formation. They noted that optineurin and the protein sequestosome are recruited to separate sites on damaged mitochondria. Damaging mutations in the protein optineurin occur in some cases of ALS and in wide-angle glaucoma.

It is important to emphasize that there are contacts between mitochondria and the endoplasmic reticulum and that damaged mitochondria may impact endoplasmic reticulum functions.

10.10.2 *The lysosomal-endosomal system in neurodegenerative diseases*

A number of studies have revealed that impaired endosomal function plays a role in Alzheimer's disease pathogenesis. Small and Petsko (2015) reported that retromer dysfunction contributes to the core pathological features in Alzheimer's disease and is also implicated in pathology in Parkinson's disease. Of particular importance is the possibility that the retromers are targets for drug discovery.

Small and Petsko noted that the retromer plays key roles in endosomal processes of sorting and trafficking. The retromer is a multiprotein complex with two primary modules. The first is a trimeric complex that serves as the cargo recognition core and is composed of vacuolar protein associated sorting units VPS26, VPS29 and VPS35. The second core module is the tubular module that forms tubules that extend out of the endosome. This module is composed of sorting nexin protein SNX1, SNX2, SNX5 and SNX6. In addition TAS related protein RAB7A and GTPases and lipid phosphatidyl-inositol-3 phosphate are important for functioning. Small and Petsko reported that a recently discovered module that interacts with the retromer include protein in the WAS family. These proteins interact with the retromer and with actin.

Retromers participate in retrograde transport, including transport of receptors from the cell surface to endosomes and lysosomes. Retromers also participate in molecules to the trans-Golgi system.

One of the ligands trafficked as retromer cargo is APP. Retromer dysfunction impedes rapid transport of cargoes such as APP through the endosomes. Small and Petsko reported that stalling of APP passage in endosomes promotes its processing to beta amyloid.

Small and Petsko reported that the retromer plays a role in the transport of cathepsin D to the endosomal lysosomal system. Retromer dysfunction is therefore a form of cathepsin deficiency and cathepsin deficiency promotes tau accumulation and abnormal processing. It is likely that abnormal endosomal function is also involved in pathogenesis of Parkinson's disease.

Xiao *et al.* (2015) emphasized that in Alzheimer's disease there is an imbalance between amyloid beta production and removal. They noted that APP undergoes non-amyloidogenic alpha processing at the plasma membrane and amyloidogenic beta and gamma cleavage within endosomes. APP products also undergo lysosomal cleavage. They studied in mice the role of the lysosomal pathway regulator, transcription factor TFEB in reducing amyloid beta levels. They reported that transfection of adenovirus containing TFEB into hippocampus of Alzheimer's mice stimulated lysosomal biogenesis and decreased levels of APP and beta amyloid. Yu *et al.* (2014) reported that APOE4 retards the passage of APP through the endosomal system and promotes generation of amyloid beta.

Bras *et al.* (2014) reported that genetic analyses implicated APOE, synuclein and lysosomal dysfunction play roles in the etiology of DLB.

10.10.2.1 *Endosome-lysosome markers in cerebrospinal fluid in Alzheimer's disease*

Armstrong *et al.* (2014) reported that changes in the endosome-lysosomal network are early changes in Alzheimer's disease. They identified six endosome-lysosome proteins in cerebrospinal fluid in Alzheimer's patients. These included three lysosomal membrane associated proteins LAMP1, LAMP2, LAMP3, LC3 (microtubule light chain 3) and early endosomal antigen EEA1.

10.10.3 *Neuro-immune system in neurodegenerative diseases*

The TREM2 receptor protein is expressed on microglia in the central nervous system. Hickman and El Khoury (2014) reported that microglia promote phagocytosis and degradation of protein aggregates that accumulate in Alzheimer's disease. The TREM2 protein that is active in myeloid cells is also expressed on macrophages and immune system dendritic cells. TREM2 protein is known to promote phagocytosis. Hickman and El Khoury reported the microglia mononuclear phagocytes are the principal immune cells in the central nervous system. They are involved in phagocytosing amyloid beta aggregates in Alzheimer's disease. However continuous accumulation of these aggregates subsequently leads microglia to produce cytokines and chemokines that are neurotoxins.

Neuman and Daly (2013) reported that specific heterozygous variants in TREM2 increase the risk for Alzheimer's disease. In 2013, Guerreiro *et al.* and Jonsson and Stefansson reported association of a specific TREM2 variant with the late onset Alzheimer's disease. TREM2 has a docking domain for the tyrosine kinase binding protein TYROBP. Homozygous rare mutations in TREM2 or in TYROBP lead to Nasu–Hakola disease that manifests with bone dysplasia and early-onset dementia.

Heneka *et al.* (2015) emphasized that there is chronic activation of the immune system in Alzheimer's disease and extensive upregulation of expression of genes that encode innate immune system products. Associations of Alzheimer's disease with variants in immune system genes have been described. These include association with a specific variant in CD33, a sialic rich protein expressed on microglial cells in the brain. In addition they confirmed that variants in TREM2, a macrophage associated protein, have also been associated with Alzheimer's disease.

10.10.4 *Prions*

Prusiner (2013) defined prions as "proteins that acquire alternative conformations that become self-propagating. He noted further that proteins within prions tend to have beta sheet structure and to form oligomers. He proposed that although proteins in prions may be functional, prions often play roles in diseases pathogenesis and particularly in late onset neurodegenerative disease. He proposed that in these diseases prion accumulation beyond a certain threshold leads to disease manifestations.

Prion formation and prion disease pathogenesis was first described in scrapie disease in sheep. Later, prions were found to cause bovine encephalopathy and chronic wasting disease in elk. Prusiner demonstrated that the products of the PRNP prion protein encoding gene in humans exist in multiple conformations. The most abundant cellular form PrPc is appropriately folded. A different form PrPsc is misfolded and this misfolded form serves as a template to induce misfolding of the normal prion form.

Misfolded prions derived from the PRNP gene occur in diseases such as kuru, CJD and fatal insomnia (Gerstmann–Straussler–Scheinker disease).

Prusiner reported that 10–20% of patients with CJD have mutations in the PRNP gene and in these patients the disease follows a familial inheritance pattern. Pathogenic PrPsc prions forms occur in 90% of cases of familial fatal insomnia. He noted that the PrPsc forms might result from aberrant post-translational modification. Pathogenic prion isoforms form oligomers but may also form amyloid fibrils. Prusiner noted that 40 different mutations have been identified in cases of Gerstmann–Straussler–Scheinker disease.

In rare familial forms of CJD expanded oligonucleotide repeats occur in the PRPN gene.

Proof of the prion concept was derived from experiments that revealed that the prion disease kuru could be transmitted from human brain extracts that contained no nucleic acid to apes and monkeys (Gajdusek, 1967). In humans kuru disease was transmitted between individuals under very unusual circumstances (eating brain tissue of deceased affected individuals).

Jaunmuktane *et al.* (2015) reported results of autopsy studies on 8 individuals between 26 and 51 years of age who died of iatrogenic CJD, likely through transmission of prions in injections of cadaver derived human growth hormone they had received during childhood. At autopsy these individuals manifested brain pathology of Alzheimer's disease including amyloid beta deposition in gray matter and in blood vessels. They also manifested amyloid deposition in pituitary glands. The authors noted that these findings promoted concerns about iatrogenic prion transmission of amyloid beta.

Jucker and Walker (2015) reported that amyloid beta could aggregate in brains of animals injected with minute amounts of beta amyloid. They reported further that when amyloid beta seeds were injected into mice abdomens there was subsequent demonstration of amyloid in cerebral blood vessels and plaques.

10.10.4.1 *Common age-related neurodegenerative diseases*

Prusiner reported that neurodegenerative diseases such as Alzheimer's disease, Parkinson's disease and ALS shared characteristics with CJD in that all these disease are of late-age onset and are associated with occurrence of abnormally folded proteins. He proposed that the mis-folded proteins in Alzheimer's disease and Parkinson's disease behave as prions. Prusiner noted that protein degradation pathways can clear low levels of prions but that at higher concentrations prion forms act as templates for the generation of additional prions, i.e. they promote self-replication. In Alzheimer's disease insoluble aggre-gates occur in plaques and in neurofibrillary tangles. There is evidence that amyloid aggregates act as prions and can spread from the entorhinal cortex to other regions of the brain.

In Parkinson's disease neurofibrillary tangles and aggregates of synuclein occur in Lewy bodies. In cases where fetal cells were grafted onto the brains of Parkinson's disease patients, the engrafted cells were shown to later shown to have accumulated synuclein and Lewy bodies.

Prusiner noted that trauma and post-traumatic stress could expedite formation of tau prions. Furthermore tau prions were shown to move between neuronal cells. Aggregates of mutant superoxide dismutase 1 were identified in familial ALS and these aggregates were shown to be self-propagating. In Huntington's disease expanded polyglutamine repeat containing huntingtin protein forms aggregates.

Prusiner concluded that early diagnosis of these neurodegenerative diseases would require the development of ligands to bind to prions and this could perhaps lead to identification of prions before symptoms develop.

Transmission of prions occurred in humans through extracts of human pituitary glands used to supply growth hormone. Prion transmission has also occurred through use of contaminated dura in neurosurgical procedures. It is important to consider that prions may contaminate instruments used in neurosurgery or electrodes used in deep brain stimulation.

Prusiner emphasized that in sporadic neurodegenerative disease wild type proteins undergo post-translational modification that generate proteins rich in beta pleated sheets and prions that undergo self-propagation as oligomers. Cleavage products are also derived from prions. Depending on the specific protein inducing prions, they may coalesce into plaques or tangles.

Jucker and Walker (2015) emphasized that strategies to stabilize proteins in their native conformation may be of therapeutic value.

Supattapone (2015) reported that host encoded cofactor molecules that may promote prion formation include phosphatidylethanolamine and single stranded RNA.

10.11 Destabilizing Protein Aggregates

Krishnan *et al.* (2014) noted that the misfolded aggregates that accumulate in neurodegenerative disease, including beta amyloid tau and alpha-synuclein, have a canonical amyloid fold. They predicted that agents that bind to intermediates of these protein aggregates may inhibit further assembly and that such agents would also destabilize formed aggregates.

Krishnan *et al.* determined that the g3p protein, a capsid protein of the filamentous bacteriophage M13 binds to amyloid folds. They demonstrated this through studies on beta amyloid fibers. Their studies revealed that g3p contains a unique amyloid targeting motif designated GAIM. These investigators postulated that this motif has potential to serve as the basis for generation of novel therapeutics for neurodegenerative disease.

11. Cancer §

"The primordial tumorigenic cell, as I propose to call it in what follows, is, according to my hypothesis a cell that harbours a specific faulty assembly of chromosomes as a consequence of an abnormal event.

The unrestrained proliferation of malignant tumour cells would then be due to a permanent excess of these stimulatory chromosomes."

T. Boveri, 1914 (translated by H. Harris, 2008)

It has been known for many decades, more than a century, that cancer is associated with abnormal cell proliferation and abnormal nuclear morphology. In addition early studies revealed that exposure to certain chemicals (e.g. coal tar) led to mutations and to cancer, providing clear evidence for the important roles of somatic mutations in cancer etiology.

Abnormalities in chromosome numbers and chromosome rearrangements in malignant tumors were clearly documented through karyotype analyses and chromosome banding techniques developed in the 1970s (Caspersson et al., 1970; Seabright, 1972). The availability of fluorescence in situ hybridization (FISH) techniques in the 1970s enabled labeling of specific genes and led to the discovery of rearrangements in chromosomes and the formation of fusion genes in some cases. In chronic myeloid leukemia, Rowley (1973) reported that fusion of the BCR gene on chromosome 22 and ABL on chromosome 9 occurred. This was subsequently shown to lead to constitutive activation of ABL tyrosine kinase.

Following the availability of the human reference genome sequence in 2001 (Lander *et al.*) and the development of high throughput next generation sequencing (NGS) methodologies, comprehensive data on the extent of somatic mutations in cancer has been obtained.

Profiles of mutations in specific tumors are useful in analysis of cancer types and are of relevance in determining cancer pathogenesis and in identifying key pathways in cancer development. Possibilities for targeted therapies may be revealed through analysis of specific driver mutations in tumors. However possibilities for targeted therapies are complicated by the large numbers of different mutations present in certain tumors.

11.1 Tumor Sequencing Data and Database Resources

The Catalogue of Somatic Mutations in Cancer (COSMIC) from the Sanger Institute in the UK (http://cancer.sanger.ac.uk/cosmic) provides comprehensive data on tumors analyzed by a number of different techniques including DNA sequencing. The Cancer Genome Atlas (TCGA) data portal (http://cancergenome.nih.gov) provides a platform for researchers to search, download, and analyze data sets of genome sequencing of tumors. Among the valuable publications that have emerged from analyses of TCGA data is a publication by Kandoth *et al.* (2013). They analyzed damaging point mutations, small insertions and deletions in 3,291 tumors across 12 tumor cells types. They identified 127 significantly mutated genes and determined that the majority of these genes functioned in 20 cellular processes. For 108 of these genes they defined the chief cell process and pathways in which each gene functioned. Cell pathways and processes in which genes mutated in cancer related genes function are listed in Table 11.1.

Kandoth *et al.* classified TP53 as in the genome integrity pathway; however it could also be classified in the transcription regulation pathway.

They reported that the number of significantly mutated genes varied across different cancers from two to six per tumor. The lowest

Table 11.1. Cell processes and pathways in which functional gene mutations were identified in tumors.

Transcription factors and transcription regulation
Cell cycle
Genome integrity
Histone modification (chromatin remodeling)
Receptor tyrosine kinase (RTK) signaling
MAPK (mitogen activated protein kinase) signaling
PI3K (phosphoinositide-3-kinase) signaling
TGFB (transforming growth factor beta) signaling
WNT signaling
Histone structure
Proteolysis
Splicing
Hippo signaling
DNA methylation
NFFEL2 Redox sensitive transcription activation
Protein phosphatases
Ribosome formation
TOR signaling

number of mutations occurred in pediatric tumors and in acute myeloid leukemia (AML). The highest number of mutations occurred in lung cancer, including lung adenocarcinoma and lung squamous carcinoma. High numbers of mutations also occurred in bladder cancer.

It is important to note that the most significantly mutated gene in their cohort was TP53 that was mutated in 42% of tumors overall. However TP53 was mutated in 95% of ovarian cancer samples and in 89% of endometrial cancer specimens.

PIK3CA (phosphatidylinositol-4,5-bisphosphate 3-kinase catalytic sub-unit alpha) was mutated in approximately 10% of cancers. Kandoth *et al.* noted that tumors that lacked mutations in PIK3CA often had mutations in PIK3R1 (phosphoinositide-3 kinase regulatory sub-unit). In the PI3K signaling pathway mutations in PTEN (dephosphorylates phosphoinositol kinase) were also relatively common.

Many cancers had mutations in chromatin remodeling genes, including lysine methyltransferases MLL2 (KMT2D), MLL3

(KMT2C), MLL4 (KMT2B) and in ARID1A, ARID5A and histone lysine demethylase (KMD5C).

Kandoth *et al.* classified NRAS and BRAF functions in the MAPK signaling pathway. They noted that mutation in NRAS and BRAF mutations were mutually exclusive in a particular tumor.

They noted that mutations in some genes occurred primarily in one tumor type. For example the von Hippel Lindau gene (VHL) and PBRM1 that encodes a subunit of an ATP dependent chromatin remodeling complex occurred primarily in kidney tumors, particularly in renal clear cell tumors. In colon and rectal tumors mutations were common in the adenomatous polyposis coli (APC) gene. In the WNT catenin signaling pathway and in KRAS mutations were common.

In studying clinical correlation of specific mutant genes Kandoth *et al.* noted that tumors with TP53 mutations generally had unfavorable prognoses and patients had a shorter time from diagnosis to death. Other gene mutations found to be particularly detrimental were those in the BAP1 (BRCA1 binding) in kidney cancer and chromatin modifying genes KDM5C and DNMT3A (DNA 5′-methylcytosine demethylase).

They emphasized that although for each cancer type there was a common set of driver genes, the combinations of drivers active in a particular tumor and in subclones of that tumor differed in different patients. Kandoth *et al.* postulated that a candidate gene panel for each tumor type could be developed.

They emphasized that characterization of specific gene mutations in a specific tumor facilitated assessment of targets for therapy.

11.1.1 *Exome sequencing on tumor/normal tissue pairs*

Lawrence *et al.* (2014) reported results of whole exome sequencing on paired tumors and normal tissues from 4,742 patients with 21 different tumor types. They noted that mutation frequencies vary significantly over cancer tumor types, from 0.03 mutations per megabase to 7,000 per megabase. Mutation frequencies also varied within tumor types.

They defined significant genes as genes that undergo mutations, including nucleotide substitutions, insertions and deletions at significant rates or in specific patterns in cancers. Lawrence *et al.* defined 23 genes that were significantly mutated in 3 or more tumor types. These included 12 genes that were mutated in 4 or more tumor types. These significantly mutated genes in cancer types are listed in Table 11.2.

By merging lists of cancer genes that were significantly mutated in single cancer types Lawrence *et al.* identified 254 cancer genes.

Table 11.2.   Significantly mutated genes.

(a) Mutated in four or more tumor types (Lawrence *et al.*, 2014)

TP53	(tumor protein 53, transcription factor, cell cycle arrest factor and RNA repair protein)
PIK3CA	(phosphatidyl-4.5-biphosphate 3-kinase catalytic factor)
PTEN	(phosphatidyl and tensin homolog dephosphorylates phosphoinositides)
RB1	(retinoblastoma 1, negative regulator of cell cycle)
KRAS	(Kirsten RAS oncogene), GTPase
BRAF	RAF family serine-threonine kinase regulates map kinase (MAPK/ERK) signaling
CDKN2A	(cyclin dependent kinase inhibitor) stabilizes TP53
FBXW7	F-box repeat protein, targets cyclins for ubiquitin degradation
ARID1A	ATP rich domain, present in ATP dependent chromatin remodeling complex
MLL2	(KMT2D) histone lysine methyl transferase
STAG2	(stromal antigen 2 cohesin complex), involved in chromatid separation in cell division

(b) Mutated in three tumor types

ATM	serine threonine kinase cell cycle checkpoint signaling and DNA damage response
CASP8	encodes caspase 8 involved in cell apoptosis
CTCF	transcription regulator
ERBB3	EGF family receptor tyrosine kinase (RTK)
HLA	A histocompatibility immune response
IDH1	isocitrate dehydrogenase 1
NF1	neurofibromin
NFEL2	redox sensitive transcription related
PIK3R1	(phosphoinositide-3-kinase regulatory sub-unit)

Significantly mutated genes in specific cancer types are listed below (note gene abbreviations are spelled out if not mentioned above):

Acute myeloblastic leukemia (AML)

FLT3 (hematopoiesis regulator tyrosine kinase, DNMT3, IDH1, IDH2, TP53, NPM1 (nucleophosmin nucleolar phosphoprotein, TET2 (methylcytosine dioxygenase), WT1 (Wilms tumor 1 transcription factor), NRAS, RUNX1 (heteromeric transcription factor), PTPN11 (non-receptor protein tyrosine phosphatase) and U2AF1 (splicing factor).

Bladder cancer

TP53, KDM6A (histone lysine demethylase) RB1, PIK3CA and ARID1A.

Colorectal cancer

APC (adenomatous polyposis coli WNT antagonist), TP53, FBXW7, NRAS BRAF, SMAD4, SMAD2 (SMAD family transcription factors activated by transmembrane tyrosine kinase signaling) and TCF7L2 (transcription factor in WNT signaling pathway).

Head and neck cancers

TP53, CDKN2A (cyclin dependent kinase inhibitor), CASP8, PIK3CA, MLL2, NSD1 (nuclear receptor transcription factor), NFEL2 (redox sensitive protein) NOTCH1 (transmembrane intercellular signaling) and FAT (atypical cadherin).

Endometrial cancer

PTEN, PIK3CA, TP53, KRAS, FBXW7, ARID1A, CTCF (transcriptional regulator) and ARHGAP35 (glucocorticoid receptor binding GTPase).

Lung adenocarcinoma

TP53, CDKN2A KRAS, KEAP1 (redox sensitive protein), STK11 (serine threonine kinase), EGFR (epidermal growth factor receptor) and SMARCA4 (SWI/SNF related chromatin remodeler).

Lung squamous carcinoma

TP53, CDKN2A, MLL2, KEAP and PIK3CA.

Lawrence *et al.* emphasized the importance of understanding the pathway levels of genes mutated in cancer with frequencies between 2 and 20%.

They noted that loss of function mutations in three specific genes helped tumors evade immune attack. These mutations occurred in HLA and TAP1 genes that process peptides for immune presentation and CD1D that presents lipid antigens to killer cells.

11.1.2 *Driver genes and driver mutations*

Vogelstein *et al.* (2013) noted that although specific genes are designated as driver genes not all mutations in that gene represent driver mutations. An example they presented was the APC gene where only mutations that truncate or disrupt the N-terminal region of APC constitute drivers. A further distinction they proposed was between mut drivers, gene mutations that served as drivers, and epi drivers, genes altered through changes in methylation modifications.

In their 2013 review, Vogelstein *et al.* listed 140 driver genes and noted that these function to provide selective growth advantage through 12 different mechanisms and these included 3 core mechanisms: Cell survival, cell fate determination and genomic maintenance.

Genes in the cell survival sub-category function in pathways which include TGFbeta, MAPK, STAT (signal transducer and activator), PI3K, RAS pathways and in apoptosis pathways.

Genes and functions that impact cell fate include genes in NOTCH, HEDGEHOG, APC pathways and genes involved in chromatin modification and in regulation of transcription.

Genes that particularly impact genomic maintenance function are involved in DNA damage control pathways.

Martincorena and Campbell (2015) reported that driver mutations that are positively selected in tumors have a higher rate of mutation and tend to have recurrent mutations. They noted that the current cancer census reported 572 recurrently mutated genes in cancer. They noted however that only three genes were recurrently mutated across different tumors: TP53 mutations occurred in 36%

of tumors, PIK3CA mutations occurred in 14% of tumors and BRAF mutations occurred in 10% of tumors.

11.1.3 *Hypermutation foci in tumors*

Alexandrov and Stratton (2014) also investigated the frequency of hypermutation foci in cancer. They referred to this hypermutation phenomenon as kataegis (Greek for thundershowers). Foci of kataegis occurred often in regions of genomic rearrangement and included up to several thousand mutations. These regions occurred in B cell lymphomas and in subsets of breast, lung and pancreatic cancer. In B cell lymphomas, kataegis frequently occurred in the proximity of immunoglobulin loci. There is evidence that kataegis is induced by AID/APOBEC-catalyzed cytidine deamination (Taylor *et al.*, 2013).

11.2 Chromosome Rearrangements in Tumors

Garraway and Lander (2013) noted that chromosome rearrangements are also pervasive in cancer and that delineating their impact is often difficult. They noted that some rearrangements might be passengers; however in some cases, specific rearrangements and gene fusions significantly impact function.

11.2.1 *Chromothripsis*

Chromothripsis occurs in specific cells in at least 2% of cancers and is characterized by shattering, rearrangement and increases in copy numbers of chromosome segments. Chromothripsis may impact one or a few chromosomes. Zhang *et al.* (2013) hypothesized that physical isolation of one or two chromosomes in a micronucleus might lead to chromothripsis. In 2015, Zhang *et al.* reported studies in which they used live cell imaging and single cell genome sequencing to define mechanisms involved in chromothripsis. They demonstrated that the capture of one or two chromosomes in a micronucleus could generate extensive breakage of chromosomes and genomic rearrangements of the segments.

Knouse and Amon (2015) noted that when an individual chromosome fails to attach to the mitotic spindle, a structure forms to isolate that chromosome. Subsequently on completion of cell division the segregated chromosome is frequently not incorporated into a nucleus but is included in a micronucleus. They noted further that micronuclear membranes are prone to rupture and that the chromosomal DNA is then exposed to the cytoplasm and DNA damage results. Even within the micronuclei the chromosomes may undergo rearrangement. The micronuclei may sometimes be incorporated into the main nucleus.

Leibowitz *et al.* (2015) emphasized that mutations are generally thought to accumulate gradually during cell divisions of cancer cells. However in the case of chromothripsis extensive genome damage accumulates in an isolated event.

11.3 Hereditary Cancers: Cancers that Follow Mendelian Inheritance Patterns

There are a number of conditions, each primarily determined by mutations in a specific gene, which are associated with the occurrence of benign and sometimes malignant tumors. It is also important to consider that certain tumors in these conditions may not be malignant in histological terms but may be malignant based on their position and damage they cause to normal functions. Particular disorders, e.g. neurofibromatosis 1 and tuberous sclerosis, may also be associated with non-tumor related phenotypic abnormalities, including, for example, pigmentary changes, neurological and neurobehavioral problems.

Databases and resources differ in their lists of hereditary tumor disorders. For example, Cancer Net (http://www.cancer.net) includes tuberous sclerosis and neurofibromatosis in their list while these condition are not included in the list from the National Cancer Institute at NIH (see Table 11.3).

It is important to note that genetic disorders associated with abnormal sensitivity to DNA damage or impaired repair of DNA damage are also associated with increased incidence of tumors, including malignant tumors.

Table 11.3. Familial Cancer syndromes included in the National Cancer Institute and Genetics Home Reference (http://ghr.nlm.nih.gov).

Basal cell nevus syndrome, Gorlin syndrome, Gorlin–Goltz syndrome, or nevoid basal cell carcinoma syndrome, autosomal dominant gene, PTCH1 (patched, receptor for signaling protein SHH1); chromosome 9q22.1

Birt–Hogg–Dubé syndrome autosomal dominant gene, FLCN (folliculin, function unknown)

Bloom syndrome autosomal recessive BLM gene (encodes a RecQ helicase that unwinds DNA necessary for replication) 15q26.1

Breast/gynecologic cancers, hereditary BRCA1/ BRCA2 (DNA damage sensors); other genes that are important include TP53, PALB2 (partner and localizer of BRCA2), ATM (cell division checkpoint kinase), CHK2 (cell cycle checkpoint kinase 2), CDH1 (cadherin1)

Carney–Stratakis Syndrome autosomal dominant gene PRKAR1A (protein kinase regulatory sub-unit, 2p16)

Cowden syndrome autosomal dominant genes PTEN, mainly also SDHB SDHD (succinate dehydrogenases), KLLN (killin, apoptosis factor transcription of this gene is controlled by TP53)

Colon cancer, hereditary nonpolyposis or Lynch syndrome autosomal dominant genes MLH1, MSH1, MSH2, MSH6 and PMS2, all involved in DNA mismatch repair; EPCAM (epithelial adhesion molecule)

Fanconi anemia, autosomal recessive genes FANCA, FANCC, FANCG, part of core complex involved in DNA repair

Hyperparathyroidism, familial, autosomal dominant, genes MEN1 (menin), CDC73 (parafibromin, CACR (calcium sensing receptor).

Li–Fraumeni syndrome autosomal dominant include breast cancers and osteosarcoma gene TP53, chromosome 17p13.2) CHEK2 (cell division checkpoint kinase chromosome 22q11).

Multiple endocrine neoplasia type 1 autosomal dominant genes MEN1 (11q13), RET (growth factor, 10q11.2) and CDKN1 (Cell cycle regulation, 12p12.2). Mutations in these genes are also predisposed to familial medullary thyroid cancer

Multiple endocrine neoplasia type 2A, 2B (Sipple syndrome) RET

Paraganglioma pheochromocytomas syndrome, hereditary, autosomal dominant, succinate dehydrogenase genes SDHB (1p35-q36), SDHC (1q23.3), SDHD (11q23), SDHAF2 (11q12.3)

Peutz–Jeghers syndrome autosomal dominant inheritance, gene STK11 (LKB1) control of rate of cell division

Polyposis, familial adenomatous and attenuated familial adenomatous polyposis autosomal recessive, APC (5q21-q22) beta catenin associated control of cell proliferation; MUTYH (1p34.1) encodes a glycosylase involved in DNA damage repair

(Continued)

Table 11.3. (*Continued*)

Polyposis, familial juvenile autosomal dominant, polyps are usually benign, occasionally they become malignant Genes BMPR1A (10q22.3) and SMAD4 (18q21.1), both genes encode products that transmit signals to nucleus to control expression of genes involved in cell growth and proliferation

Prostate cancer, hereditary many different gene mutations predispose to prostate cancer however in a minority of cases specific mutations in BRCA1, BRCA2 (important for DNA damage response) and HOXB13 (transcription factor) predispose men to prostate cancer and have autosomal dominant effect.

Renal cell cancer, hereditary with uterine leiomyomas autosomal dominant, specific gene mutations in fumarate dehydratase (FH) metabolic enzyme (1q24.1)

Renal cell cancer, hereditary papillary 5 different genes are known to carry mutations that predispose tis cancer, RNF139 (ubiquitin ligase in endoplasmic reticulum; PBRM (ATP dependent chromatin remodeling); OGG1 8-oxyguanine DNA glycosylase excision of damaged DNA bases; SETD2 histone methyltransferase important in transcription; RET encodes nuclear transmembrane protein transmits signals for growth

Von Hippel–Lindau (VHL) autosomal dominant syndrome caused by defects in the VHL gene (3p25.3). VHL encoded protein is a component of the VCB-CUL2 complex that detects and targets damaged proteins for destruction

In a number of conditions associated defects in detection of DNA damage and conditions associated with defects in DNA damage repair, there is an increased tendency to tumor and cancer development.

11.3.1 *Additional important facts regarding cancers in autosomal dominant conditions described above*

It is important to note that not all mutations in these cancer predisposing genes cause cancer. For conditions labeled as autosomal dominant, a second mutation or genomic change needs to occur in the normal homolog of the predisposing gene in order for cancer to manifest.

11.4 Common Cancers, Susceptibility and Familial Factors

Frank *et al.* (2015) reviewed a large data set collected in Sweden on 8,148,737 individuals that included information gathered between 1958 and 2010. They noted that across cancer types the individual risk for a specific cancer was increased if a parent or sibling of an individual had that specific cancer. They determined that if a parent or siblings were affected with a specific cancer, the risk in an individual was twofold compared with the general population risk. However they showed that risks differed with specific types of cancers, e.g. familial risk was much higher than twofold for Hodgkin's lymphoma.

Frank *et al.* emphasized that in addition to the high penetrance genes, there are low penetrance cancer-risk genes, many of which remain undiscovered.

Hindorff *et al.* (2011) reported that genome-wide association studies identified common variants associated with cancer. Many of these increased the risk less than 2.5 fold; however the degrees of risk increase varied with different markers and different forms of cancer.

11.5 Mutational Signature Patterns and Insights into Cancer Predisposing Factors

There is now consensus that cancers originate from single cells that acquire somatic mutations. Alexandrov and Stratton summarized information from sequencing projects that revealed that specific patterns of mutations occur in response to specific cancer inducing factors. The first evidence of an agent that induced specific cancer predisposing mutations was from Sanger sequencing. Ultra-violet (UV) irradiation led to generation of pyrimidine dimers, CC to TT, in DNA. Patients with skin cancer manifested C>T and CC>TT mutations even in the p53 gene. In smokers with lung cancer tumors, sequencing revealed predominance of C>A substitutions. In breast cancer patients C>A mutations predominated. In bladder cancer C>G mutations predominated.

Alexandrov and Stratton noted that somatic mutations arise throughout life due to exogenous and endogenous mutational processes.

As NGS projects have provided more extensive sequencing data, mathematical modeling computational programs have been developed and applied to analyze the mutational patterns C>A, C>G,

C > T, T > A, T > C and T > G. In addition information on bases immediately 5-prime and 3-prime to the mutated bases were taken into account. This yielded the 96 possible mutation classification patterns. Some cancers showed predominance of one or two substitution patterns while in other cancers there was evidence of multiple patterns of mutations.

Signature pattern 1 involved C > T substitutions at NpCpG trinucleotides (non-promoter trinucleotides). Alexandrov and Stratton reported that this pattern likely results from spontaneous deamination at 5-methylcytosine sites. Theses investigators have proposed that aging is a precipitating factor in these changes.

In the signature pattern 2, C > T and C > G patterns predominate. These changes likely represent the activity of APOBEC cytidine deaminases that convert cytidine to uracil and must be followed on replication by base excision repair (APOBEC cytidine deaminase will be discussed further later). In cervical cancer cells, the signature pattern 2 APOBEC predominated.

Alexandrov and Stratton reported that in most cancers more than one mutational signature pattern predominated. In lung adenocarcinoma, the signature pattern 4 associated with smoking predominated; however there was evidence for the aging related mutational pattern 1 and some evidence of signature pattern 2 (APOBEC). Mutation patterns in chronic lymphocytic leukemia were predominantly 1B aging and 2 APOBEC.

Another aspect of mutation signals taken into account was strand bias and whether mutations occurred predominantly on the transcribed strand or the untranscribed strand. Alexandrov and Stratton reported that T > C mutations on the transcribed strand occurred in hepatocellular carcinomas. The C > T mutations in melanoma occur predominantly on the untranscribed antisense strand.

In addition to nucleotide signal changes, the presence of indels (insertions, deletions and duplications) were also examined. Alexandrov and Stratton reported that in signature pattern 6, small indels occurred at nucleotide repeats. This pattern was consistent with microsatellite instability that is associated with mismatch repair deficiency. This signature was predominant in colorectal cancer.

Larger deletions up to 50 nucleotides were features of signature pattern 3 and this signature pattern occurred in breast cancer with BRCA1 or BRCA2 mutations and also in pancreatic cancer.

Martincorena and Campbell (2015) reported that certain herbal medications that contain aristocholic acid predispose to bladder cancer and to an increased frequency of A > T mutations.

These investigators also emphasized that it is important to note that somatic mutations are ongoing throughout life. Cancer risk below 20 is 2%; beyond 80 years of age it is 50%. They also noted that 10% of healthy patients older than 60 years of age carry a clonal population of cells with characteristics of leukemia.

11.6 Processes Leading to Cancer: DNA Damage and Impaired Repair

11.6.1 *Types of DNA damage*

Types of damage that occur include damage to the amine groups in DNA, hydrolysis of bases (including deamination, depurination or depyrimidation) and alkylation with oxidation of bases. A major alkylation mutation involves the formation of O-alkylguanosine. Breaks in the backbone of DNA can be single stranded or double stranded. Other forms of DNA damage include the formation of abnormal cross linkages between bases in the same strand or on opposite strands (Helleday *et al.*, 2014).

11.6.2 *Repair mechanisms*

Jackson and Bartek (2009) reviewed DNA damage repair mechanisms. Direct repair of alkylating damage can be achieved through the enzyme O6-methylguanine methyltransferase that is encoded by the gene MGMT. Base excision repair may involve the activity of monofunctional DNA glycosylases that remove a damaged nitrogenous base and leave the sugar phosphate backbone intact. Bifunctional DNA glycosylases cleave the nitrogenous base and also cleave phosphodiester bonds to create a single strand. Endonucleases can also cleave the DNA backbone to release a damaged base. DNA polymerases then synthesize a new DNA strand using the complementary strand as template. Base excision repair can also involve the activity of other proteins including XRRC1, PARP1 (poly-ADP ribose polymerase 1) and PARP2 (poly-ADP ribose polymerase 2).

Nucleotide excision repair is frequently used to repair helix distorting damage and it includes removal of a base and also removal of 12–24 adjacent bases. Damaged nucleotides and adjacent nucleotides are removed by endonucleases. DNA lesions that require nucleotide excision repair are detected by xeroderma pigmentosum (XP) class of proteins; other important proteins in the resolution of these lesions include DNA polymerases, PCNA (cofactor of polymerase), RPA (replication factor polymerase accessory factor) and ligases.

Mismatch repair mechanisms are activated when defects in the Watson–Crick base-pairing are detected. Sensing of mismatches involves MSH2-MSH6, MSH2-MSH3, and a MLH1-PMS, MLH3-PLH (microsatellite mismatch sensing proteins) and EXO1 endonuclease. Removal of the damaged region by exonuclease is followed by re-synthesis with polymerase and sealing with ligase. Correction involves activity of PCNA (cofactor of polymerase), RFC (replication factor accessory of polymerase), RPA, ligase 1, polymerase gamma and epsilon.

Strand breaks can be single stranded or double stranded. Single stranded breaks are repaired by similar mechanisms used in base excision repair and the homologous strand is used as a template. Double stranded DNA breaks are potentially very damaging to the genome. Two different processes are used for their repair. In non-homologous end-joining, specific sensors detect double-stranded breaks; these sensors include Ku protein (XRRC5/6) that bind to DNA near the break and activate DNA polynucleotide kinases, XRRC4, XLF (NHEJ1) and ligase IV, to facilitate joining. Jackson and Bartek noted that non-homologous end joining can also use MRE11-RAD50-NBS1 (MRN) complex, Artemis nuclease, aprataxin and polymerases mu and gamma for healing.

There is also a microhomology end-joining process. Non-homologous end joining can introduce mutations during repair.

Homologous recombination-based repair of double stranded DNA breaks uses information on the intact sister chromatin in the G2 phase of the cell cycle. Homologous recombination repair may also utilize duplicate copies of the gene if they are present elsewhere either within the same chromosome or elsewhere in the genome. In addition homologous recombination may utilize similar but non-identical sequences in the genome as templates for repair. Homologous recombination repair is dependent on the availability of a large number of different proteins; these include RAD51 and RAD51-related proteins.

PARP (poly ADP ribose polymerase) plays roles in single stranded and double stranded DNA repair. There is also recruitment of a specific histone h2AX to site of DNA damage.

A number of double stranded DNA repair proteins play important roles at telomeres and participate in the alternate repair of telomere processes and in processes that prevent telomere shortening.

DNA damage and repair and conditions associated with defects in repair are also discussed in Chapter 2.

11.6.3 *TERT (telomerase reverse transcriptase) promoter mutations in cancer*

The continued proliferation of cancer cells is facilitated through telomere maintenance due to continued telomerase activity or through the alternative lengthening of telomere mechanism. Heidenreich *et al.* (2014) reported that in most cancers there is up regulation of TERT that prevents loss of telomeres that would impair cell proliferation. The TERT gene maps to chromosome 5p15.33 and the region upstream of TERT contains binding sites for several transcription factors, including cMYC.

Heidenreich *et al.* reported that normal BRCA1 forms a complex with MYC and that this inhibits TERT activity. Also p53 can down regulate TERT. In contrast the ETS transcription factor can stimulate TERT transcription. Factors that increase ETS transcription include RAS and RAF and oncogenic forms of EGF.

Germline mutations in the TERT promoter were described in a family with a history of severe forms of melanoma in several members

(Horn *et al.*, 2013). Heidenreich *et al.* and Horn *et al.* reported that somatic mutations in the TERT promoter occurred in melanomas. Heidenreich *et al.* noted that TERT promoter mutations have also been described in other forms of cancer, particularly in bladder cancer where they are associated with decreased survival and increased disease recurrence following initial treatment. Additional information on DNA damage and repair is included in Chapter 2.

11.7 Cell Cycle and Relevance to Cancer

Williams and Stoeber (2012) reviewed aspects of the cell cycle. They noted that cells may be permanently withdrawn from the cell cycle and differentiated neurons are an example; cells may be reversibly withdrawn from division and be in GO phase.

The S phase is the stage of DNA replication; in the G2 phase the cell prepares to enter mitosis; in the M phase one cell divides to rise to two daughter cells; and G1 follows mitosis when division is completed.

Cell factors that drive cells through the cycle include cyclin-dependent kinase and their activators, the cyclins. S phase activities require the actions of cyclin A (CCNA2) and its partner CDK2. In the G2 phase, cyclin B and its partner CDK1 promote passage to M phase. Cyclin B (CCNB) is a regulatory protein that forms a complex with maturation promoting factor MPF. In G1, cyclin D with CDK4 and CDK6 promote completion of mitosis.

In summary, cyclins function as regulators of CDK kinases. The G2 phase requires CCNB1 and CDK1; the S phase requires CCNA2 and CDK2. G1 and transition to M require CCND1 and kinases CDK4 and CDK6.

11.7.1 Checkpoint and sensor proteins

Checkpoint and sensor proteins monitor that progression through the cell cycle is correct. In the case of DNA damage, for example, sensor proteins may trigger cell cycle arrest. DNA double stranded breaks particularly activate sensor proteins and the DNA damage response complex. Key to the DNA damage response is the MRN complex (MRE11, RAD50, and NBS1) and recruitment of the ATM kinase. CDK2 in the S phase can phosphorylate CHEK1 protein in response to double stranded DNA breaks. Williams and Stoeber emphasized that deregulation of the cell cycle can lead to uncontrolled proliferation. CDK inhibitors can also induce cell cycle arrest. CDK inhibitors are often inactivated through mutations that occur in cancer.

Retinoblastoma (Rb) protein plays a key role in regulating transition from G1 to S phase. Unphosphorylated Rb protein binds to E2F transcription factor and inhibits transcription of genes that encode proteins essential to the cell cycle. Phosphorylation of Rb leads to its dissociation from E2F so that transcription is no longer inhibited (Giacinti and Giordano, 2006). Phosphorylation of Rb may occur through a number of different mechanisms including activation of upstream signaling pathways or through cyclin D and CDK4 activation.

Many of the traditional chemotherapeutic agents used in cancer therapy have focused on the cell cycle. However the problems with many of these agents are that they damage cell cycle in tumor cells and in normal cells, leading to severe side effects due to impaired cell replication in bone marrow in the mucosa of the gastrointestinal tract or in the skin.

Williams and Stoeber proposed that G1 arrest through knockdown of CDC7 inhibitors represented a key possibility for cancer therapeutics since this knockdown was reversible and had less effect on self-renewing tissues (e.g. mucosa, bone-marrow and skin). CDC7 is a protein with kinase activity that is a key regulator at the G1 to S cell cycle transition. CDC7 phosphorylates MCM (minichromosome maintenance) proteins. MCM has helicase activity and it is activated when phosphorylated. This activation is essential for unwinding of the double stranded DNA at the origins of replication to single stranded DNA. Single stranded DNA is essential to enter and initiate replication. There is also evidence that MCM proteins can be phosphorylated by CDC2. A number of CDC7 inhibitors are in clinical trials. A key regulator in the G2 to M transition in the cell cycle is PLK1 polo-like kinase. In mitosis Aurora kinase stabilizes the mitotic spindle.

11.8 Key Genes and Pathways in Cancer

11.8.1 *P53*

The classical view of wild type p53 function was that it limited progression from G1 to S phase of the cell cycle. Bieging *et al.* (2014) reported that there is now evidence that p53 plays a role in DNA damage response. The canonical p53 associated functions for tumor suppression include cell cycle arrest, cell senescence and apoptosis and DNA repair. Bieging *et al.* note that inactivation of p53 in tumors is most commonly due to missense mutations in the DNA binding domain. However impaired p53 function can also result from variants, particularly at the carboxyl end of the protein, that impair folding and contact of p53 with interacting proteins.

The p53 protein acts as a transcription factor and the lists of genes that undergo p53 regulation of expression is continually growing. In their publication Bieging *et al.* include a list of those genes. P53 regulates expression of genes in metabolism, genes involved in control of response to oxidative stress, genes involved in autophagy. In addition there is evidence that mutant p53 promotes invasion of tumor cells into tissue and the occurrence of metastases.

Bieging *et al.* reviewed and illustrated the classical p53 activation and response to DNA damage. Acute DNA damage, particularly double stranded DNA breaks, lead to activation of ATM and ATR kinases that activate checkpoint genes CHK1 and CHK2. These kinases then phosphorylate p53, promote its release from MDM2 and MDM4 and stabilize p53. Under non-activating conditions MDM2 and MDM4 are bound to the transcription activating domain of p53. Phosphorylation of p53 also promotes its binding to transcriptional co-activators. Activated p53 then acts to promote transcription and expression of genes that inhibit G1 to S transition and genes that promote DNA repair.

Bieging *et al.* report that newer studies indicate that p53 activation can be induced by a number of factors besides DNA damage. These factors include telomere reduction, oxidative stress, hypoxia and nutrient stress.

Autophagy plays important roles in tumor suppression and there is evidence that p53 induces transcription and expression of a number of genes in the autophagy pathway. Napoli and Flores (2013) summarized functional connections between p53 and autophagy. Autophagy involves the degradation of damaged proteins, organelles and protein aggregates in a process that involves uptake of these components in double membrane structures and the subsequent fusion of these membrane structures with lysosomes. The first step in this process is generation of the pre-autophagic structures, the membranes that isolate components to be digested. Napoli and Flores report that formation of the pre-autophagic structures is dependent on the MTOR-ULK pathway.

Soussi and Wiman (2015) reported that mutations have been described in 386 of the 393 amino acids in TP53. However the mutations most frequent in tumors involve the DNA binding core domain. They noted further that more than 80% of somatic and germline mutations lead to synthesis of a stable mutant TP53 protein that accumulates in the nucleus. They suggested that most mutations led to loss of function of the wild type p53 and to the generation of a mutant form of p53 with different properties and mutations that produced oncomorphic proteins. They noted that as wild type p53 decreases cancer protection decreases.

Brachova *et al.* (2013) postulated that oncomorphic p53 mutations may lead to activation of additional transcription factors and that oncomorphic TP53 proteins may bind to additional proteins not bound by wild-type TP53.

Soussi and Wiman reported that small molecules which target P53 mutant proteins with p.R157H and p.Y220C substitutions have been developed and showed encouraging results. They noted that drugs need to be developed that target p.R273H, p.R248W and p.R175H oncomorphic p53 mutants.

11.8.2 *PIK3CA2, phosphoinositides and signaling*

Phosphoinositide-3-kinase (PI3K) phosphorylates the 3′ hydroxyl group of phosphatidylinositol. The PI3K enzymes are responsible for production of phosphatidylinositol-3-phosphate (PI3P) (PIP), phosphatidylinositol3-4-biphosphate (PI3,4P2) (PIP2) and phosphatidylinositol3-4-5-triphosphate (PI3,4,5P3) (PIP3).

Members of class one PI3K are each composed of an alpha catalytic sub-unit and a regulator sub-unit gamma. The PIK3CA2 gene produces a catalytic subunit for class 1 PI3K. The phosphatidyl inositol 3-4-5-triphosphate that forms as a result of PI3K activity activates AKT, a serine threonine kinase (also known as protein kinase B). AKT in turn activates a number of downstream pathways, including the MAP kinase ERK pathway and the MTOR pathway. Activation of the MTOR pathway enhances cell growth and proliferation. Activation of ERK (extracellular signal regulated kinase) promotes its translocation across the nuclear membrane into the nucleus where it stimulates activity of transcription factors involved in promotion of cell cycle progression (Burotto *et al.*, 2014).

Mutations that enhance production of PIK3CA2 catalytic sub-unit therefore increase PI3K activity and increase AKT, ERK and MTOR pathway signaling, promoting cell proliferation and growth.

Figure 11.1 illustrates PIK3CA2 activation and downstream signaling.

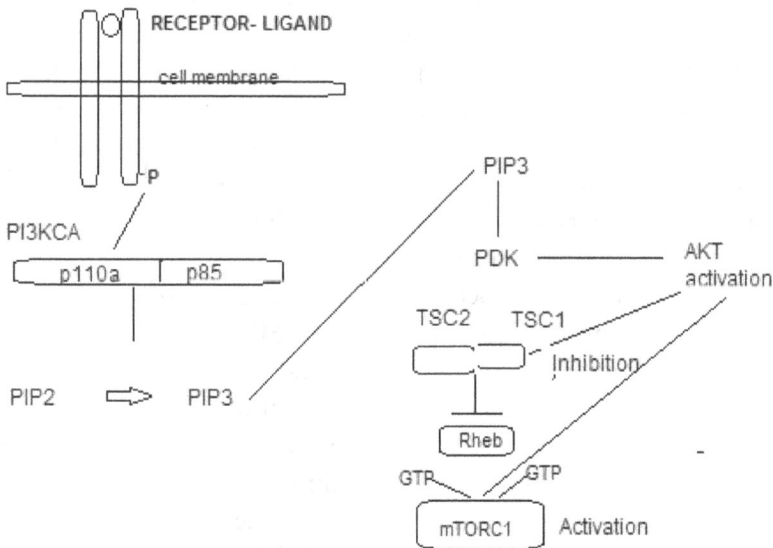

Figure 11.1. PIK3A activity following receptor tyrosine kinase (RTK) signaling and leading to activation of downstream pathways including phosphatidyl inositol, AKT and mTORC1 activation.

11.8.3 *BRAF*

BRAF is a RAS-like protein; it is activated by RAS following stimulation of receptor tyrosine kinases (RTKs) and co-activators. Activated BRAF then stimulates activity in the MAPK/ERK pathway. Constitutive activation of BRAF occurs in a number of different cancers including melanoma, small cell lung cancer, colon cancer and thyroid cancer. Drugs have been developed that target specific oncogenic BRAF mutations including the V600E and V600K mutations (Holderfield *et al.*, 2014).

Because resistance often develops when single molecule inhibitors are used, combination therapies are now recommended. These include combinations with different small molecule inhibitors or combinations with small molecule inhibitors and immune therapy or combinations with a small molecule inhibitor and conventional chemotherapy.

11.8.4 *CDKN2A*

The CDKN2A gene, located on chromosome 9p21, encodes a number of different protein isoforms. One protein isoform p16INK4a binds to and inactivates the cyclin kinases CDK4 and CDK6 that are involved in cell cycle regulation. CDKN2A also encodes another isoform, the p14ARF protein that protects TP53 from breakdown by complexing with MDM2, a ubiquitin ligase-like molecule that degrades p53. The two isoforms derived from CDKN2A both function to inhibit progression through the cell cycle.

CDKN2A mutations occur in many different types of cancer. Mutations that impair activity of the p16INK4A isoform fail to inhibit cyclin kinase CDK4 and CDK6 and therefore fail to control cell cycle activity; this leads to increased cell division and proliferation. Mutations that lead to reduced expression of p14ARF protein impair its ability to protect TP53 protein from breakdown.

Many tumors have somatic CDKN2A gene mutations. However germline mutations also occur in the CDKN2A gene and lead to increased cancer predisposition (National Cancer Institute and Genetics Home Reference, 2016).

11.8.5 *MTOR pathway*

Pourdehnad *et al.* (2013) noted that MTOR over-expression occurs in a broad range of cancers and is associated with poor prognosis. They noted that MTORC1, which impacts protein synthesis, is also often deregulated in cancer. MTORC1 phosphorylates the translation initiation factor binding protein and it phosphorylates the ribosomal protein P70S6 kinase. Unphosphorylated 4EBP1 negatively regulates EIF4E the translation initiation factor. When 4EBP1 (EIF4ABP1) is phosphorylated through MTORC1 activity, it no longer suppresses EIF4E and translation initiation then follows. EIF4E then promotes recruitment of the 40S ribosomal subunit to the 5′ cap of mRNA and translation is initiated. Protein synthesis in tumors is therefore dependent on MTOR phosphorylation of 4EBP1 (EIF4ABP1).

11.8.6 *Receptor tyrosine kinases (RTKs) and signaling*

Progress in determining mechanisms through which signaling alters cellular growth and differentiation has relevancy for therapeutics in cancer and for advances in stem cell biology and its applications. Much attention has been focused on RTK and its activation through binding of ligands to cell surface receptors.

Lemmon and Schlesinger (2010) reviewed signaling functions of RTKs. They noted that there are 20 different families of RTKs in humans and that at least 58 different genes encode RTKs. Cell surface receptors frequently undergo dimerization upon binding to activating ligands. Dimerization of the receptors is however not required for all types of receptors. Ligand binding leads to activation of the intra-cellular tyrosine kinase domain of the receptors. This activation involves phosphorylation of tyrosine. Some receptors can be activated in monomeric form, e.g. ERBB3 (HER3). ERBB2 (HER2) is however usually recruited into dimers.

The RTK families include EGFR, platelet-derived growth factor receptor (PDGFR), insulin receptor family, fibroblast growth factor receptor (FGFRs), vascular endothelial growth factor receptor (VGFR), neurotropic tyrosine kinase receptor (NTRK1), ephrin receptor and 12 others.

The activated receptor tyrosine functions as a node in a complex network that transmits information from the exterior to the interior of the cell.

Phosphorylation of RTKs is followed by formation of signaling complexes. Lemmon and Schlessinger reported that signaling molecules contain SH2 or SRC (Sarc homology) domain or PTB (phosphotyrosine binding) domains specifically bind to phosphotyrosines in the intra-cellular regions of the receptors. The GRB2 protein is an adaptor molecule that contains two SH2 domains. In some cases the signaling molecules bind indirectly to the phosphorylated receptor through adaptor molecules.

Other molecules that bind directly to phosphorylated RTKs or indirectly through interactions with SH3 or PTB linkages include phospholipase C and phosphoinositol containing molecules. The SH3 adaptor protein may bind additional proteins, including SOS (a guanine nucleotide exchange factor) that promotes activation of the signaling protein RAS. In addition to SOS other molecules, e.g. RAS activator RASGRP1 can participate in RAS activation. Figure 11.2

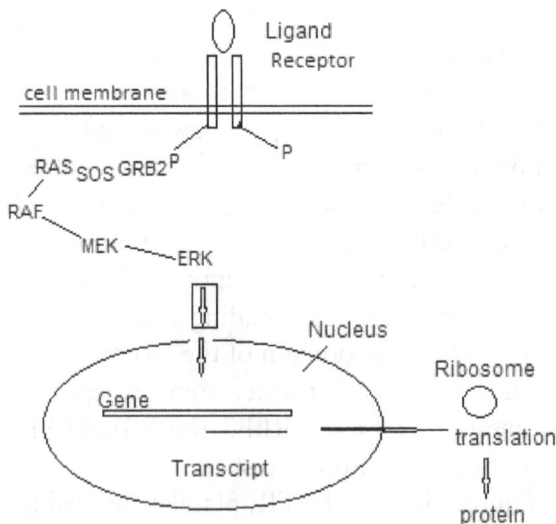

Figure 11.2.   Receptor ligand binding, receptor tyrosine kinase (RTK) signaling, activation of RAS and downstream signaling to activate gene expression, transcript generation and protein synthesis.

illustrates receptor ligand binding, phosphorylation, downstream signaling and transcription activation.

RAS activation triggers the activity of a number of downstream signaling pathways, including RAF, MEK and ERK signaling pathways. Phosphorylated ERK can enter the nucleus and trigger gene transcription. Activated RAS can also promote signaling through the PI3K and RAS pathways, ultimately triggering MEKK1 and JNK pathways. JNK can then enter the nucleus and trigger gene transcription.

11.8.7 *Relevance to cancer therapy*

Although kinase inhibitors are widely used in cancer therapy, other steps in the process of cell signaling that lead to growth promotion are being investigated as targets (Gough, 2013).

Mutations in RAS proteins are powerful cancer drivers; however they have proven to be very difficult to target. This failure is in part due to the fact that the RAS molecule has a smooth surface with few binding locations for small molecules. A number of different approaches to circumvent this problem are being explored (Ledford, 2015). One approach includes using small molecules that change the shape of RAS protein to reveal binding pockets. Other approaches involve targeting specific mutations in RAS that frequently occur in cancer, e.g. G12C mutations that replace glycine with cysteine; cysteine can be readily targeted.

Regad (2015) reported that RAS MAP kinase pathways and the RAS PI3K-AKT pathways have been identified as playing key roles in proliferation, differentiation and survival. The RTKs include receptors for growth factors, hormones, cytokines and other signaling ligands. Cancer mutations may not only impact the RTK but also genes that encode components of downstream signaling pathways.

Specific therapeutic agents currently employed include monoclonal antibodies that target extra-cellular ligand binding regions of RTKs and small molecules that target the intra-cellular tyrosine kinase domains. Monoclonal antibodies have been developed against EGFR, ERBB2 (HER2), VEGFRs and FGFRs.

Regad reported that most of the cancer related mutations in the MAP signaling pathway involve RAS and RAF genes. These include KRAS mutation in pancreatic cancer, cancers of the intestinal and biliary tract, the lung, endometrium and ovarian cancer. NRAS mutations occur in nervous system cancers, melanomas and lymphoid and thyroid cancers.

HRAS mutations are common in the urinary tract, cervix, prostate and skin. BRAF mutations occur in melanoma, and in thyroid and colon cancers. Small molecule inhibitors have been developed particularly to target BRAF.

Activation of RTKs also activate phospho-inositol-3-kinase leading to generation of PIP2 (phosphatidyl-inositol-4-5 biphosphate) and PIP3 (phosphatidyl-3-4-5 triphosphate). PIP3 in turn activates the AKT pathway and AKT phosphorylates MTOR that promotes proliferation. Activity of PTEN prevents AKT activity. AKT mutations have been identified in several cancers. Regad reported that small molecule inhibitors of the PI3K and AKT pathways have been developed.

Regad emphasized the problem of resistance to inhibitors and proposed therapies with combinations of inhibitors to circumvent resistance.

11.8.8 *Investigation of pathways that drive tumor progression through functional genomics and short inhibitory RNAs (siRNAs) based studies*

Osteosarcoma is an aggressive bone cancer that affects children, adolescents and young adults. Perry *et al.* (2014) noted that there are TP53 germline mutations that predispose to osteosarcoma and sporadic TP53 mutations often occur in these tumors. They noted however that genome analysis of osteosarcomas revealed very high frequencies of mutations, copy number changes and genomic rearrangements.

In experiments designed to identify potential pathway targets for therapies, Perry *et al.* carried out gene set enrichment analyses (GSEA). In these analyses mutations found in tumor samples are compared with canonical biological standards to identify differences in pathways. Using the GSEA approach Perry *et al.* determined that there were alterations in 22 pathways in osteosarcomas and that 22 of these pathways included the TP53; in 8 pathways MTOR was included.

Perry *et al.* then carried out a functional genomics screen using pooled screen of murine osteosarcoma cells with siRNAs against 8,400 genes. The siRNAs were transfected into primary osteosarcoma cell lines. They determined that siRNAs against PIK3CA (catalytic subunit of PI3K) and MTOR were significantly depleted by the tumor cells. Furthermore knockdown of the PIK3CA and MTOR genes caused growth arrest of the osteosarcoma cells.

Perry *et al.* proposed that the PIK3 and MTOR pathways represent likely targets for therapy of osteosarcomas.

11.9 Epigenetics, Chromatin, Regulators, Transcription and Cancer

Suvà *et al.* (2013) noted that shared mechanisms which involve transcription factors, chromatin regulators and epigenetic states play key roles in determination of cell renewal, cell differentiation and in cancer.

They noted that there is now evidence that transcription factors act not only at gene promoter regions but also at distal enhancer elements in the genome. Furthermore transcription factors act in combinatorial patterns. Transcription factor binding is dependent on underlying chromatin structure; however transcription factors also recruit chromatin regulators to modify chromatin structure.

In recent years the concept of pioneer transcription factors has emerged. Iwafuchi-Do and Zaret (2014) described features of pioneer transcription factors and noted that they can bind to target sites in closed chromatin that is in DNA nuclease resistant chromatin. Binding of these pioneer transcription factors increases accessibility to chromatin remodelers, chromatin modifiers and specific histone variants.

In 2006, Takahashi and Yamanaka demonstrated that specific transcription factors could induce reprogramming of fibroblasts to pluripotent cells. The transcription factors they utilized were OCT4, SOX2, KLF4 and MYC. Suvà *et al.* noted that studies since 2006 have demonstrated that OCT4 and SOX2 are strictly required to introduce pluripotency and that other transcription factors (e.g. Nano and Lin28) can substitute for KLF4 and MYC. They emphasized that reprogramming requires the right combination of transcription factors and that it induces changes to the epigenetic landscape, including chromatin modification and remodeling.

Suvà *et al.* emphasized the parallelism between oncogenic transformation and cellular reprogramming. They noted that several reprogramming transcription factors are oncogenes and epigenetic modifications occur in oncogenesis and in reprogramming. SOX2, MYC and KIF4 are important transcription factors in human cancers. SOX2 is amplified in squamous cell carcinoma of the lung and esophagus and in small cell lung cancer. KIF4 expression promotes certain breast cancers and skin cancer. MYC and NANOG are expressed in multiple malignancies.

Changes in patterns of DNA methylation and histone methylation occur both in cellular reprogramming and oncogenesis. Repressive chromatin states are marked by histone H3 lysine 9 trimethylation (H3K9me3), and H3K27me and binding of Polycomb repressive complexes. Actively expressed genes are associated with H3K4 methylation. In transcription initiation, an H3K79 methyltransferase DOT1L is an important as a regulator and this methyltransferase shows altered expression in cellular reprogramming and oncogenesis.

Suvà *et al.* listed 13 chromatin regulators that have been shown to be mutated or abnormally expressed in certain cancers (see Table 11.4). They expressed the opinion that studies in cellular reprogramming have expanded functional insights into tumor biology.

11.9.1 *Recurrent somatic mutations in regulatory elements in cancer*

Melton *et al.* (2015) analyzed whole genome sequencing data on cancers from 436 patients with 8 cancer subtypes. They discovered mutations in non-protein coding DNA that contained regulatory

Table 11.4. Chromatin regulators involved in certain cancers.

SUV39H1 Histone methyltransferase trimethylates H3K9
SETDB1 Histone methyltransferase transcription repression
G9A (EHMT2) Demethylates histone, represses transcription
UTX (KDM6A) demethylates trimethylated H3
PRC2 Polycomb repressive complex
ARID1A regulates transcription through chromatin remodeling
MLL1 (KMT2A) H3K4 methyltransferase
MLL2 (KMT2D) H3K4 methyltransferase
MLL3 (KMT2) methyltransferase and transcription co-activator
LSD1 (KDM1A) histone deacetylase gene silencer
DOT1L K3K79 methyltransferase
KDM2B lysine specific demethylases
DNMT3A/B DNA cytosine 5 methyltransferases

Table 11.5. Mutations in non-protein coding DNA segments act as regulatory sites for the following cancer related genes.

TERT (telomerase reverse transcriptase)
GNAS complex locus, imprinted gene expression
INPP4B Inositolpolyphosphate-4-phosphatase type IIB
MAP2K2 Mitogen activated protein kinase 2
BCL11B Zinc finger protein related to BCLII lymphocyte protein
NEDD4LNEDD4 family of ubiquitin ligases
ANKRD11 Ankyrin repeat protein inhibits transcription
TRPM Calcium permeable cation channel
P2RY8 Purinergic receptor

---

elements involved in the control of expression of specific cancer genes and these are listed in Table 11.5.

Melton *et al.* noted that TRPM and INBB4B were described as tumor suppressor genes. GNAS, MAP2K2, BCL11B and P2RY8 are included in the Cancer Gene Census. A number of the regulatory sites identified are transcription factor binding sites.

11.10 Metabolic Reprogramming in Cancer

Warburg (1925) first reported that cancer cells preferentially converted glucose to lactate even under oxygen rich conditions, while non-cancerous cells under oxygen rich conditions preferably utilize glucose to fuel metabolic processes that produce CO_2. The modern application of the Warburg effect is the use of ^{18}F-2-deoxyglucose and positive emission tomography (PET imaging) to localize cancerous tumors based on their metabolic difference from normal tissues.

Ward and Thompson (2012) reviewed advances during the past decade in the analysis of cancer metabolism. They emphasized that the altered metabolism in cancer cells is under complex regulatory control primarily directed by growth factor signaling. They noted that elucidation of the entry into the tricarboxylic acid (TCA) cycle and delineation of ATP production in mitochondria represented one of the great discoveries of the 20th century. Non-proliferating cells metabolize glucose primarily through oxidative phosphorylation to produce ATP. They noted that when mitochondrial function is adequate in proliferating cancer cells, these cells utilize mitochondrial enzymes for the synthesis of anabolic precursors.

Ward and Thompson provided evidence that the phosphoinositide adenylate kinase activation and PI3K/AKT/mTORC1 pathway drive metabolic changes including mitochondrial reprogramming. There is evidence that activated PI3K/AKT is associated with increased transport of glucose into cells and activation of hexokinase to process glucose that enters cells. Levels of hexokinase II, the enzyme that phosphorylates glucose and generates glucose-6-phosphate, is elevated in certain tumors (e.g. hepatomas).

Activation of PI3K adenylate pathway promotes glucose-6-phosphate metabolism and synthesis of acetyl co-enzyme A (acetylCoA) which is required for synthesis of lipids that are important for membrane synthesis. AcetylCoA generation is dependent on activity of pyruvate dehydrogenase followed by metabolism through the TCA cycle in mitochondria. Citrate generated in this cycle is transported from mitochondria to the cytoplasm and is metabolized through the activity of ATP citrate lyase (ACL). Ward and Thompson reported that knockdown or pharmacologic inhibition of ACL was effective in reducing cellular proliferation.

MTORC1 activity enhances synthesis of proteins from aminoacids. Ward and Thompson emphasized that MTORC1 functions are intertwined with mitochondrial metabolism. Alpha-ketoglutarate generated in the TCA cycle in mitochondria is transaminated to produce glutamate; oxaloacetate generated in the citric acid cycle is converted to aspartate and then to asparagine. Early studies demonstrated that certain leukemia cells are very sensitive to asparagine deprivation and these leukemias were successfully treated with L-asparaginase. They also reported that the MYC oncogene acts as a transcription factor that enhances mitochondrial gene expression and mitochondrial biogenesis. Interestingly cells that express oncogenic MYC are addicted to glutamine and are dependent on glutamine as a carbon source for mitochondrial metabolism. Removal of glutamine or inhibition of glutamine generation leads to cell death.

There is evidence that loss of wild type p53 activity leads to metabolic reprogramming in tumor cells and to increased reliance on mitochondrial function. The drug Metformin1 that inhibits mitochondrial complex 1 was shown to be toxic to p53 depleted cells (Buzzai et al., 2005). Weinberg et al. (2010) reported that cells with activated RAS signaling are dependent on efficient mitochondrial function.

Düvel et al. (2010) reported that MTORC1 activity promotes lipogenesis mediated through the transcription factor SREBP and that this is essential for cholesterol and lipid synthesis.

11.10.1 *Alternate enzyme isoforms in cancer cells*

The PKM (pyruvate kinase M) gene on chromosome 15q22 gives rise to two transcripts from which two proteins PKM1 and PKM2 are generated. Mazurek et al. (2007) reported that embryonic cells and proliferating cells in tumors preferentially express the PKM2 isoform; in most tissues the M1 form is expressed. Goldberg and Sharp (2012) used short inhibitory RNAs (siRNAs) to knockdown PKM2 transcripts in tumor cells. They reported that PKM2 knockdown led to increased apoptosis of cancer cells.

Phosphoglycerate dehydrogenase (PHGDH) is often overexpressed in tumor cells. This enzyme oxidizes 3-phosphoglycerate to 3-phosphohydroxypyruvate that gives rise in successive reactions to serine. Serine is used as a building block in the synthesis of proteins, nucleotides and sphingosine. There is evidence that suppression of PHGDH expression is useful in therapy of tumors that manifest elevated expression of this enzyme; these include certain breast cancer tumors and melanoma (Possemato *et al.*, 2011).

11.11 Oncometabolites in Specific Cancers

In 2012, Jin *et al.* reported that a specific mutation at amino acid 132 in the isocitrate dehydrogenase 1 (IDH1) was present in 70% of grade 1 astrocytomas and oligodendrocytomas. Mutations in the corresponding amino acid R140 in the IDH2 enzyme were found in a number of other tumors, including bone tumors, liver tumors and cases of leukemia.

There is evidence that these mutant forms of isocitrate dehydrogenase have new activity; they convert alpha-ketoglutarate to D2-hydroxyglutaric acid. This metabolite is not produced by wild type IDH1 or IDH2. There is evidence that D2-hydroxyglutaric acid impairs cellular differentiation. Ward and Thompson noted that magnetic resonance studies that detect D2-hydroxyglutaric acid might be useful in clinical detection of gliomas. In addition screening tumors and body fluids for D2-hydroxyglutaric acid might be useful in screening for tumors.

Jin *et al.* reported that D2-hydroxyglutaric acid inhibits activity of the telomerase enzyme TERT that is involved in telomere maintenance. *In vitro* studies demonstrated that knockdown of mutant IDH1 or of mutant IDH2 with short hairpin RNAs led to significant reduction of proliferation in fibrosarcomas.

Ward and Thompson (2012) reported evidence that D2-hydroxyglutaric acid impacted chromatin and had particular effect on histone methylation, demethylation and epigenetic regulation.

Germline loss of function mutations in the TCA cycle enzymes succinate dehydrogenase and fumarate dehydrogenase lead to tumors such as pheochromocytomas, gangliomas, leiomyomas and renal cell carcinomas (Hoekstra and Bayley, 2012; Sanz-Ortega *et al.*, 2013). These germline succinate dehydrogenase and fumarate dehydrogenase mutations lead to increased levels of succinate and fumarate within tumors.

11.12 Approaches to Therapy Including Targeting of Mutations, Fusion Gene Products, Exploiting Synthetic Lethality and Use of Immune-Based Therapy

11.12.1 *Cancer epigenetics and drug discovery*

Primary efforts in this area concentrated on the discovery of drugs that impacted DNA and histone modifications, including methyltransferase, demethylases, histone acetyltransferase and histone deacetylases. Drugs that impact DNA methylation through inhibition of DNA methyltransferases (DNMTs) include azacytidine decitabine and disulfiram. Inhibitors of lysine acetylation (HDAC inhibitors) include vorinostat.

Campbell and Tummino (2014) presented information on efforts that target tumor somatic alterations which impacted epigenetic functions. They presented summaries of reports of activating mutations or function altering mutations in the histone methyltransferase EZH2 in cases of diffuse large B cell lymphomas (DLBCL) and in follicular lymphomas. In these cases EZH2 became an oncogenic driver. The EZH2 mutations in these cases lead to increased activity and effective EZH2 inhibitors (e.g. GSK126) which have been developed to target the increased oncogenic activity.

Aberrant increased activity of DOT1L methyltransferase occurs in certain AMLs and in certain acute lymphoblastic leukemias. Small molecule inhibitors of DOT1L have proven useful in treatment of these leukemias.

Overexpression of the histone methyltransferase MMSET in t(4;14)+ translocation positive multiple myeloma patients was reported to be the driving factor in the pathogenesis of a specific form of myeloma. This enzyme may present a target for therapeutic development. Campbell and Tummino noted that somatic alterations and modifications occur in many tumor types in the SWI/SNF complex that is involved in nucleosome remodeling and transcription and present other possible targets.

Campbell and Tummino noted that potential epigenetic targets often exist in large complexes. In these complexes one protein may be substituted for another so that the function of the complex is retained. They reported that treatments have been developed to target specific readers of epigenetic modifications, including specific bromodomain proteins, and they presented an example. A specific translocation t(15; 19) (q14:p31) leads to fusion between a bromo-domain protein, either BRD3 or BRD4, and a nuclear protein. This fusion serves as a driver mutation in certain testis cancers. A specific small molecule BET was developed to target the fusion protein and was found to be useful in treatment of the tumor.

11.12.2 *Synthetic lethality*

Synthetic lethality in the context of cancer was succinctly explained by Kaelin (2005). He wrote:

> "Two genes are synthetically lethal if mutation in either alone is compatible with viability but mutations of both lead to death."

Kaelin emphasized that although many chemicals have been identified that kill cancer cells, the bottleneck in therapy is caused by the fact that many chemicals kill not only cancer cells but also normal dividing cells in the body, e.g. cells in the bone marrow, gastrointestinal mucosa and epithelial cells in the skin. In addition many cancer drugs harm cells in the heart and lungs and cerebellum. Kaelin noted that the concept of synthetic lethality could be extended to situations where mutations in two different genes worked together to impair cellular fitness, a situation he referred to as "synthetic sick".

One of the most effective examples of the application of synthetic lethality concepts in cancer therapy involves the use of PARP1 inhibitors to treat breast cancer associated with mutations in the BRCA genes. Farmer *et al.* (2005) noted that normal BRCA1 and BRCA2 both play roles in the repair of double stranded breaks and that they achieve this through promotion of repair by homologous recombination. BRCA1 normally functions as a scaffolding protein onto which DNA damage repair proteins assemble. These complexes include histone H2AX.

Farmer *et al.* also noted that the enzyme PARP1 is involved in repair of single stranded DNA breaks. They then demonstrated that cells with BRCA1 or BRCA2 dysfunction were highly sensitive to treatment with PARP1 inhibitor. PARP1 inhibitor treatment of BRCA1 or BRCA2 mutant cells led to chromosome instability, cell cycle arrest and apoptosis. Farmer *et al.* and Bryant *et al.* (2005) proposed that the loss of PARP1 activity that resulted from the use of PARP1 inhibitors triggered foci of DNA deletions that required repair. Therefore inhibition of PARP1 in the presence of BRCA1 or BRCA2 deficiency led to the inability to repair DNA breaks. Defects in two different pathways that repair DNA breaks therefore promoted death of cancer cells.

There is also evidence that over-expression of PARP1 plays a role in the resistance of cancer cells to chemotherapy or to radiotherapy.

Utilization of concepts of synthetic lethalities in cancer treatment has greatly expanded. McLornan *et al.* (2014) reviewed recent applications. They noted that the oncogenic drivers previously considered "undruggable", including MYC and KRAS, can be targeted with synthetic lethality approaches. Synthetic lethality can be achieved using combinations of drugs directed against different genomic changes in a specific cancer. Furthermore changes in cell cycle oncogenes, cell metabolism, stromal characteristics and cell proteome can be targeted. Also specific chemotherapeutic agents and specific cancer causing mutations in particular genes may constitute synthetic lethal combinations.

Martin *et al.* (2009) reported that the MSH2 gene which is mutated in colon cancer is involved in the removal of base mismatches introduced during the repair of damaged DNA. They noted that tumors with MSH2 mutations were particularly sensitive to the drug methotrexate. Methotrexate leads to oxidative lesions in DNA, e.g. 8-hydroxy-2-deoxyguanosine. These lesions can be cleared if MSH2 function is intact. DNA lesions such as those introduced by methotrexate cannot be cleared in the presence of mutations in the mismatch repair genes MSH2 and MSH1.

McLornan *et al.* reported that a number of different PARP inhibitors have been developed. In addition it is possible that PARP1 inhibitors have effects beyond preventing repair of DNA defects. Weaver *et al.* (2015) reported that functioning PARP1 also impacts cellular energetics and cell death pathways. They reported evidence that PARP1 inhibitors can inhibit cancer cell growth and metastases. PARP1 inhibitors are used in conjunction with alkylating agents in treatment of tumors.

It is however important to note that resistance to PARP1 inhibitors can arise through several mechanisms. One mechanism involves increased extrusion of the drugs from cells through over-expression of P-glycoprotein transporters. P-glycoprotein transporter inhibitors (e.g. verapamil) may restore sensitivity to inhibitors.

There is evidence that malignant cells are more sensitive than normal cells to proteasome inhibitors. Crawford *et al.* (2011) proposed that cancer cells regulate anti-apoptotic pathways. They demonstrated increased proteasome activity in leukemia cells and increased sensitivity to proteasome inhibitors. Increased sensitivity to proteasome inhibitors was also demonstrated in myeloma cells.

There is evidence that loss of the transcription factor GATA in tumors with mutant KRAS may impair tumor growth. Loss of GATA also negatively impacts the RHO signaling pathway. Boidot *et al.* (2010) proposed that GATA1 could be a target for treatment of breast cancer.

Steckel *et al.* (2012) reported that the combination of therapy with cell cycle inhibitors (e.g. gemcitabine) that block cell cycle at the G1 to S transition and proteasome inhibitors (e.g. bortezomib) provided benefit. They also suggested that the combination of drugs which target the proteasome and topoisomerase inhibitors that disrupt DNA replication were useful.

11.12.3 *Targeting the mTOR pathway*

Inhibitors of the mTOR pathway have been approved for treatment of a number of different cancers (Fasolo and Sessa, 2012). MYC over-expression occurs in a broad range of cancers and is associated with poor prognosis. There is evidence that MYC modulates protein synthesis in tumors and that this is essential to the MYC oncogenic effect. Pourdehnad *et al.* (2013) reported that MYC and MTOR converge on a common mode of protein synthesis and that MTOR inhibition may be useful in treatment of MYC driven cancers.

11.12.4 *Aspects of targeted therapy resistance*

Nathanson *et al.* (2014) reported that the mechanisms of resistance that cancer cells develop to targeted therapies include target suppression and activation of alternative kinases to maintain downstream signaling. In addition substantial genomic heterogeneity may exist in tumors so that targeted therapy only succeeds in suppressing a sub-population of the tumor cells.

These investigators analyzed EGFR that is frequently mutated in gliomas. They noted that in gliomas the mutated EGFR is frequently present in extra-chromosomal elements. Nathanson *et al.* demonstrated that in tumor cells which acquired resistance to therapies that targeted mutated EGFR, the mutant EGFR was initially eliminated. However mutant EGFR on extra-chromosomal elements reappeared after drug withdrawal.

11.12.5 *Immunotherapy and cancer treatment*

Immunotherapy of cancer involves at least three different strategies: Immune checkpoint therapy, adoptive T cell therapy and personalized cancer vaccines (Mueller, 2015). These approaches are discussed in further detail in Chapter 6.

In second generation personalized medicine trials, therapies that target more than one genomic alteration in a specific tumor are used in combination. In some cases targeted therapies are used in combination with immunotherapies (Arnedos *et al.*, 2014).

11.12.6 *Genomic analysis and relevance to assessment of sensitivity of tumors to PD1 blockade*

Rizvi *et al.* (2015) used whole exome sequencing to analyze non-small cell lung tumors in patients treated with the immune checkpoint inhibitor that targeted PD1 (programmed cell death receptor). They reported that patients who had tumors with higher non-synonymous mutation burdens showed improved response. In different tumors the frequency on non-synonymous exonic mutations ranges between 45 and 1,732. They also noted that tumors with a molecular signature indicating smoking damaging and tumors with evidence of DNA repair pathway mutations showed improved response.

11.13 Liquid Biopsies and Circulating Extra-Cellular Vesicles, Cancer Diagnosis and Monitoring of Response

11.13.1 *Liquid biopsies and analysis of cancers resistant to therapy*

Liquid biopsies, the analysis of cell free DNA in plasma, have been successfully used to reveal the mechanisms by which specific cancers become resistant to targeted therapies. Meador and Lovly (2015) reported that non-small cell lung cancer tumors that harbor activating mutation in the tyrosine kinase domain of EGFR could be successfully treated with EGFR tyrosine kinase inhibitors. However resistance to treatment developed within 9 to 12 months. Resistance in approximately 50% of tumors was found to be due to a specific secondary mutation in EGFR, EGFR 90 T>M and that mutation could be successfully treated with an additional tyrosine kinase inhibitor. Other tumors acquired different EGFR mutations. There was evidence that the range of different EGFR mutations could be more successfully evaluated in cell free tumor DNA than in individual tumor biopsies.

In colon cancer tumor resistance with EGFR specific monoclonal antibodies were investigated by Siravegna *et al.* (2015). Specific factors leading to therapeutic resistance included mutations in NRAS, KRAS, MET or ERB2. Siravegna *et al.* reported that mutations leading to therapeutic resistance were more readily detected in cell free circulating DNA than in additional tumor biopsies.

Meador and Lovly emphasized the advantages of cell free circulating DNA in enabling longitudinal analyses of the impact of tumor mutations and of the cellular mechanisms involved in therapeutic resistance. These studies also facilitated the design of combinatorial therapies.

11.13.2 *Detection of brain tumor DNA in cerebrospinal fluid*

Pan *et al.* (2015) utilized cerebrospinal fluid (CSF) to detect evidence of the presence of tumor specific cell free DNA in patients with primary or metastatic brain tumors. In some cases they analyzed DNA in CSF for genes with known driver mutations. In other cases they characterized global genomic aberrations in CSF DNA. For seven patients where prior studies on tumors had revealed cancer specific gene mutations, PCR of CSF DNA generated the corresponding amplicons. In other patients tumor genomic aberrations were identified through DNA analyses. This second strategy then corresponded to liquid biopsy and in six out of seven patients studied they identified tumor mutations in patients with solid brain tumors.

Wang *et al.* (2015) carried out studies in CSF to detect tumor specific DNA. They emphasized the importance of CSF DNA studies for disease monitoring in cases with brain tumors. They noted that monitoring of treatment efficacy through imaging or even through repeat tumor resection was sub-optimal since treatment effects, including scarring and necrosis, impair interpretation of imaging results and may also impair accurate histological tumor analyses. They noted further that patients are sometimes kept on ineffective treatments for long periods due to ineffective tumor monitoring procedures.

Wang *et al.* studied CSF DNA form 35 patients and they used information on previously established tumor markers in those patients to facilitate CSF DNA analyses. In four cases where no prior information on tumor DNA markers was available, they utilized CSF DNA for whole exome sequencing and were able to identify tumor specific mutations in two cases. Wang *et al.* emphasized that the CSF DNA studies revealed tumor specific DNA in cases where tumors were adjacent to cortical surfaces or adjacent to ventricles, to basal brain regions, adjacent to cisternae or present in the spinal cord.

They identified tumor specific sequences in DNA from cases with gliomas, ependymomas or medulloblastomas.

11.13.3 *Circulating extra-cellular vesicles in the diagnosis of pancreatic cancer*

Exosomes are secreted membrane enclosed vesicles derived from cellular endosomes. Exosomes are released into the extra-cellular space and then into the circulation. There are reports that exosomes released from tumors into the circulation could be isolated by ultra-centrifugation of relatively small quantities of serum (Théry, 2015).

Melo *et al.* (2015) isolated exosomes from blood of patients with pancreatic cancer. They then used mass spectrometry analysis to identify components of exosomes. They determined that in pancreatic cancer patients circulating exosomes were positive for the proteoglycan glypican 1.

In mouse model systems they determined that the percentage of glypican 1 positive exosomes increased with tumor size. They then carried out studies to evaluate whether glypican 1 levels and circulating exosome levels could be used to distinguish pancreatic cancer patients from healthy donors. They reported that glypican 1 levels and circulated exosomes could be detected in cases with early pancreatic cancer.

Other biomarkers for pancreatic cancer including carcinoembryonic antigen CA150, DUPAN 2 and alpha fetoprotein are increased not only in pancreatic cancer patients but also in patients with chronic pancreatitis.

Melo *et al.* proposed that glypican 1 expression can also be used as a target to isolate pancreatic specific exosomes.

11.14 Pediatric Cancers and Therapies

Although survival gains have been made for hematologic malignancies in children, treatments for solid tumors are less successful. Glade Bender *et al.* (2015) reported that clinical tumor genomic profiling has yet to demonstrate clinical utility.

There is evidence that pediatric cancers show high frequencies of epigenetic changes. Frequently mutated genes are those that encode histone proteins H3.3 and H3.1, and chromatin modifier encoding genes, e.g. ATRX, SMARCA4, CHD7 and methyltransferases (Huether *et al.*, 2014). However specific epigenetic therapies for solid tumors are not yet in clinical use.

Mackall *et al.* (2014) reported that immune based therapies are being considered for therapy of tumors in children. Allogeneic hematopoietic stem cell therapy has been used to treat pediatric patients with leukemia when response to chemotherapy was inadequate or when recurrences recurred. Monoclonal antibodies against disialoganglioside have proven useful in patients with refractory neuroblastoma.

Progress in the discovery of genomic defect in neuroblastoma reported by Peifer *et al.* (2015) has direct relevance for treatment. Neuroblastoma is a pediatric tumor of the central nervous system. Some of these tumors regress but those that fail to regress constitute high risk neuroblastomas that have a poor prognosis and are difficult to treat.

Peifer *et al.* reported that in 31% of high risk neuroblastomas, recurrent genomic rearrangements occur on chromosome 5p15.33 proximal to the TERT telomerase locus. These genomic changes are associated with a significant increase in telomerase levels. Other genes that have defects in neuroblastoma include MYCN (neuroblastoma MYC homolog) and ATRX. Peifer *et al.* reported that TERT, MYCN and ATRX all impact telomere lengthening. They propose that telomerase inhibitors are likely useful agents for the treatment of high risk neuroblastomas.

12. Epilogue §

"Today it is quite clear that Mendelian inheritance of traits, including diseases, is the exception not the rule. Nevertheless the entire language of genetics is in terms of individual genes for individual phenotypes with one function rather than the ensemble and emergent properties of the genome."

A. Chakravarti et al. (2013)

In attempting to unravel complexity in disorders in which genetic and genomic factors play roles in pathogenesis, we reveal more clearly aspects of individual variation. Individual variation does not only include the variation in the primary gene defects that lead to a specific disease but also individual variant factors that modify disease presentations and disease outcomes.

Determinations of average disease manifestations, average prognoses and average therapeutic responses have limited relevance seen through the lens of individual variation.

In considering individual variation and disease, we are taken both back in time to concepts of the mid-20th century and forward in time to the goals of the Personalized Precision Medicine Initiative of 2015.

In 1966, Harry Harris wrote:

"There is thus ample indication of inborn diversity in the biochemical make-up of individuals. This may be reflected in differences in the patterns of metabolic processes or in differences in the structures of macromolecules. A particular kind of variant may be rare or common. It may result in pathological consequences for the individual or is may lead to no obvious effect of viability or biological fitness."

Francis Collins and the USA National Institutes of Health have defined the Precision Medicine Initiative (https://www.nih.gov/precision-medicine-initiative-cohort-program):

"Precision medicine is an emerging approach for disease treatment and prevention that takes into account individual variability in genes, environment and lifestyle for each person."

It is clear then that precision medicine will involve a strong focus on genes, environments and gene — environment interactions. Combinations of gene and environmental effects are clearly most relevant to complex common disorders; however environmental factors also have relevance to outcomes of less common disorders with a genetic basis.

Much work needs to be done in determining the mechanisms through which environments impact genome stability and gene functions. There is evidence that epigenetics and chromatin modification constitute intermediary mechanisms through which environments impact gene expression. The environments that need to be considered include both the endogenous environment created by physiological and biochemical processes in the individual and exogenous environments. There are a number of interesting studies relevant to gene environment interactions (Benayoun *et al.*, 2015).

12.1 Genetic and Environmental Factors in Stress and Anxiety Disorders

There is evidence that early life stress, including anxiety disorders in the mother, are associated with long term changes in the regulation of the stress hormone system (Klengel and Binder, 2015). Stress activates the glucocorticoid receptor (GR). This receptor is expressed in different isoforms.

Cortisol levels increase during stress. Cortisol binds to the GR. The activated receptor enters the nucleus where it binds to specific

DNA sequence elements and glucocorticoid response elements (GRE), thus impacting expression of a number of different genes.

In the cytoplasm in the non-activated state, the GR is bound to heat shock proteins and other proteins. Release from this binding is required for cortisol activation of the GR. One protein that binds to GR and decreases its ability to bind to cortisol is FKBP5 (FK506 binding protein). FKBP5 therefore increases resistance to cortisol and glucocorticoids, and there is evidence that it modulates response to other hormone receptors.

A specific series of polymorphisms and a specific haplotype across the FKBP5 encoding gene that includes snp rs1360780 impairs the binding of FKBP5 to the GR and is associated with increased response to cortisol. There is evidence that the T allele in rs1360780 is associated with increased attention to threat (Klengel and Binder, 2015).

12.2 Personalized Responses to Defined Nutritional Intake

Zeevi *et al.* (2015) measured 46,898 post-prandial glucose responses (PPGR) in 800 individuals fed a defined identical diet. The PPGR is an important parameter in control of glycaemia. Zeevi *et al.* demonstrated that the PPGR levels were associated with body mass index (BMI) and with wake up glucose levels. There was also association between PPGR and microbiome composition. However they concluded that there were likely to be as yet unknown associations with particular biosynthetic pathways and transport and secretion systems.

Zeevi *et al.* then initiated individualized dietary intervention based on multiple aspects of glucose metabolism including PPGR. They reported that with these interventions the fluctuations in blood glucose levels were less extreme.

12.3 Individual Variations, in Physiological Interactions and Cancer Risk

There is evidence that excess weight may increase the risk of developing breast cancer in BRCA mutation carriers. (Velasquez-Manoff, 2015). In addition there is evidence that physical exercise can lower breast cancer risk. Studies being carried out indicate that the basis for this may be that exercise increases the level of BRCA gene expression from the normal BRCA1 gene, thus compensating in part for deficiency activity of the mutant BRCA1 gene.

12.4 Endogenous and Exogenous Factors, DNA Damage and Chromatin

Environmental factors play roles in DNA damage. DNA damage changes chromatin structure, epigenetic modifications and leads to altered transcription, which is a key factor in aging (Burgess *et al.*, 2012). In addition defective repair following DNA damage predisposes to mutations, genomic imbalance and cancer.

Burgess *et al.* emphasized that studies on premature aging syndromes have revealed that defects in DNA damage repair are key factors in aging. Further insight into root causes of genomic, genetic and epigenetic disruption will promote preventive measures and decrease the burdens of age-related conditions.

12.5 Disease Classifications and Relationships

One of the goals of the Precision Medicine Initiative is to develop new disease classifications and define disease relationships. In reviewing material for this book, it becomes clear that diagnoses-based classifications that rely solely on specific symptoms and constellations of manifestations prevent us, in many cases, from determining the underlying root causes and pathogenesis. Furthermore it is clear that for a number of diseases discoveries of the root cause are very valuable for therapeutic design and for clarification of preventive measures.

12.6 Individual Variation in Drug Response

Individual variations in drug responses have been studied for a number of years. Early studies included defining variations in the incidence of adverse responses to the antimalarial drug primaquine and the discovery that deficiency of glucose-6-phosphae dehydrogenase was the main contributing factor in causing these side effects (Beutler *et al.*,

1968). Subsequently aberrant drug response was shown to be due, in many cases, to defects in enzymes involved in the metabolism of drugs, particularly the cytochrome P450 family of enzymes. More recently, allelic variations at the HLA-B locus were found to be responsible for hypersensitivity to floxacillin and allelic variation at a different site in HLA-B led to sensitivity to the anti-retroviral drug abacavir. HLA-B variants have also been shown to lead to increased sensitivity to the anti-epileptic drug carbamazepine (Wang *et al.*, 2011).

However much work remains to be done in understanding how individual variations not only lead to drug hypersensitivity but also how they impact drug effectiveness.

12.7 Understanding Tumor Heterogeneity to Improve Targeted Therapy in Cancer

Studies on tumor heterogeneity and target therapy failures provide evidence for the emergence of new driver mutations in sub-clones. McGranahan *et al.* (2015) noted that branched evolution occurs in tumors and this gives rise to subclones that differ from the initiating tumor clones. They proposed that targeted therapies need to be stratified according to the proportion of tumor cells in which a particular driver mutation is identified.

12.8 Need for Increased Understanding of Gene Regulation

Although important new information on gene regulation has been gathered through activities in the ENCODE project (http://www.genome.gov/encode/), it is evident that we have yet much to learn about aspects of gene expression regulation and about how environmental factors impact gene regulation and expression.

12.9 Expanding Concepts of Forms of Genetic and Genomic Variants, their Penetrance and their Interactions

There is abundant evidence that rare variants, variants of intermediate frequency and common variants all contribute to disease (Manolio *et al.*, 2009). Further investigations of the penetrance of specific variants are required. Penetrance can be defined as the fraction of the phenotype due to a particular genotypic variant (Antonarakis *et al.*, 2010). It will be necessary to determine to what extent penetrance is impacted by epigenetic and environmental factors. In addition inter-action between variants (epistasis) will require analysis.

12.10 Epistasis

Epistasis is the phenomenon of gene-gene interactions, the impact of variants in one gene on the functioning of another gene, and the joint impacts of variants in genes that operate in specific pathways or specific physiological processes.

A number of investigators have emphasized that single nucleo-tide polymorphisms are usually modeled as having additive cumula-tive effects on phenotypes rather than interactive effects (Hemani *et al.*, 2014). The effective analysis of interactive effects will require development of increasingly sophisticated models and bioinformatic resources. Continued progress in elucidation of gene networks and pathways will facilitate modeling for analyses of epistasis.

12.11 Mechanisms of Disease and Downstream Effects of Gene Mutations

Continued progress in the analysis of disease mechanisms will be required for the development of effective therapeutic approaches to many diseases in which genetic and genomic factors play roles.

Insights into specific disease mechanisms, even for disease due to causative mutations of high effect in single genes, frequently require detailed analyses of the downstream effects of the mutation for development of effective therapies. One example is Marfan disease, due to mutations in fibrillin 1 (FBN1), where the downstream effects include increased bioavailability of transforming growth factor TGF beta due to its defective binding by the mutant FBN1. Effective therapy requires medications that inhibit the effects of the increased bioavailability of TGF beta and its impact on extra-cellular matrix (Doyle *et al.*, 2012).

References §

Ahmadian M, Suh JM, Hah N *et al*. PPARγ signaling and metabolism: the good, the bad and the future. *Nat Med*. 2013 May;19(5):557–566. doi: 10.1038/nm.3159. Review. PMID:23652116.

Akers AL, Johnson E, Steinberg GK *et al*. Biallelic somatic and germline mutations in cerebral cavernous malformations (CCMs): evidence for a two-hit mechanism of CCM pathogenesis. *Hum Mol Genet*. 2009 Mar 1;18(5):919–930. doi: 10.1093/hmg/ddn430. PMID:19088123.

Alexandrov LB, Stratton MR. Mutational signatures: the patterns of somatic mutations hidden in cancer genomes. *Curr Opin Genet Dev*. 2014 Feb;24:52–60. doi: 10.1016/j.gde.2013.11.014. PMID:24657537.

Allou L, Lambert L, Amsallem D *et al*. 14q12 and severe Rett-like phenotypes: new clinical insights and physical mapping of FOXG1-regulatory elements. *Eur J Hum Genet*. 2012 Dec;20(12):1216–1223. doi: 10.1038/ejhg.2012.127. PMID:22739344.

Ament SA, Szelinger S, Glusman G, Ashworth J. Rare variants in neuronal excitability genes influence risk for bipolar disorder. *Proc Natl Acad Sci USA*. 2015 Mar 17;112(11):3576–3581. doi:10.1073/pnas.1424958112. PMID:25730879.

Amin V, Harris RA, Onuchic V *et al*. Epigenomic footprints across 111 reference epigenomes reveal tissue-specific epigenetic regulation of lincRNAs. *Nat Commun*. 2015 Feb 18;6:6370. doi: 10.1038/ncomms7370. PMID:25691256.

Amm I, Sommer T, Wolf DH. Protein quality control and elimination of protein waste: the role of the ubiquitin-proteasome system. *Biochim Biophys Acta*. 2014 Jan;1843(1):182–196. doi: 10.1016/j.bbamcr.2013.06.031. PMID:23850760.

345

Amo T, Saiki S, Sawayama T *et al.* Detailed analysis of mitochondrial respiratory chain defects caused by loss of PINK1. *Neurosci Lett.* 2014 Sep 19;580:37–40. doi 10.1016/j.neulet.2014.07.045. PMID: 25092611.

Antonarakis SE, Chakravarti A, Cohen JC, Hardy J. Mendelian disorders and multifactorial traits: the big divide or one for all? *Nat Rev Genet.* 2010 May;11(5):380–384. doi: 10.1038/nrg2793. PMID:20395971.

AOCS American Oil Chemists lipid Library. Accessed November 2015. http://lipidlibrary.aocs.org/

Ardisson Korat AV, Willett WC, Hu FB. Diet, lifestyle, and genetic risk factors for type 2 diabetes: a review from the Nurses' Health Study, Nurses' Health Study 2, and Health Professionals' Follow-up Study. *Curr Nutr Rep.* 2014 Dec 1;3(4):345–354.

Armanios M, Blackburn EH. The telomere syndromes. *Nat Rev Genet.* 2012 Oct;13(10):693–704. doi: 10.1038/nrg3246. PMID:22965356.

Armanios M. Telomerase and idiopathic pulmonary fibrosis. *Mutat Res.* 2012 Feb 1;730(1–2):52–58. PMID:22079513.

Armanios M. Telomeres and age-related disease: how telomere biology informs clinical paradigms. *J Clin Invest.* 2013 Mar;123(3):996–1002. doi: 10.1172/JCI66370. PMID:23454763.

Armstrong A, Mattsson N, Appelqvist H *et al.* Lysosomal network proteins as potential novel CSF biomarkers for Alzheimer's disease. *Neuromolecular Med.* 2014 Mar;16(1):150–160. doi: 10.1007/s12017-013-8269-3. PMID:24101586.

Arnedos M, Vielh P, Soria JC *et al.* The genetic complexity of common cancers and the promise of personalized medicine: is there any hope? *J Pathol.* 2014 Jan;232(2):274–282. doi: 10.1002/path.4276. PMID:24114621.

Artis D, Spits H. The biology of innate lymphoid cells. *Nature.* 2015 Jan 15;517(7534):293–301. doi: 10.1038/nature14189. PMID:25592534.

Ash PE, Vanderweyde TE, Youmans KL *et al.* Pathological stress granules in Alzheimer's disease. *Brain Res.* 2014 Oct 10;1584:52–58. doi: 10.1016/j.brainres.2014.05.052. PMID:25108040.

Autry AE, Monteggia LM. Brain-derived neurotrophic factor and neuropsychiatric disorders. *Pharmacol Rev.* 2012 Apr;64(2):238–258. doi: 10.1124/pr.111.005108. Review. PMID:22407616.

Bachmann-Gagescu R, Dempsey JC, Phelps IG *et al.* Joubert syndrome: a model for untangling recessive disorders with extreme genetic heterogeneity. *J Med Genet.* 2015 Aug;52(8):514–522. doi: 10.1136/jmedgenet-2015-103087. PMID:26092869.

Bahi-Buisson N, Nectoux J, Rosas-Vargas H *et al.* Key clinical features to identify girls with CDKL5 mutations. *Brain.* 2008 Oct;131(Pt 10): 2647–2661. doi: 10.1093/brain/awn197. PMID:18790821.

Bahi-Bussion N, Poirier K, Fourniol F *et al.* The wide spectrum of tubulinopathies: what are the key features for the diagnosis? *Brain.* 2014 Jun;137 (Pt 6): 1676–1700. doi: 10.1093/brain/awu082. PMID:24860126.

Bailey A, Luthert P, Dean A *et al.* A clinicopathological study of autism. *Brain.* 1998 May;121 (Pt 5):889–905. PMID:9619192.

Ball AR Jr, Chen YY, Yokomori K. Mechanisms of cohesin-mediated gene regulation and lessons learned from cohesinopathies. *Biochim Biophys Acta.* 2014 Mar;1839(3):191–202. doi: 10.1016/j.bbagrm.2013.11.002. PMID:24269489.

Bannister AJ, Kouzarides T. Regulation of chromatin by histone modifications. *Cell Res.* 2011 Mar;21(3):381–395. doi: 10.1038/cr.2011.22. Epub 2011 Feb 15. PMID:21321607.

Barash Y, Calarco JA, Gao W *et al.* Deciphering the splicing code. *Nature.* 2010 May 6; 465(7294):53–59. doi: 10.1038/nature09000. PMID:20445623.

Barr ML, Bertram EG. A morphological distinction between neurones of the male and female, and the behaviour of the nucleolar satellite during accelerated nucleoprotein synthesis. *Nature.* 1949 Apr 30;163(4148): 676. PMID:18120749.

Bauer DE, Orkin SH. Hemoglobin switching's surprise: the versatile transcription factor BCL11A is a master repressor of fetal hemoglobin. *Curr Opin Genet Dev.* 2015 Sep 13;33:62–70. doi: 10.1016/j. gde.2015.08.001. Review. PMID:26375765.

Beaulieu JM, Espinoza S, Gainetdinov RR. Dopamine receptors— IUPHAR Review 13. *Br J Pharmacol.* 2015 Jan;172(1):1–23. PMID: 25671228.

Benayoun BA, Caburet S, Veitia RA. Forkhead transcription factors: Key players in health and disease. *Trends Genet.* 2011 Jun;27(6):224–232. doi: 10.1016/j.tig.2011.03.003. PMID:21507500.

Benayoun BA, Pollina EA, Brunet A. Epigenetic regulation of ageing: linking environmental inputs to genomic stability. *Nat Rev Mol Cell Biol.* 2015 Oct;16(10):593–610. doi: 10.1038/nrm4048. PMID: 26373265.

Ben-David E, Shifman S. Combined analysis of exome sequencing points toward a major role for transcription regulation during brain development in autism. *Mol Psychiatry.* 2013 Oct;18(10):1054–1056. doi: 10.1038/mp.2012.148. PMID:23147383.

Berg AT, Dobyns WB. Progress in autism and related disorders of brain development. *Lancet Neurol.* 2015 Nov;14(11):1069–1070. doi: 10.1016/S1474-4422(15)00048-4. 16. PMID:25891008.

Berlow RB, Dyson HJ, Wright PE. Functional advantages of dynamic protein disorder. *FEBS Lett.* 2015 Sep 14;589(19 Pt A):2433–2440. doi: 10.1016/j.febslet.2015.06.003. PMID:26073260.

Berndsen CE, Wolberger C. New insights into ubiquitin E3 ligase mechanism. *Nat Struct Mol Biol.* 2014 Apr;21(4):301–307. doi: 10.1038/nsmb.2780. PMID:24699078.

Beutler B, Jiang Z, George l P *et al.* Genetic analysis of host resistance: Toll-like receptor signaling and immunity at large. *Annu Rev Immunol.* 2006;24:353–89. Review. PMID:16551253.

Beutler E, Mathai CK, Smith JE. Biochemical variants of glucose-6-phosphate dehydrogenase giving rise to congenital nonspherocytic hemolytic disease. *Blood.* 1968 Feb;31(2):131–150. PMID:5643703.

Bieging KT, Mello SS, Attardi LD. Unravelling mechanisms of p53-mediated tumour suppression. *Nat Rev Cancer.* 2014 May;14(5):359–370. doi: 10.1038/nrc3711. PMID:24739573.

Björklund A, Dunnett SB. Dopamine neuron systems in the brain: an update. *Trends Neurosci.* 2007 May;30(5):194–202. Review. PMID:17408759.

Blasco MA. Telomeres and human disease: ageing, cancer and beyond. *Nat Rev Genet.* 2005 Aug;6(8):611–22. Review. PMID:16136653.

Bloom GS. Amyloid-β and tau: the trigger and bullet in Alzheimer disease pathogenesis. *JAMA Neurol.* 2014 Apr;71(4):505–508. doi: 10.1001/jamaneurol.2013.5847. Review. PMID:24493463.

Blume-Jensen P, Hunter T. Oncogenic kinase signalling. *Nature.* 2001 May 17;411(6835):355–365. PMID:11357143.

Bode A, Lynch JW. The impact of human hyperekplexia mutations on glycine receptor structure and function. *Mol Brain.* 2014 Jan 9;7:2. doi: 10.1186/1756-6606-7-2. PMID:24405574.

Boettcher M, McManus MT. Choosing the right tool for the job: RNAi, TALEN, or CRISPR. *Mol Cell.* 2015 May 21;58(4):575–585. doi: 10.1016/j.molcel.2015.04.028. PMID:26000843.

Bogenhagen DF. Mitochondrial DNA nucleoid structure. *Biochim Biophys Acta.* 2012 Sep–Oct;1819(9–10):914–920. doi: 10.1016/j.bbagrm.2011.11.005. PMID:22142616.

Boidot R, Végran F, Jacob D, *et al.* The transcription factor GATA-1 is overexpressed in breast carcinomas and contributes to surviving upregulation via a promoter polymorphism. *Oncogene.* 2010 Apr 29;29(17):2577–2584. doi: 10.1038/onc.2009.525. PMID:20101202.

Boland CR, Thibodeau SN, Hamilton SR *et al.* A National Cancer Institute workshop on microsatellite instability for cancer detection and familial predisposition: Development of international criteria for the determination of microsatellite instability in colorectal cancer. *Cancer Res.* 1998 Nov 15;58(22):5248–57. PMID:9823339.

Borrelli E, Nestler EJ, Allis CD *et al.* Decoding the epigenetic language of neuronal plasticity. *Neuron.* 2008 Dec 26;60(6):961–974. doi: 10.1016/j.neuron.2008.10.012. Review. PMID:19109904.

Boveri T. Zur Frage der Entstehung maligner Tumoren. Fischer, Jena. 1914. Referred to in publication by Harris H, PMID:18089651

Brachova P, Thiel KW, Leslie KK. The consequence of oncomorphic TP53 mutations in ovarian cancer. *Int J Mol Sci.* 2013 Sep 23;14(9): 19257–75. doi: 10.3390/ijms140919257. PMID:24065105.

Brancaccio M, Pivetta C, Granzotto M *et al.* Emx2 and Foxg1 inhibit gliogenesis and promote neuronogenesis. *Stem Cells.* 2010 Jul;28(7): 1206–1218. doi: 10.1002/stem.443. PMID:20506244.

Brandler WM, Sebat J. From *de novo* mutations to personalized therapeutic interventions in autism. *Annu Rev Med.* 2015;66:487–507. doi: 10.1146/annurev-med-091113-024550. PMID:2558765.

Bras J, Guerreiro R, Darwent L *et al.* Genetic analysis implicates APOE, SNCA and suggests lysosomal dysfunction in the etiology of dementia with Lewy bodies. *Hum Mol Genet.* 2014 Dec 1;23(23):6139–6146. doi: 10.1093/hmg/ddu334. PMID:24973356.

Brewer JW. Regulatory crosstalk within the mammalian unfolded protein response. *Cell Mol Life Sci.* 2014 Mar;71(6):1067–1079. doi: 10.1007/s00018-013-1490-2. PMID:24135849.

Broere F, Apasov SG, Sitkovsky MV *et al.* T cell subsets and T cell-mediated immunity. In Nijkamp FP and Parnham MJ (eds), *Principles of Immunopharmacology.* 15–27, 2011 Springer.

Brouwers MC, van Greevenbroek MM, Stehouwer CD *et al.* The genetics of familial combined hyperlipidaemia. *Nat Rev Endocrinol.* 2012 Feb 14;8(6):352–362. doi: 10.1038/nrendo.2012.15. PMID:22330738.

Brown CJ, Lafreniere RG, Powers VE *et al.* Localization of the X inactivation centre on the human X chromosome in Xq13. *Nature.* 1991 Jan 3;349(6304):82–4. PMID

Brown MS, Goldstein JL. Receptor-mediated endocytosis: insights from the lipoprotein receptor system. *Proc Natl Acad Sci USA.* 1979 Jul;76(7): 3330–3337. Review. PMID:226968.

Brownlee PM, Meisenberg C, Downs JA. The SWI/SNF chromatin remodelling complex: Its role in maintaining genome stability and preventing tumourigenesis. *DNA Repair (Amst)*. 2015 Aug;32:127–33. doi: 10.1016/j.dnarep.2015.04.023. PMID:25981841.

Brownlie RJ, Zamoyska R. T cell receptor signalling networks: branched, diversified and bounded. *Nat Rev Immunol*. 2013 Apr;13(4):257–269. doi: 10.1038/nri3403. PMID:23524462.

Brück D, Wenning GK, Stefanova N *et al*. Glia and alpha-synuclein in neurodegeneration: A complex interaction. *Neurobiol Dis*. 2015 Mar 10. pii: S0969-9961(15)00064-9. doi: 10.1016/j.nbd.2015.03.003. PMID: 25766679.

Bryant HE, Schultz N, Thomas HD *et al*. Specific killing of BRCA2-deficient tumours with inhibitors of poly(ADP-ribose) polymerase. *Nature*. 2005 Apr 14;434(7035):913–7. Erratum in: *Nature*. 2007 May 17;447(7142):346. PMID:15829966.

Buckholtz JW, Meyer-Lindenberg A. Psychopathology and the human connectome: toward a transdiagnostic model of risk for mental illness. *Neuron*. 2012 Jun 21;74(6):990–1004. doi: 10.1016/j.neuron.2012.06.002. PMID:22726830.

Buckingham SD, Jones AK, Brown LA, Sattelle DB. Nicotinic acetylcholine receptor signalling: roles in Alzheimer's disease and amyloid neuroprotection. *Pharmacol Rev*. 2009 Mar;61(1):39–61. doi: 10.1124/pr.108.000562. PMID:19293145.

Burgess RC, Misteli T, Oberdoerffer P. DNA damage, chromatin, and transcription: the trinity of aging. *Curr Opin Cell Biol*. 2012 Dec;24(6):724–730. doi: 10.1016/j.ceb.2012.07.005. PMID:22902297.

Burotto M, Chiou VL, Lee JM *et al*. The MAPK pathway across different malignancies: a new perspective. Cancer. 2014 Nov 15;120(22):3446–56. doi: 10.1002/cncr.28864. Review. PMID:24948110.

Buzzai M, Bauer DE, Jones RG *et al*. The glucose dependence of Akt-transformed cells can be reversed by pharmacologic activation of fatty acid beta-oxidation. *Oncogene*. 2005 Jun 16;24(26):4165–73. PMID:15806154.

Cabianca DS, Casa V, Bodega B *et al*. A long ncRNA links copy number variation to a polycomb/trithorax epigenetic switch in FSHD muscular dystrophy. *Cell*. 2012 May 11;149(4):819–31. doi: 10.1016/j.cell.2012.03.035. PMID:22541069.

Cai N, Chang S, Li Y *et al*. Molecular signatures of major depression. *Curr Biol*. 2015 May 4;25(9):1146–56. doi: 10.1016/j.cub.2015.03.008. PMID: 25913401.

Caldecott KW. Single-strand break repair and genetic disease. *Nat Rev Genet.* 2008 Aug;9(8):619–31. doi: 10.1038/nrg2380. Review. PMID: 18626472.

Campbell IM, Shaw CA, Stankiewicz P, Lupski JR. Somatic mosaicism: implications for disease and transmission genetics. *Trends Genet.* 2015 Jul;31(7):382–392. doi: 10.1016/j.tig.2015.03.013. Review. PMID: 25910407.

Campbell RM, Tummino PJ. Cancer epigenetics drug discovery and development: the challenge of hitting the mark. *J Clin Invest.* 2014 Jan;124(1):64–9. doi: 10.1172/JCI71605. PMID:24382391.

Cancer Net. Accessed November 2015. http://www.cancer.net

Capell BC, Erdos MR, Madigan JP *et al.* Inhibiting farnesylation of progerin prevents the characteristic nuclear blebbing of Hutchinson-Gilford progeria syndrome. *Proc Natl Acad Sci.* USA S A. 2005 Sep 6;102(36):12879–84. PMID:16129833.

Carreno BM, Magrini V, Becker-Hapak M *et al.* Cancer immunotherapy. A dendritic cell vaccine increases the breadth and diversity of melanoma neoantigen-specific T cells. *Science.* 2015 May 15;348(6236): 803–808. doi: 10.1126/science.aaa3828. PMID:25837513.

Caspersson T, Zech L, Johansson C. Analysis of human metaphase chromosome set by aid of DNA-binding fluorescent agents. *Exp Cell Res.* 1970 Oct;62(2):490–2. available. PMID:5495462.

Catalogue of Somatic Mutations in Cancer (COSMIC). Accessed November 2015. http://cancer.sanger.ac.uk/cosmic

Cautain B, Hill R, de Pedro N *et al.* Components and regulation of nuclear transport processes. *FEBS J.* 2015 Feb;282(3):445–462. doi: 10.1111/febs.13163. PMID:25429850.

Chakravarti A, Clark AG, Mootha VK Distilling pathophysiology from complex disease genetics. *Cell.* 2013 Sep 26;155(1):21–6. doi: 10.1016/j.cell.2013.09.001. PMID:24074858.

Chandler DJ, Lamperski CS, Waterhouse BD. Identification and distribution of projections from monoaminergic and cholinergic nuclei to functionally differentiated subregions of prefrontal cortex. *Brain Res.* 2013 Jul 19;1522:38–58. doi: 10.1016/j.brainres.2013.04.057. PMID:23665053.

Chen JA, Peñagarikano O, Belgard TG *et al.* The emerging picture of autism spectrum disorder: genetics and pathology. *Annu Rev Pathol.* 2015; 10:111–44. doi: 10.1146/annurev-pathol-012414-040405. Review. PMID:25621659.

Chen L, Chen K, Lavery LA *et al.* MeCP2 binds to non-CG methylated DNA as neurons mature, influencing transcription and the timing of onset for Rett syndrome. *Proc Natl Acad Sci USA.* 2015 Apr 28; 112(17):5509–14. doi: 10.1073/pnas.1505909112. Erratum in: *Proc Natl Acad Sci USA.* 2015 Jun 2;112(22):E2982. PMID:25870282.

Chen LY, Redon S, Lingner J. The human CST complex is a terminator of telomerase activity. *Nature.* 2012 Aug 23;488(7412):540–4. doi: 10.1038/nature11269. PMID:22763445.

Chen M, Manley JL. Mechanisms of alternative splicing regulation: insights from molecular and genomics approaches. *Nat Rev Mol Cell Biol.* 2009 Nov;10(11):741–54. doi: 10.1038/nrm2777 Review. PMID: 19773805.

Choi AM, Ryter SW, Levine B. Autophagy in human health and disease. *N Engl J Med.* 2013 May 9;368(19):1845–6. doi: 10.1056/NEJMc130 3158. PMID:23656658.

Chow ML, Pramparo T, Winn ME *et al.* Age-dependent brain gene expression and copy number anomalies in autism suggest distinct pathological processes at young versus mature ages. *PLoS Genet.* 2012;8(3):e1002592. doi: 10.1371/journal.pgen.1002592. PMID:22457638.

Ciccia A, Elledge SJ. The DNA damage response: making it safe to play with knives. *Mol Cell.* 2010 Oct 22;40(2):179–204. doi: 10.1016/j.molcel.2010.09.019. Review. PMID:20965415.

Cirulli ET, Lasseigne BN, Petrovski S *et al.* Exome sequencing in amyotrophic lateral sclerosis identifies risk genes and pathways. *Science.* 2015 Mar 27;347(6229):1436–41. doi: 10.1126/science.aaa3650. PMID:25700176.

Claussnitzer M, Dankel SN, Kim KH *et al.* FTO obesity variant circuitry and adipocyte browning in humans. *N Engl J Med.* 2015 Sep 3; 373(10):895–907. doi: 10.1056/NEJMoa1502214. PMID:26287746.

Clavaguera F, Akatsu H, Fraser G *et al.* Brain homogenates from human tauopathies induce tau inclusions in mouse brain. *Proc Natl Acad Sci USA.* 2013 Jun 4;110(23):9535–40. doi: 10.1073/pnas.1301175110. PMID:23690619.

Coe BP, Girirajan S, Eichler EE. A genetic model for neurodevelopmental disease. *Curr Opin Neurobiol.* 2012 Oct;22(5):829–36. doi: 10.1016/j.conb.2012.04.007. Review. PMID:22560351.

Cohen JC, Boerwinkle E, Mosley TH Jr, Hobbs HH. Sequence variations in PCSK9, low LDL, and protection against coronary heart disease. *N Engl J Med.* 2006 Mar 23;354(12):1264–1272. PMID:16554528.

Cohen OS, Chapman J, Korczyn AD *et al.* Familial Creutzfeldt-Jakob disease with the E200K mutation: longitudinal neuroimaging from asymptomatic to symptomatic CJD. *J Neurol.* 2015 Mar;262(3):604–13. doi: 10.1007/s00415-014-7615-1. PMID:25522698.

Colvert E, Tick B, McEwen F *et al.* Heritability of Autism Spectrum Disorder in a UK Population-Based Twin Sample. *JAMA Psychiatry.* 2015 May;72(5):415–23. doi: 10.1001/jamapsychiatry.2014.3028.

Cong L, Ran FA, Cox D *et al. Science.* 2013 Feb 15;339(6121):819–23. doi: 10.1126/science.1231143. PMID:23287718.

Conley ME, Dobbs AK, Farmer DM *et al.* Primary B cell immunodeficiencies: comparisons and contrasts. *Annu Rev Immunol.* 2009;27:199–227. doi: 10.1146/annurev.immunol.021908.132649. Review. PMID: 19302039.

Cooper GM, Coe BP, Girirajan S *et al.* A copy number variation morbidity map of developmental delay. *Nat Genet.* 2011 Aug 14;43(9):838–46. doi: 10.1038/ng.909. PMID:21841781.

Couch FB, Bansbach CE, Driscoll R *et al.* ATR phosphorylates SMARCAL1 to prevent replication fork collapse. *Genes Dev.* 2013 Jul 15;27(14): 1610–23. doi: 10.1101/gad.214080.113. PMID:23873943.

Crawford LJ, Walker B, Irvine AE. Proteasome inhibitors in cancer therapy. *J Cell Commun Signal.* 2011 Jun;5(2):101–10. doi: 10.1007/s12079-011-0121-7. PMID:21484190.

Crino PB, Aronica E, Baltuch G, *et al.* Biallelic TSC gene inactivation in tuberous sclerosis complex. *Neurology.* 2010 May 25;74(21):1716–23. doi: 10.1212/WNL.0b013e3181e04325. PMID:20498439.

Cross-Disorder Group of the Psychiatric Genomics Consortium. Identification of risk loci with shared effects on five major psychiatric disorders: a genome-wide analysis. *Lancet.* 2013 Apr 20;381(9875): 1371–1379. doi: 10.1016/S0140-6736(12)62129-1. Erratum in: *Lancet.* 2013 Apr 20;381(9875):1360. PMID:23453885.

Croteau DL, Popuri V, Opresko PL, Bohr VA. Human RecQ helicases in DNA repair, recombination, and replication. *Annu Rev Biochem.* 2014;83: 519–52. doi: 10.1146/annurev-biochem-060713-035428. Review. PMID: 24606147.

Cuervo AM, Macian F. Autophagy and the immune function in aging. *Curr Opin Immunol.* 2014 Aug;29:97–104. doi: 10.1016/j.coi.2014.05.006. Review. PMID:24929664.

Curr Top Microbiol Immunol. 1967;40:59–63. PMID:6069978.

Cusanelli E, Romero CA, Chartrand P. Telomeric noncoding RNA TERRA is induced by telomere shortening to nucleate telomerase molecules at short telomeres. *Mol Cell.* 2013 Sep 26;51(6):780–91. doi: 10.1016/j.molcel.2013.08.029. PMID:24074956.

Cuthbert BN, Insel TR. Toward the future of psychiatric diagnosis: the seven pillars of RDoC. *BMC Med.* 2013 May 14;11:126. doi: 10.1186/1741-7015-11-126. PMID:23672542.

da Rocha ST, Boeva V, Escamilla-Del-Arenal M *et al.* Jarid2 Is implicated in the initial Xist-induced targeting of PRC2 to the inactive X chromosome. *Mol Cell.* 2014 Jan 23;53(2):301–16. doi: 10.1016/j.molcel.2014.01.002. PMID:24462204.

Darnell JC, Klann E. The translation of translational control by FMRP: therapeutic targets for FXS. *Nat Neurosci.* 2013 Nov;16(11):1530–6. doi: 10.1038/nn.3379. PMID:23584741.

Dasika GK, Lin SC, Zhao S, Sung P, Tomkinson A, Lee EY. DNA damage-induced cell cycle checkpoints and DNA strand break repair in development and tumorigenesis. *Oncogene.* 1999 Dec 20;18(55):7883–99. Review. PMID:10630641.

Day C, Shepherd JD. Arc: building a bridge from viruses to memory. *Biochem J.* 2015 Jul 1;469(1):e1–3. doi: 10.1042/BJ20150487. PMID 26173260.

De Jager PL, Srivastava G, Lunnon K, Burgess J *et al.* Alzheimer's disease: early alterations in brain DNA methylation at ANK1, BIN1, RHBDF2 and other loci. *Nat Neurosci.* 2014 Sep;17(9):1156–63. doi: 10.1038/nn.3786. PMID:25129075.

De Rubeis S, Buxbaum JD. Genetics and genomics of autism spectrum disorder: embracing complexity. *Hum Mol Genet.* 2015 Oct 15;24(R1): R24–31. doi: 10.1093/hmg/ddv273. Review. PMID:26188008.

De Rubeis S, He X, Goldberg AP *et al.* Synaptic, transcriptional and chromatin genes disrupted in autism. *Nature.* 2014 Nov 13;515(7526): 209–15. doi: 10.1038/nature13772. PMID:25363760.

De Sandre-Giovannoli A, Bernard R, Cau P *et al.* Lamin a truncation in Hutchinson-Gilford progeria. *Science.* 2003 Jun 27;300(5628):2055. PMID:12702809.

Deantonio C, Cotella D, Macor P *et al.* Phage display technology for human monoclonal antibodies. *Methods Mol Biol.* 2014;1060:277–95. doi: 10.1007/978-1-62703-586-6_14. PMID 24037846.

DeKosky ST, Scheff SW, Styren SD. Structural correlates of cognition in dementia: quantification and assessment of synapse change. Neurodegeneration. 1996 Dec;5(4):417–21. PMID:9117556.

Deng Z, Glousker G, Molczan A *et al*. Inherited mutations in the helicase RTEL1 cause telomere dysfunction and Hoyeraal-Hreidarsson syndrome. *Proc Natl Acad Sci USA*. 2013 Sep 3;110(36):E3408–16. doi: 10.1073/pnas.1300600110. PMID:23959892.

Dennis EL, Thompson PM. Typical and atypical brain development: a review of neuroimaging studies. *Dialogues Clin Neurosci*. 2013 Sep; 15(3):359–84. Review. PMID:24174907.

Derrien T, Johnson R, Bussotti G *et al*. The GENCODE v7 catalog of human long noncoding RNAs: analysis of their gene structure, evolution, and expression. *Genome Res*. 2012 Sep;22(9):1775–89. doi: 10.1101/gr.132159.111. PMID:22955988.

Derrien T, Johnson R, Bussotti G *et al*. The GENCODE v7 catalog of human long noncoding RNAs: analysis of their gene structure, evolution, and expression. *Genome Res*. 2012 Sep;22(9):1775–89. doi: 10.1101/gr.132159.111. PMID:22955988.

DiGiovanna JJ, Kraemer KH. Shining a light on xeroderma pigmentosum. *J Invest Dermatol*. 2012 Mar;132(3 Pt 2):785–96. doi: 10.1038/jid.2011.426. Review. PMID:22217736.

Dijk M, Typas D, Mullenders L *et al*. Insight in the multilevel regulation of NER. *Exp Cell Res*. 2014 Nov 15;329(1):116–23. doi: 10.1016/j.yexcr.2014.08.010. PMID:25128816.

Dinopoulos A, Matsubara Y, Kure S. Atypical variants of nonketotic hyperglycinemia. *Mol Genet Metab*. 2005 Sep–Oct;86(1–2):61–9. Review. PMID:16157495.

Dittmer TA, Sahni N, Kubben N *et al*. Systematic identification of pathological lamin A interactors. *Mol Biol Cell*. 2014 May;25(9):1493–510. doi: 10.1091/mbc.E14-02-0733. PMID:24623722.

Djebali S, Davis CA, Merkel A, Dobin A *et al*. Landscape of transcription in human cells. *Nature*. 2012 Sep 6;489(7414):101–8. doi: 10.1038/nature11233. PMID:22955620.

Do R, Stitziel NO, Won HH *et al*. Exome sequencing identifies rare LDLR and APOA5 alleles conferring risk for myocardial infarction. *Nature*. 2015 Feb 5;518(7537):102–6. doi:10.1038/nature13917. PMID:25487149.

Docherty LE, Rezwan FI, Poole RL *et al*. Mutations in NLRP5 are associated with reproductive wastage and multilocus imprinting disorders in humans. *Nat Commun*. 2015 Sep 1;6:8086. doi: 10.1038/ncomms9086. PMID:26323243.

Dominissini D, He C. Cancer: Damage prevention targeted. *Nature* 508; 191–192:(10 April 2014)doi:10.1038/nature13221. PMID:24695227.

Doyle JJ, Gerber EE, Dietz HC. Matrix-dependent perturbation of TGFβ signaling and disease. *FEBS Lett.* 2012 Jul 4;586(14):2003–2015. doi: 10.1016/j.febslet.2012.05.027. PMID:22641039.

Driver JA, Zhou XZ, Lu KP. Regulation of protein conformation by Pin1 offers novel disease mechanisms and therapeutic approaches in Alzheimer's disease. *Discov Med.* 2014.

Duggal G, Wang H, Kingsford C. Higher-order chromatin domains link eQTLs with the expression of far-away genes. *Nucleic Acids Res.* 2014 Jan;42(1):87–96. doi: 10.1093/nar/gkt857. Epub 2013 Oct 1. PMID: 24089144.

Düvel K, Yecies JL, Menon S, Raman P. Activation of a metabolic gene regulatory network downstream of mTOR complex. *Mol Cell.* 2010 Jul 30;39(2):171–83. doi: 10.1016/j.molcel.2010.06.022. PMID: 20670887.

Ebisch SJ, Gallese V, Willems RM *et al.* Altered intrinsic functional connectivity of anterior and posterior insula regions in high-functioning participants with autism spectrum disorder. *Hum Brain Mapp.* 2011 Jul;32(7):1013–28. doi: 10.1002/hbm.21085. PMID:20645311.

Eggermann T, Heilsberg AK, Bens S, Siebert R, Beygo J, Buiting K, Begemann M, Soellner L. Additional molecular findings in 11p15-associated imprinting disorders: an urgent need for multi-locus testing. *Nature.* 2015 Feb 19;518(7539):365–9. doi: 10.1038/nature14252.

Eletr ZM, Wilkinson KD. Regulation of proteolysis by human deubiquitinating enzymes. *Biochim Biophys Acta.* 2014 Jan;1843(1):114–28. doi: 10.1016/j.bbamcr.2013.06.027. PMID:23845989.

Elledge SJ. Accidents and Damage Control. *Cell.* 2015 Sep 10;162(6): 1196–200. doi: 10.1016/j.cell.2015.08.042. PMID:26359977.

Enns GM, Shashi V, Bainbridge M *et al.* Mutations in NGLY1 cause an inherited disorder of the endoplasmic reticulum-associated degradation pathway. *Genet Med.* 2014 Oct;16(10):751-8. doi: 10.1038/gim.2014.22. PMID:24651605.

Epel ES, Blackburn EH, Lin J, Dhabhar FS *et al.* Accelerated telomere shortening in response to life stress. *Proc Natl Acad Sci USA.* 2004 Dec 7;101(49):17312–5. PMID:15574496.

EPICURE Consortium, Genome-wide association analysis of genetic generalized epilepsies implicates susceptibility loci at 1q43, 2p16.1, 2q22.3 and 17q21.32. *Hum Mol Genet.* 2012 Dec 15;21(24):5359–72. doi: 10.1093/hmg/dds373. PMID:22949513.

Episkopou H, Draskovic I, Van Beneden A *et al.* Alternative Lengthening of Telomeres is characterized by reduced compaction of telomeric chromatin. *Nucleic Acids Res.* 2014 Apr;42(7):4391–405. doi: 10.1093/nar/gku114. PMID:24500201.

Eriksson M, Brown WT, Gordon LB *et al.* Recurrent de novo point mutations in lamin A cause Hutchinson-Gilford progeria syndrome. *Nature.* 2003 May 15;423(6937):293–8. Epub 2003 Apr 25. PMID: 12714972.

Estes ML, McAllister AK. Immune mediators in the brain and peripheral tissues in autism spectrum disorder. *Nat Rev Neurosci.* 2015 Aug; 16(8):469–86. doi: 10.1038/nrn3978. PMID:26189694.

Exner N, Lutz AK, Haass C, *et al.* Mitochondrial dysfunction in Parkinson's disease: molecular mechanisms and pathophysiological consequences. *EMBO J.* 2012 Jun 26;31(14):3038–62. doi: 10.1038/emboj.2012.170. PMID:22735187.

Fall T, Ingelsson E. Genome-wide association studies of obesity and metabolic syndrome. *Mol Cell Endocrinol.* 2014 Jan 25;382(1): 740–57. doi: 10.1016/j.mce.2012.08.018. PMID:22963884.

Farh KK, Marson A, Zhu J *et al.* Genetic and epigenetic fine mapping of causal autoimmune disease variants. *Nature.* 2015 Feb 19;518(7539): 337–43. doi: 10.1038/nature13835. PMID:25363779.

Farmer H, McCabe N, Lord CJ, *et al.* Targeting the DNA repair defect in BRCA mutant cells as a therapeutic strategy. *Nature.* 2005 Apr 14; 434(7035):917–21. PMID:15829967.

Fasolo A, Sessa C. Targeting mTOR pathways in human malignancies. *Curr Phar Des.* 2012;18(19):2766–77.

Fassone E, Rahman S. Complex I deficiency: clinical features, biochemistry and molecular genetics. *J Med Genet.* 2012 Sep;49(9):578–90. doi: 10.1136/jmedgenet-2012-101159. Review. Erratum in: *J Med Genet.* 2012 Oct;49(10):668. PMID:22972949.

Fedor MJ. Alternative splicing minireview series: combinatorial control facilitates splicing regulation of gene expression and enhances genome diversity. *J Biol Chem.* 2008 Jan 18;283(3):1209–10. PMID: 1802442.

Feoktistova K, Tuvshintogs E, Do A *et al.* Human eIF4E promotes mRNA restructuring by stimulating eIF4A helicase activity. *Proc Natl Acad Sci USA.* 2013 Aug 13;110(33):13339–44. doi: 10.1073/pnas.1303781110. PMID:23901100.

Ferguson-Smith MA. Karyotype-Phenotype correlations in gonadal dysgenesis and their bearing on the pathogenesis of malformations. *J Med Genet.* 1965 Jun;2(2):142–55. Review. PMID:14295659.

Fire A, Xu S, Montgomery MK, Kostas SA *et al.* Potent and specific genetic interference by double-stranded RNA in Caenorhabditis elegans. *Nature.* 1998 Feb 19;391(6669):806–11. PMID:9486653.

Fishel R, Lescoe MK, Rao MR *et al.* The human mutator gene homolog MSH2 and its association with hereditary nonpolyposis colon cancer. *Cell.* 1993 Dec 3;75(5):1027-38. PMID:8252616.

Fishel R. Mismatch Repair. *J Biol Chem.* 2015 Sep 9. pii: jbc.R115.660142. PMID:26354434.

Florio M, Albert M, Taverna E *et al.* Human-specific gene ARHGAP11B promotes basal progenitor amplification and neocortex expansion. *Science.* 2015 Mar 27;347(6229):1465–70. doi: 10.1126/science. aaa1975. PMID:25721503.

Fogel BL, Wexler E, Wahnich A *et al.* RBFOX1 regulates both splicing and transcriptional networks in human neuronal development. *Hum Mol Genet.* 2012 Oct 1;21(19):4171–86. doi: 10.1093/hmg/dds240. PMID: 22730494.

Fox BW, Tibbetts RS. Neurodegeneration: Problems at the nuclear pore. *Nature.* 2015 Sep 3;525(7567):36–7. doi: 10.1038/nature15208. PMID:26308896.

Frank C, Fallah M, Sundquist J *et al.* Population Landscape of familial cancer. *Sci Rep.* 2015 Aug 10;5:12891. doi: 10.1038/srep12891. PMID: 26256549.

Freeze HH, Chong JX, Bamshad MJ *et al.* Solving glycosylation disorders: fundamental approaches reveal complicated pathways. *Am J Hum Genet.* 2014 Feb 6;94(2):161–75. doi: 10.1016/j.ajhg.2013.10.024. Review. PMID:24507773.

Freibaum BD, Lu Y, Lopez-Gonzalez R et al. GGGGCC repeat expansion in C9orf72 compromises nucleocytoplasmic transport. *Nature.* 2015 Sep 3;525(7567):129–33. doi: 10.1038/nature14974. PMID:26308899.

Fromer M, Pocklington AJ, Kavanagh DH, Williams HJ *et al. De novo* mutations in schizophrenia implicate synaptic networks. *Nature.* 2014 Feb 13;506(7487):179–84. doi:10.1038/nature12929. PMID:24463507.

Fujita M, Kinoshita T. GPI-anchor remodeling: potential functions of GPI-anchors in intracellular trafficking and membrane dynamics. *Biochim Biophys Acta.* 2012 Aug;1821(8):1050–8. doi: 10.1016/j.bbalip.2012. 01.004. Review. PMID:22265715.

Gabel HW, Kinde B, Stroud H *et al.* Disruption of DNA-methylation-dependent long gene repression in Rett syndrome. *Nature.* 2015 Jun 4;522(7554):89–93. doi: 10.1038/nature14319. PMID:25762136.

Gad H, Koolmeister T, Jemth AS *et al.* MTH1 inhibition eradicates cancer by preventing sanitation of the dNTP pool. *Nature.* 2014 Apr 10;508(7495):215–21. doi: 10.1038/nature13181. PMID:24695224.

Gaj T, Gersbach CA, Barbas CF 3rd. ZFN, TALEN, and CRISPR/Cas-based methods for genome engineering. *Trends Biotechnol.* 2013 Jul;31(7): 397–405. doi: 10.1016/j.tibtech.2013.04.004. PMID:23664777.

Gajdusek C. Discussion on kuru, scrapie and the experimental kuru-like syndrome in chimpanzees. *Curr Top Microbiol Immunol.* 1967;40:59–63. PMID: 6069978.

Galligan JT, Martinez-Noël G, Arndt V *et al.* Proteomic analysis and identification of cellular interactors of the giant ubiquitin ligase HERC2. *J Proteome Res.* 2015 Feb 6;14(2):953–66. doi: 10.1021/pr501005v. PMID:25476789.

Gao AW, Cantó C, Houtkooper RH. Mitochondrial response to nutrient availability and its role in metabolic disease. *EMBO Mol Med.* 2014 May 1;6(5):580–9. doi: 10.1002/emmm.201303782. Review. PMID: 24623376.

Garraway LA, Lander ES. Lessons from the cancer genome. *Cell.* 2013 Mar 28;153(1):17–37. doi: 10.1016/j.cell.2013.03.002. Review. PMID: 23540688.

Gaspar HB, Qasim W, Davies EG *et al.* How I treat severe combined immunodeficiency. *Blood.* 2013 Nov 28;122(23):3749–58. doi: 10.1182/blood-2013-02-380105. PMID:24113871.

Gécz J, Shoubridge C, Corbett M. The genetic landscape of intellectual disability arising from chromosome X. *Trends Genet.* 2009 Jul;25(7):308–16. doi: 10.1016/j.tig.2009.05.002. Review. PMID:19556021.

Gendrel AV, Heard E. Noncoding RNAs and epigenetic mechanisms during X-chromosome inactivation. *Annu Rev Cell Dev Biol.* 2014;30: 561–80. doi: 10.1146/annurev-cellbio-101512-122415. Epub 2014 Jun 27. Review. PMID:25000994.

Genetic Modifiers of Huntington's Disease (GeM-HD) Consortium.

Genetics Home Reference. Accessed November 2015. http://ghr.nlm.nih.gov

Geschwind DH, Flint J. Genetics and genomics of psychiatric disease. *Science.* 2015 Sep 25;349(6255):1489–94. doi: 10.1126/science.aaa8954. PMID:26404826.

Geschwind DH, State MW. Gene hunting in autism spectrum disorder: on the path to precision medicine. *Lancet Neurol.* 2015 Nov;14(11): 1109–20. doi: 10.1016/S1474-4422(15)00044-7. PMID:25891009.

Giacinti C, Giordano A. RB and cell cycle progression. *Oncogene.* 2006 Aug 28;25(38):5220–7. Review. PMID:16936740.

Gidalevitz T, Stevens F, Argon Y. Orchestration of secretory protein folding by ER chaperones. *Biochim Biophys Acta.* 2013 Nov;1833(11): 2410–24. doi: 10.1016/j.bbamcr.2013.03.007. PMID:23507200.

Gilissen C, Hehir-Kwa JY, Thung DT *et al.* Genome sequencing identifies major causes of severe intellectual disability. *Nature.* 2014 Jul 17;511(7509):344–7. doi: 10.1038/nature13394. PMID:24896178.

Gil-Rodríguez MC, Deardorff MA, Ansari M, Tan CA *et al. De Novo* heterozygous mutations in SMC3 cause a range of Cornelia de Lange syndrome-overlapping phenotypes. *Hum Mutat.* 2015 Apr;36(4): 454–62. doi: 10.1002/humu.22761. PMID:25655089.

Girdea M, Dumitriu S, Fiume M, et al. PhenoTips: patient phenotyping software for clinical and research use. *Hum Mutat.* 2013 Aug;34(8): 1057–65. doi: 10.1002/humu.22347. PMID:23636887.

Gjoneska E, Pfenning AR, Mathys H *et al.* PMID:25693568 Conserved epigenomic signals in mice and humans reveal immune basis of Alzheimer's disease. *Nature.* 2015 Feb 19;518(7539):365–9. doi: 10.1038/nature 14252.

Glade Bender J, Verma A, Schiffman JD. Translating genomic discoveries to the clinic in pediatric oncology. *Curr Opin Pediatr.* 2015 Feb;27(1): 34–43. doi:10.1097/MOP.0000000000000172. Review. PMID:25502895.

Goedert M, Jakes R. Mutations causing neurodegenerative tauopathies. *Biochim Biophys Acta.* 2005 Jan 3;1739(2–3):240–50. Review. PMID:15615642.

Goedert M. Neurodegeneration. Alzheimer's and Parkinson's diseases: The prion concept in relation to assembled Aβ, tau, and α-synuclein. *Science.* 2015 Aug 7;349(6248):1255555. doi: 10.1126/science.1255555. Review. PMID:26250687.

Goldberg MS, Sharp PA. Pyruvate kinase M2-specific siRNA induces apoptosis and tumor regression. *J Exp Med.* 2012 Feb 13;209(2):217–24. doi: 10.1084/jem.20111487.

Goldstein JL, Brown MS. A century of cholesterol and coronaries: from plaques to genes to statins. *Cell.* 2015 Mar 26;161(1):161–72. doi: 10.1016/j.cell.2015.01.036. Review. PMID:25815993.

Gomes AV. Genetics of proteasome diseases. *Scientifica (Cairo)*. 2013;2013:637629. doi: 10.1155/2013/637629. PMID:24490108.

Gordon LB, Massaro J, D'Agostino RB, Sr, *et al.* Progeria Clinical Trials Collaborative. Impact of farnesylation inhibitors on survival in Hutchinson-Gilford progeria syndrome. *Circulation.* 2014 Jul 1; 130(1):27–34. doi:10.1161/CIRCULATIONAHA.113.008285 PMID: 24795390.

Gorinski N, Ponimaskin E. Palmitoylation of serotonin receptors. *Biochem Soc Trans.* 2013 Feb 1;41(1):89–94. doi: 10.1042/BST20120235. Review. PMID:23356264.

Goris A, Liston A. The immunogenetic architecture of autoimmune disease Cold Spring Harb Perspect Biol. 2012 Mar 1;4(3). pii: a007260. doi: 10.1101/cshperspect.a007260. PMID:22383754.

Gough NR. Focus issue: From genomic mutations to oncogenic pathways. *Sci Signal.* 2013 Mar 26;6(268):eg3. doi: 10.1126/scisignal.2004149. PMID:23532330.

Graham JM Jr, Schwartz CE. MED12 related disorders. *Am J Med Genet A.* 2013 Nov;161A(11):2734–40. doi: 10.1002/ajmg.a.36183. PMID: 24123922.

Gray SP, Di Marco E, Okabe J *et al.* NADPH oxidase 1 plays a key role in diabetes mellitus-accelerated atherosclerosis. *Circulation.* 2013 May 7;127(18):1888–902. doi: 10.1161/CIRCULATIONAHA.112.132159.

Green AJ, Sepp T, Yates JR. Clonality of tuberous sclerosis harmatomas shown by non-random X-chromosome inactivation. *Hum Genet.* 1996 Feb;97(2):240–3. PMID:8566961.

Green AJ, Smith M, Yates JR. Loss of heterozygosity on chromosome 16p13.3 in hamartomas from tuberous sclerosis patients. *Nat Genet.* 1994 Feb;6(2):193–6. PMID:8162074.

Green EK, Rees E, Walters JT, Smith KG *et al.* Copy number variation in bipolar disorder. *Mol Psychiatry.* 2015 Jan 6. doi: 10.1038/mp.2014. 174.l. PMID:25560756.

Greene CS, Krishnan A, Wong AK *et al.* Understanding multicellular function and disease with human tissue-specific networks. *Nat Genet.* 2015 Jun;47(6):569–76. doi: 10.1038/ng.3259. PMID:25915600.

Gregory AP, Dendrou CA, Attfield KE *et al.* TNF receptor 1 genetic risk mirrors outcome of anti-TNF therapy in multiple sclerosis. *Nature.* 2012 Aug 23;488(7412):508–11. doi: 10.1038/nature11307. PMID: 22801493.

Gross AM, Ideker T. Molecular networks in context. *Nat Biotechnol*. 2015 Jul;33(7):720–1. doi: 10.1038/nbt.3283. PMID:26154012.

Gruber AR, Martin G, Keller W *et al*. Means to an end: mechanisms of alternative polyadenylation of messenger RNA precursors. *Wiley Interdiscip Rev RNA*. 2014 Mar–Apr;5(2):183–96. doi: 10.1002/wrna.1206. Review. PMID:24243805.

Gruenbaum Y, Margalit A, Goldman RD *et al*. The nuclear lamina comes of age. *Nat Rev Mol Cell Biol*. 2005 Jan;6(1):21–31. Review. PMID: 15688064.

Grumach AS, Kirschfink M. Are complement deficiencies really rare? Overview on prevalence, clinical importance and modern diagnostic approach. *Mol Immunol*. 2014 Oct;61(2):110–7. doi: 10.1016/j.molimm.2014.06.030. Review. PMID:25037634.

GTEx Consortium. Human genomics. The Genotype-Tissue Expression (GTEx) pilot analysis: multitissue gene regulation in humans. *Science*. 2015 May 8;348(6235):648–60. doi: 10.1126/science.1262110. PMID:25954001.

Gubin MM, Artyomov MN, Mardis ER *et al*. Tumor neoantigens: building a framework for personalized cancer immunotherapy. *J Clin Invest*. 2015 Sep 1;125(9):3413–21. doi: 10.1172/JCI80008. PMID: 26258412.

Guerreiro R, Wojtas A, Bras J *et al*. TREM2 variants in Alzheimer's disease. *N Engl J Med*. 2013 Jan 10;368(2):117–27. doi: 10.1056/NEJMoa1211851. PMID:23150934.

Guerrini R, Dobyns WB. Malformations of cortical development: clinical features and genetic causes. *Lancet Neurol*. 2014 Jul;13(7):710–26. doi: 10.1016/S1474-4422(14)70040-7.Review. PMID:24932993.

Gupta S, Ellis SE, Ashar FN *et al*. Transcriptome analysis reveals dysregulation of innate immune response genes and neuronal activity-dependent genes in autism. *Nat Commun*. 2014 Dec 10;5:5748. doi: 10.1038/ncomms6748. PMID:25494366.

Gurel PS, Hatch AL, Higgs HN. Connecting the cytoskeleton to the endoplasmic reticulum and Golgi. *Curr Biol*. 2014 Jul 21;24(14):R660–72. doi: 10.1016/j.cub.2014.05.033. PMID:25050967.

Haeusler AR, Donnelly CJ, Periz G *et al*. C9orf72 nucleotide repeat structures initiate molecular cascades of disease. *Nature*. 2014 Mar 13;507(7491):195–200. doi: 10.1038/nature13124. PMID:24598541.

Hahamy A, Behrmann M, Malach R. The idiosyncratic brain: distortion of spontaneous connectivity patterns in autism spectrum disorder. *Nat Neurosci*.2015Feb;18(2):302–9.doi:10.1038/nn.3919.PMID:25599222.

Hall J, Trent S, Thomas KL, O'Donovan MC. Genetic risk for schizophrenia: convergence on synaptic pathways involved in plasticity. *Biol Psychiatry*. 2015 Jan 1;77(1):52–8. doi: 10.1016/j.biopsych.2014.07. 011. PMID:25152434.

Hall JA, Dominy JE, Lee Y *et al.* The sirtuin family's role in aging and age-associated pathologies. *J Clin Invest*. 2013 Mar;123(3):973–9. doi: 10.1172/JCI64094. PMID:23454760.

Halliday M, Mallucci GR. Targeting the unfolded protein response in neurodegeneration: A new approach to therapy. *Neuropharmacology*. 2014 Jan;76 (Pt A):169–74. doi: 10.1016/j.neuropharm.2013.08.034. PMID:24035917.

Hara M, Ohba C, Yamashita Y, *et al.* De novo SHANK3 mutation causes Rett syndrome-like phenotype in a female patient. *Am J Med Genet A*. 2015 Jul;167(7):1593–6. doi: 10.1002/ajmg.a.36775. PMID:25931020.

Hargreaves DC, Crabtree GR. ATP-dependent chromatin remodeling: genetics, genomics and mechanisms. *Cell Res*. 2011 Mar;21(3): 396–420. doi: 10.1038/cr.2011.32. PMID:21358755.

Harms M, Seale P. Brown and beige fat: development, function and therapeutic potential. *Nat Med*. 2013 Oct;19(10):1252–63. doi: 10.1038/ nm.3361. PMID:24100998.

Harris H. Chapter 1 Introduction, *Human Biochemical Genetics*. Page 8, 1996, Cambridge University Press.

Harris H. Concerning the origin of malignant tumours by Theodor Boveri. Translated and annotated by Henry Harris. Preface. *J Cell Sci*. 2008 Jan;121 Suppl 1:v-vi. doi: 10.1242/jcs.025759. PMID:18089651.

Harrow J, Frankish A, Gonzalez JM *et al.* GENCODE: the reference human genome annotation for The ENCODE Project. *Genome Res*. 2012 Sep;22(9):1760–74. doi: 10.1101/gr.135350.111. PMID:22955987.

Haslbeck M, Vierling E. A first line of stress defense: small heat shock proteins and their function in protein homeostasis. *J Mol Biol*. 2015 Apr 10;427(7):1537–48. doi: 10.1016/j.jmb.2015.02.002. PMID:25681016.

Hawkes CA, Härtig W, Kacza J *et al.* Perivascular drainage of solutes is impaired in the ageing mouse brain and in the presence of cerebral amyloid angiopathy. *Acta Neuropathol*. 2011 Apr;121(4):431–43. doi: 10.1007/s00401-011-0801-7. PMID:21259015.

Hawrylycz MJ, Lein ES, Guillozet-Bongaarts AL *et al.* An anatomically comprehensive atlas of the adult human brain transcriptome. *Nature*. 2012 Sep 20;489(7416):391–9. doi: 10.1038/nature11405. PMID 22996553.

Hazenberg MD, Verschuren MC, Hamann D *et al.* T cell receptor excision circles as markers for recent thymic emigrants: basic aspects, technical approach, and guidelines for interpretation. *J Mol Med* (Berl). 2001 Nov;79(11):631–40. PMID:11715066.

He Y, Ecker JR. Non-CG Methylation in the Human Genome. *Annu Rev Genomics Hum Genet.* 2015 Aug 24;16:55–77. doi: 10.1146/annurev-genom-090413-025437. PMID:26077819.

Heidari N, Phanstiel DH, He C *et al.* Genome-wide map of regulatory interactions in the human genome. *Genome Res.* 2014 Dec;24(12):1905–17. doi: 10.1101/gr.176586.114. PMID:25228660.

Heidenreich B, Rachakonda PS, Hemminki K *et al.* TERT promoter mutations in cancer development. *Curr Opin Genet Dev.* 2014 Feb;24:30–7. doi: 10.1016/j.gde.2013.11.005. Review. PMID:24657534.

Helleday T, Eshtad S, Nik-Zainal S. Mechanisms underlying mutational signatures in human cancers. *Nat Rev Genet.* 2014 Sep;15(9):585–98. doi: 10.1038/nrg3729. PMID:24981601.

Hemani G, Shakhbazov K, Westra HJ *et al.* Detection and replication of epistasis influencing transcription in humans. *Nature.* 2014 Apr 10;508(7495):249–53. doi: 10.1038/nature13005. PMID:24572353.

Heneka MT, Carson MJ, El Khoury J *et al.* Neuroinflammation in Alzheimer's disease. *Lancet Neurol.* 2015 Apr;14(4):388–405. doi: 10.1016/S1474-4422(15)70016-5. Review. PMID:25792098.

Heneka MT, Golenbock DT, Latz E. Innate immunity in Alzheimer's disease. *Nat Immunol.* 2015 Mar;16(3):229–36. doi: 10.1038/ni.3102. Review. PMID:25689443.

Heride C, Urbé S, Clague MJ. *Curr Biol.* 2014 Mar 17;24(6):R215–20. doi: 10.1016/j.cub.2014.02.002. Review. PMID:24650902.

Hickman SE, El Khoury J. TREM2 and the neuroimmunology of Alzheimer's disease. *Biochem Pharmacol.* 2014 Apr 15;88(4):495–8. doi: 10.1016/j.bcp.2013.11.021. PMID:24355566.

Higginbotham HR, Gleeson JG. The centrosome in neuronal development. *Trends Neurosci.* 2007 Jun;30(6):276–83. PMID:17420058.

. Higgs DR, Goodbourn SE, Lamb J *et al.* Alpha-thalassaemia caused by a polyadenylation signal mutation. *Nature.* 1983 Nov 24–30;306 (5941):398–400. PMID:6646217.

Hindorff LA, Gillanders EM, Manolio TA. Genetic architecture of cancer and other complex diseases: lessons learned and future directions. *Carcinogenesis.* 2011 Jul;32(7):945–54. doi: 10.1093/carcin/bgr056. Review. PMID:21459759.

Hipp MS, Park SH, Hartl FU. Proteostasis impairment in protein-misfolding and -aggregation diseases. *Trends Cell Biol.* 2014 Sep;24(9):506–14. doi: 10.1016/j.tcb.2014.05.003. Review. PMID:24946960.

Hoekstra AS, Bayley JP. The role of complex II in disease. Biochim Biophys Acta. 2013 May;1827(5):543–51. doi: 10.1016/j.bbabio.2012.11.005. PMID:23174333.

Holderfield M, Deuker MM, McCormick F *et al.* Targeting RAF kinases for cancer therapy: BRAF-mutated melanoma and beyond. *Nat Rev Cancer.* 2014 Jul;14(7):455–67. doi: 10.1038/nrc3760. Review. PMID:24957944.

Holdt LM, Teupser D. Long noncoding RNA-MicroRNA pathway controlling nuclear factor IA, a novel atherosclerosis modifier gene. *Arterioscler Thromb Vasc Biol.* 2015 Jan;35(1):7–8. doi: 10.1161/ATVBAHA. 114.304485. PMID:25520519.

Holmström KM, Finkel T. Cellular mechanisms and physiological consequences of redox-dependent signalling. *Nat Rev Mol Cell Biol.* 2014 Jun;15(6):411–21. doi: 10.1038/nrm3801. Review. PMID: 24854789.

Holoch D, Moazed D. RNA-mediated epigenetic regulation of gene expression. *Nat Rev Genet.* 2015 Feb;16(2):71–84. doi: 10.1038/nrg3863. Review. PMID:25554358.

Holohan B, Wright WE, Shay JW. Cell biology of disease: Telomeropathies: an emerging spectrum disorder. *J Cell Biol.* 2014 May 12;205(3): 289–99. doi: 10.1083/jcb.201401012. PMID:24821837.

Hoppins S, Nunnari J. Cell Biology. Mitochondrial dynamics and apoptosis — the ER connection. *Science.* 2012 Aug 31;337(6098): 1052–4. doi: 10.1126/science.1224709. PMID:22936767.

Horn S, Figl A, Rachakonda PS *et al.* TERT promoter mutations in familial and sporadic melanoma. *Science.* 2013 Feb 22;339(6122):959–61. doi: 10.1126/science.1230062. PMID:23348503.

Horsthemke B. Mechanisms of imprint dysregulation. *Am J Med Genet C Semin Med Genet.* 2010 Aug 15;154C(3):321–8. doi: 10.1002/ ajmg.c.30269. Review. PMID:20803654.

Hottman DA, Chernick D, Cheng S et al. HDL and cognition in neurodegenerative disorders. *Neurobiol Dis.* 2014 Dec;72 Pt A:22–36. doi: 10.1016/j.nbd.2014.07.015. PMID:25131449.

Hovatta I. Genetics: dynamic cellular aging markers associated with major depression. *Curr Biol.* 2015 May 18;25(10):R409–11. doi: 10.1016/j. cub.2015.03.036. PMID:25989078.

Hsu PD, Lander ES, Zhang F. Development and applications of CRISPR-Cas9 for genome engineering. *Cell*. 2014 Jun 5;157(6):1262–78. doi: 10.1016/j.cell.2014.05.010. Review. PMID:24906146.

Hu H, Haas SA, Chelly J, Van Esch H *et al*. X-exome sequencing of 405 unresolved families identifies seven novel intellectual disability genes. *Mol Psychiatry*. 2015 Feb 3. doi: 10.1038/mp.2014.193. PMID:25644381.

Hu WF, Chahrour MH, Walsh CA. The diverse genetic landscape of neurodevelopmental disorders. *Annu Rev Genomics Hum Genet*. 2014; 15:195–213. doi: 10.1146/annurev-genom-090413-025600. Review. PMID:25184530.

Huang Y, Mahley RW. Apolipoprotein E: structure and function in lipid metabolism, neurobiology, and Alzheimer's diseases. *Neurobiol Dis*. 2014 Dec;72 Pt A:3–12. doi: 10.1016/j.nbd.2014.08.025. PMID:25173806.

Huether R, Dong L, Chen X, Wu G, Parker M. The landscape of somatic mutations in epigenetic regulators across 1,000 paediatric cancer genomes. *Nat Commun*. 2014 Apr 8;5:3630. doi: 10.1038/ncomms4630. PMID:24710217.

Hunter T. Tyrosine phosphorylation: thirty years and counting. *Curr Opin Cell Biol*. 2009 Apr;21(2):140–6. doi: 10.1016/j.ceb.2009.01.028. Review. PMID:19269802.

Identification of Genetic Factors that Modify Clinical Onset of Huntington's Disease. *Cell*. 2015 Jul 30;162(3):516–26. doi: 10.1016/j.cell.2015.07.003. PMID:26232222.

Inobe T, Matouschek A. Paradigms of protein degradation by the proteasome. *Curr Opin Struct Biol*. 2014 Feb;24:156–64. doi: 10.1016/j.sbi.2014.02.002.

Irimia M, Weatheritt RJ, Ellis JD *et al*. A highly conserved program of neuronal microexons is misregulated in autistic brains. *Cell*. 2014 Dec 18;159(7):1511–23. doi: 10.1016/j.cell.2014.11.035. PMID:25525873.

Ishida T, Ishida M, Tashiro S *et al*., Role of DNA damage in cardiovascular disease. *Circ J*. 2014;78(1):42–50. Review. PMID:24334614.

Iwafuchi-Doi M, Zaret KS. Pioneer transcription factors in cell reprogramming. *Genes Dev*. 2014 Dec 15;28(24):2679–92. doi: 10.1101/gad.253443.114. Review. PMID:25512556.

Jackson S, Xiong Y. CRL4s: the CUL4-RING E3 ubiquitin ligases. *Trends Biochem Sci*. 2009 Nov;34(11):562–70. doi: 10.1016/j.tibs.2009.07.002. 7. Review. PMID:19818632.

Jackson SP, Bartek J. The DNA-damage response in human biology and disease. *Nature.* 2009 Oct 22;461(7267):1071–8. doi: 10.1038/nature08467. Review. PMID:19847258.

Jaeken J. Congenital disorders of glycosylation. *Handb Clin Neurol.* 2013;113:1737–43. doi: 10.1016/B978-0-444-59565-2.00044-7. Review. PMID:23622397.

Jaffe AE, Shin J, Collado-Torres L *et al.* Developmental regulation of human cortex transcription and its clinical relevance at single base resolution. *Nat Neurosci.* 2015 Jan;18(1):154–61. doi: 10.1038/nn.3898. PMID:25501035.

Jarome TJ, Thomas JS, Lubin FD. The epigenetic basis of memory formation and storage. *Prog Mol Biol Transl Sci.* 2014;128:1–27. doi: 10.1016/B978-0-12-800977-2.00001-2. PMID:25410539.

Jaunmuktane Z, Mead S, Ellis M *et al.* Evidence for human transmission of amyloid-β pathology and cerebral amyloid angiopathy. *Nature.* 2015 Sep 10;525(7568):247–50. doi: 10.1038/nature15369. PMID: 26354483.

Javle M, Curtin NJ. The role of PARP in DNA repair and its therapeutic exploitation. *Br J Cancer.* 2011 Oct 11;105(8):1114–22. doi: 10.1038/bjc.2011.382. PMID:21989215.

Jedrychowski MP, Wrann CD, Paulo JA *et al.* Detection and Quantitation of Circulating Human Irisin by Tandem Mass Spectrometry. *Cell Metab.* 2015 Oct 6;22(4):734–40. doi: 10.1016/j.cmet.2015.08.001. PMID:26278051.

Ji Y, Eichler EE, Schwartz S *et al.* Structure of chromosomal duplicons and their role in mediating human genomic disorders. *Genome Res.* 2000 May;10(5):597–610. Review. PMID:10810082.

Ji Z, Luo W, Li W, Hoque M *et al.* Transcriptional activity regulates alternative cleavage and polyadenylation. *Mol Syst Biol.* 2011 Sep 27;7:534. doi: 10.1038/msb.2011.69. PMID:21952137.

Jin G, Pirozzi CJ, Chen LH *et al.* Mutant IDH1 is required for IDH1 mutated tumor cell growth. *Oncotarget.* 2012 Aug;3(8):774–82. PMID:22885298.

Johansen T, Lamark T. Selective autophagy goes exclusive. *Nat Cell Biol.* 2014 May;16(5):395–7. PMID:24914435.

Jonsson T, Stefansson K. TREM2 and neurodegenerative disease. *N Engl J Med.* 2013 Oct 17;369(16):1568–9. doi: 10.1056/NEJMc1306509. PMID:24131183.

Jørgensen AB, Frikke-Schmidt R, Nordestgaard BG *et al.* Loss-of-function mutations in APOC3 and risk of ischemic vascular disease. *N Engl J Med.* 2014 Jul 3;371(1):32–41. doi:10.1056/NEJMoa1308027.PMID:24941082.

Jorgenson LA, Newsome WT, Anderson DJ *et al.* The BRAIN Initiative: developing technology to catalyse neuroscience discovery. *Philos Trans R Soc Lond B Biol Sci.* 2015 May 19;370(1668). pii: 20140164. doi: 10.1098/rstb.2014.0164. PMID:25823863.

Joshi G, Chi Y, Huang Z, Wang Y. Aβ-induced Golgi fragmentation in Alzheimer's disease enhances Aβ production. *Proc Natl Acad Sci USA.* 2014 Apr 1;111(13):E1230–9. doi: 10.1073/pnas.1320192111. PMID: 24639524.

Jovičić A, Mertens J, Boeynaems S *et al.* Modifiers of C9orf72 dipeptide repeat toxicity connect nucleocytoplasmic transport defects to FTD/ALS. *Nat Neurosci.* 2015 Aug 26;18(9):1226–9. doi: 10.1038/nn.4085. PMID:26308983.

Jucker M, Walker LC. Neurodegeneration: Amyloid-β pathology induced in humans. *Nature.* 2015 Sep 10;525(7568):193–4. doi: 10.1038/525193a. PMID:26354478.

Kaelin WG Jr. The concept of synthetic lethality in the context of anticancer therapy. *Nat Rev Cancer.* 2005 Sep;5(9):689–98. Review. PMID:16110319.

Kamiya Y, Satoh T, Kato K. Molecular and structural basis for N-glycan-dependent determination of glycoprotein fates in cells. *Biochim Biophys Acta.* 2012 Sep;1820(9):1327–37. doi: 10.1016/j.bbagen.2011.12.017. Review. PMID:22240168.

Kandel ER. The molecular biology of memory storage: a dialogue between genes and synapses. *Science.* 2001 Nov 2;294(5544):1030–8. Review. PMID:11691980.

Kandoth C, McLellan MD, Vandin F *et al.* Mutational landscape and significance across 12 major cancer types. *Nature.* 2013 Oct 17; 502(7471):333–9. doi: 10.1038/nature12634. PMID:24132290.

Kanduri C. Long noncoding RNAs: Lessons from genomic imprinting. *Biochim Biophys Acta.* 2015 May 22. pii: S1874–9399(15)00104–2. doi: 10.1016/j.bbagrm.2015.05.006. [Epub ahead of print] Review. PMID:26004516.

Karch CM, Goate AM. Alzheimer's disease risk genes and mechanisms of disease pathogenesis. *Biol Psychiatry.* 2015 Jan 1;77(1):43–51. doi: 10.1016/j.biopsych.2014.05.006. PMID:4951455.

Kassner U, Salewsky B, Wühle-Demuth M *et al.* Severe hypertriglyceride-mia in a patient heterozygous for a lipoprotein lipase gene allele with two novel missense variants. *Eur J Hum Genet.* 2015 Sep;23(9): 1259–61. doi: 10.1038/ejhg.2014.295. PMID:25585702.

Katahira J. mRNA export and the TREX complex. *Biochim Biophys Acta.* 2012 Jun;1819(6):507–13. doi: 10.1016/j.bbagrm.2011.12.001. Review. PMID:22178508.

Kathiresan S, Willer CJ, Peloso GM *et al.* Common variants at 30 loci contribute to polygenic dyslipidemia. *Nat Genet.* 2009 Jan;41(1):56–65. doi: 10.1038/ng.291. PMID:19060906.

Kato M, Das S, Petras K *et al.* Polyalanine expansion of ARX associated with cryptogenic West syndrome. *Neurology.* 2003 Jul 22;61(2): 267–76. PMID:12874418.

Kavanagh DH, Tansey KE, O'Donovan MC, Owen MJ. Schizophrenia genetics: emerging themes for a complex disorder. *Mol Psychiatry.* 2015 Feb;20(1):72–6. doi: 10.1038/mp.2014.148. PMID:25385368.

Kemper TL, Bauman ML. The contribution of neuropathologic studies to the understanding of autism. *Neurol Clin.* 1993 Feb;11(1):175–87. PMID: 8441369.

Kermanizadeh A, Chauché C, Brown DM *et al.* The role of intracellular redox imbalance in nanomaterial induced cellular damage and genotoxicity: a review. *Environ Mol Mutagen.* 2015 Mar;56(2):111–24. doi: 10.1002/em.21926. Review. PMID:25427446.

Khoury GA, Baliban RC, Floudas CA. Proteome-wide post-translational modification statistics: frequency analysis and curation of the swissprot database. *Sci Rep.* 2011 Sep 13;1. pii: srep00090. PMID: 22034591.

Kim CA, Berg JM. A 2.2 A resolution crystal structure of a designed zinc finger protein bound to DNA. *Nat Struct Biol.* 1996 Nov;3(11):940–5. PMID:8901872.

Kim H, D'Andrea AD. Regulation of DNA cross-link repair by the Fanconi anemia/BRCA pathway. *Genes Dev.* 2012 Jul 1;26(13):1393–408. doi: 10.1101/gad.195248.112. Review. PMID:22751496.

Kim MS, Pinto SM, Getnet D *et al.* A draft map of the human proteome. *Nature.* 2014 May 29;509(7502):575–81. doi: 10.1038/nature13302. PMID:24870542.

Kim YE, Hipp MS, Bracher A *et al.* Molecular chaperone functions in protein folding and proteostasis. *Annu Rev Biochem.* 2013;82:323–55. doi: 10.1146/annurev-biochem-060208-092442. PMID:23746257.

Kim YJ, Wilson DM 3rd. Overview of base excision repair biochemistry. *Curr Mol Pharmacol.* 2012 Jan;5(1):3–13. PMID:22122461.

Kinde B, Gabel HW, Gilbert CS, *et al.* Reading the unique DNA methylation landscape of the brain: Non-CpG methylation, hydroxymethylation, and MeCP2. *Proc Natl Acad Sci USA.* 2015 Jun 2;112(22):6800–6. doi: 10.1073/pnas.1411269112. 2015. PMID:25739960.

King IF, Yandava CN, Mabb AM *et al.* Topoisomerases facilitate transcription of long genes linked to autism. *Nature.* 2013 Sep 5;501(7465): 58–62. doi: 10.1038/nature12504. PMID:23995680.

Kirov G. CNVs in neuropsychiatric disorders. *Hum Mol Genet.* 2015 Oct 15;24(R1):R45–9. doi: 10.1093/hmg/ddv253. PMID:26130694.

Kleefstra T, Schenck A, Kramer JM *et al.* The genetics of cognitive epigenetics. *Neuropharmacology.* 2014 May;80:83–94. doi: 10.1016/j. neuropharm.2013.12.025. Epub 2014 Jan 13. PMID:24434855.

Klei L, Sanders SJ, Murtha MT *et al.* Common genetic variants, acting additively, are a major source of risk for autism. *Mol Autism.* 2012 Oct 15;3(1):9. doi: 10.1186/2040-2392-3-9. PMID:23067556.

Klengel T, Binder EB. Epigenetics of stress-related psychiatric disorders and gene × environment interactions. *Neuron.* 2015 Jun 17;86(6):1343–57. doi: 10.1016/j.neuron.2015.05.036. Review. PMID:26087162.

Knight JC. Approaches for establishing the function of regulatory genetic variants involved in disease. *Genome Med.* 2014 Oct 31;6(10):92. doi: 10.1186/s13073-014-0092-4. eCollection 2014. PMID:25473428.

Knight JC. Genomic modulators of the immune response. *Trends Genet.* 2013 Feb;29(2):74–83. doi: 10.1016/j.tig.2012.10.006. Review. PMID: 23122694.

Knouse KA, Amon A. Cell biology: the micronucleus gets its big break. *Nature.* 2015 Jun 11;522(7555):162–3. doi: 10.1038/nature14528. PMID:26017308.

Kohli RM, Zhang Y. TET enzymes, TDG and the dynamics of DNA demethylation. *Nature.* 2013 Oct 24;502(7472):472–9. doi: 10.1038/ nature12750. Review. PMID:24153300.

Komander D. The emerging complexity of protein ubiquitination. *Biochem Soc Trans.* 2009 Oct;37(Pt 5):937–53. doi: 10.1042/BST0370937. Review. PMID:19754430.

Kondo A, Shahpasand K, Mannix R *et al.* Antibody against early driver of neurodegeneration cis P-tau blocks brain injury and tauopathy. *Nature.* 2015 Jul 23;523(7561):431–6. doi:10.1038/nature14658. PMID:26176913.

Kong CM, Lee XW, Wang X. Telomere shortening in human diseases. *FEBS J.* 2013 Jul;280(14):3180–93. doi: 10.1111/febs.12326. PMID: 23647631.

Kosho T, Miyake N Carey JC. Coffin-Siris syndrome and related disorders involving components of the BAF (mSWI/SNF) complex: historical review and recent advances using next generation sequencing. *Am J Med Genet C Semin Med Genet.* 2014 Sep;166C(3):241–51. doi: 10.1002/ajmg.c.31415. PMID:25169878.

Kovacs GG, Rozemuller AJ, van Swieten JC, Gelpi E *et al.* Neuropathology of the hippocampus in FTLD-tau with Pick bodies: a study of the BrainNet Europe Consortium. *Neuropathol Appl Neurobiol.* 2013 Feb;39(2):166–78. doi: 10.1111/j.1365-2990.2012.01272.x. PMID: 22471883.

Koyama R, Ikegaya Y. Microglia in the pathogenesis of autism spectrum disorders. *Neurosci Res.* 2015 Jun 25. pii: S0168-0102(15)00162-5. doi: 10.1016/j.neures.2015.06.005. Review. PMID:26116891.

Krishnan R, Tsubery H, Proschitsky MY *et al.* A bacteriophage capsid protein provides a general amyloid interaction motif (GAIM) that binds and remodels misfolded protein assemblies. *J Mol Biol.* 2014 Jun 26;426(13):2500–19. doi:10.1016/j.jmb.2014.04.015.PMID:24768993.

Kropski JA, Lawson WE, Young LR *et al.* Genetic studies provide clues on the pathogenesis of idiopathic pulmonary fibrosis. *Dis Model Mech.* 2013 Jan;6(1):9–17. doi: 10.1242/dmm.010736. Review. PMID: 23268535.

Kumaran R, Cookson MR. Pathways to Parkinsonism redux: convergent pathobiological mechanisms in genetics of Parkinson's disease. *Hum Mol Genet.* 2015 Jun 22. pii: ddv236. PMID:26101198.

Kupfer GM. Fanconi anemia: a signal transduction and DNA repair pathway. *Yale J Biol Med.* 2013 Dec 13;86(4):491–7. Review. PMID:24348213.

Kurian MA, Gissen P, Smith M *et al.* The monoamine neurotransmitter disorders: an expanding range of neurological syndromes. *Lancet Neurol.* 2011 Aug;10(8):721–33. doi: 10.1016/S1474-4422(11)70141–7. Review. PMID:21777827.

Kwan A, Puck JM. History and current status of newborn screening for severe combined immunodeficiency. *Semin Perinatol.* 2015 Apr;39(3):194–205. doi: 10.1053/j.semperi.2015.03.004 PMID:25937517.

Labrecque N, Baldwin T, Lesage S. Molecular and genetic parameters defining T-cell clonal selection. *Immunol Cell Biol.* 2011 Jan;89(1): 16–26. doi: 10.1038/icb.2010.119. PMID:20956988.

Lander ES *et al.* Initial sequencing and analysis of the human genome. *Nature.* 2001 Feb 15;409(6822):860–921. PMID:11237011.

Lange LA, Hu Y, Zhang H *et al.* Whole-exome sequencing identifies rare and low-frequency coding variants associated with LDL cholesterol. *Am J Hum Genet.* 2014 Feb 6;94(2):233–45. doi: 10.1016/j.ajhg.2014.01.010. PMID:24507775.

Längst G, Manelyte L. Chromatin remodelers: From function to dysfunction. *Genes (Basel).* 2015 Jun 12;6(2):299–324. doi: 10.3390/genes6020299. PMID:26075616.

Lanzer P, Boehm M, Sorribas V *et al.* Medial vascular calcification revisited: review and perspectives. *Eur Heart J.* 2014 Jun 14;35(23):1515–25. doi: 10.1093/eurheartj/ehu163. PMID:24740885.

Larance M, Lamond AI. Multidimensional proteomics for cell biology. *Nat Rev Mol Cell Biol.* 2015 May;16(5):269–80. doi: 10.1038/nrm3970. PMID:25857810.

Lawrence MS, Stojanov P, Mermel CH *et al.* Discovery and saturation analysis of cancer genes across 21 tumour types. *Nature.* 2014 Jan 23;505(7484):495–501. doi: 10.1038/nature12912. PMID:24390350.

Leach FS, Nicolaides NC, Papadopoulos N *et al.* Mutations of a mutS homolog in hereditary nonpolyposis colorectal cancer. *Cell.* 1993 Dec 17;75(6):1215–25. PMID:8261515.

Leal G, Comprido D, Duarte CB. BDNF-induced local protein synthesis and synaptic plasticity. *Neuropharmacology.* 2014 Jan;76 Pt C: 639–56. doi: 10.1016/j.neuropharm.2013.04.005. PMID:23602987.

Leder P. Discontinuous genes. *N Engl J Med.* 1978 May 11;298(19): 1079–81. PMID:643015.

Ledford H. Cancer: The Ras renaissance. *Nature.* 2015 Apr 16; 520(7547):278–80. doi: 10.1038/520278a. PMID:25877186.

Lee JT, Bartolomei MS. X-inactivation, imprinting, and long noncoding RNAs in health and disease. *Cell.* 2013 Mar 14;152(6):1308–23. doi: 10.1016/j.cell.2013.02.016. PMID:23498939.

Leibowitz ML, Zhang CZ, Pellman D. Chromothripsis: a new mechanism for rapid karyotype evolution. *Annu Rev Genet.* 2015 Oct 6. PMID:26442848.

Lejeune F, Maquat LE. Mechanistic links between nonsense-mediated mRNA decay and pre-mRNA splicing in mammalian cells. *Curr Opin Cell Biol.* 2005 Jun;17(3):309–15. Review. PMID:15901502.

Lemmon MA, Schlessinger J. Cell signaling by receptor tyrosine kinases. *Cell.* 2010 Jun 25;141(7):1117–34. doi: 10.1016/j.cell.2010.06.011. Review. PMID:20602996.

Leslie M. Cleanup crew. *Science.* 2015 Mar 6;347(6226):1058–9, 1061. doi: 10.1126/science.347.6226.1058. PMID:25745143.

Levine B, Packer M, Codogno P. Development of autophagy inducers in clinical medicine. *J Clin Invest*. 2015 Jan;125(1):14–24. doi: 10.1172/JCI73938. PMID:25654546.

Levinthal, Cyrus. Are there pathways for protein folding? *Journal de Chimie Physique et de Physico-Chimie Biologique* 1968;65: 44–45.

Li D, Guo B, Wu H *et al*. TET Family of Dioxygenases: Crucial Roles and Underlying Mechanisms. *Cytogenet Genome Res*. 2015. PMID: 26302812.

Li GM. Mechanisms and functions of DNA mismatch repair. *Cell Res*. 2008 Jan;18(1):85–98. Review. PMID:18157157.

Li W, Lee MH, Henderson L *et al*. Human endogenous retrovirus-K contributes to motor neuron disease. *Sci Transl Med*. 2015 Sep 30;7(307):307ra153. doi: 10.1126/scitranslmed.aac8201. PMID:26424568.

Li X, Battle A, Karczewski KJ, Zappala Z *et al*. Transcriptome sequencing of a large human family identifies the impact of rare noncoding variants. *Am J Hum Genet*. 2014 Sep 4;95(3):245–56. doi: 10.1016/j.ajhg.2014.08.004. PMID:25192044.

Li YI, Sanchez-Pulido L, Haerty W *et al*. RBFOX and PTBP1 proteins regulate the alternative splicing of micro-exons in human brain transcripts. *Genome Res*. 2015 Jan;25(1):1–13. doi: 10.1101/gr.181990.114. PMID:25524026.

Licatalosi DD, Darnell RB. RNA processing and its regulation: global insights into biological networks. *Nat Rev Genet*. 2010 Jan;11(1): 75–87. doi: 10.1038/nrg2673. Review. PMID:20019688.

Lidell ME, Betz MJ, Enerbäck S. Two types of brown adipose tissue in humans. *Adipocyte*. 2014 Jan 1;3(1):63–6. doi: 10.4161/adip.26896. PMID:24575372.

Lieber MR. The mechanism of double-strand DNA break repair by the nonhomologous DNA end-joining pathway. *Annu Rev Biochem*. 2010; 79:181–211. doi: 10.1146/annurev.biochem.052308.093131. PMID: 20192759.

Lindhurst MJ, Parker VE, Payne F *et al*. Mosaic overgrowth with fibroadipose hyperplasia is caused by somatic activating mutations in PIK3CA. *Nat Genet*. 2012 Jun 24;44(8):928–33. doi: 10.1038/ng.2332. PMID 22729222.

Lindhurst MJ, Wang JA, Bloomhardt HM *et al*. AKT1 gene mutation levels are correlated with the type of dermatologic lesions in patients with Proteus syndrome. *J Invest Dermatol*. 2014 Feb;134(2):543–6. doi: 10.1038/jid.2013.312. PMID:23884311.

Lister R, Mukamel EA, Nery JR *et al.* Global epigenomic reconfiguration during mammalian brain development. *Science.* 2013 Aug 9;341(6146): 1237905. doi: 10.1126/science.1237905. PMID:23828890.

Liston A, Gray DH. Homeostatic control of regulatory T cell diversity. *Nat Rev Immunol.* 2014 Mar;14(3):154–65. doi: 10.1038/nri3605. Review. PMID:24481337.

Litchfield DW, Shilton BH, Brandl CJ *et al.* Pin1: Intimate involvement with the regulatory protein kinase networks in the global phosphorylation landscape. *Biochim Biophys Acta.* 2015 Mar 10. pii: S0304-4165(15)00081-1. doi: 10.1016/j.bbagen. 2015.02.018. Review. PMID:25766872.

Lodato MA, Woodworth MB, Lee S, *et al.* Somatic mutation in single human neurons tracks developmental and transcriptional history. *Science.* 2015 Oct 2;350(6256):94–8. doi: 10.1126/science.aab1785. PMID:26430121.

Long J, Tokhunts R, Old WM *et al.* Identification of a family of fatty-acid-speciated sonic hedgehog proteins, whose members display differential biological properties. *Cell Rep.* 2015 Mar 3;10(8):1280–7. doi: 10.1016/j.celrep.2015.01.058. PMID:25732819.

Lubs HA, Stevenson RE, Schwartz CE. Fragile X and X-linked intellectual disability: four decades of discovery. *Am J Hum Genet.* 2012 Apr 6; 90(4):579–90. doi: 10.1016/j.ajhg.2012.02.018. PMID:22482801.

Lunnon K, Smith R, Hannon E *et al.* Methylomic profiling implicates cortical deregulation of ANK1 in Alzheimer's disease. *Nat Neurosci.* 2014 Sep;17(9):1164–70. doi: 10.1038/nn.3782. PMID:25129077.

Lyon MF. Gene action in the X-chromosome of the mouse (*Mus Musculus* L). *Nature.* 1961 Apr 22;190:372–3. PMID:13764598.

MacArthur DG, Manolio TA, Dimmock DP *et al.* Guidelines for investigating causality of sequence variants in human disease. *Nature.* 2014 Apr 24;508(7497):469–76. doi: 10.1038/nature13127. PMID:24759409.

Mackall CL, Merchant MS, Fry TJ. Immune-based therapies for childhood cancer. *Nat Rev Clin Oncol.* 2014 Dec;11(12):693–703. doi: 10.1038/ nrclinonc.2014.177. Epub 2014 Oct 28. Review. PMID:25348789.

Mahadevan S, Wen S, Wan YW *et al.* NLRP7 affects trophoblast lineage differentiation, binds to overexpressed YY1 and alters CpG methylation. *Hum Mol Genet.* 2014 Feb 1;23(3):706–16. doi: 10.1093/hmg/ddt457. PMID:24105472.

Maity T, Fuse N, Beachy PA. Molecular mechanisms of Sonic hedgehog mutant effects in holoprosencephaly. *Proc Natl Acad Sci USA.* 2005 Nov 22;102(47):17026–31. PMID:16282375.

Malek TR, Bayer AL. Tolerance, not immunity, crucially depends on IL-2. *Nat Rev Immunol.* 2004 Sep;4(9):665–74. PMID:15343366.

Manolio TA, Collins FS, Cox NJ *et al.* Finding the missing heritability of complex diseases. *Nature.* 2009 Oct 8;461(7265):747–53. doi: 10.1038/nature08494. PMID:19812666.

Mapstone M, Cheema AK, Fiandaca MS *et al.* Plasma phospholipids identify antecedent memory impairment in older adults. *Nat Med.* 2014 Apr;20(4):415–8. doi: 10.1038/nm.3466. Epub 2014 Mar 9. PMID:24608097.

Martin SA, McCarthy A, Barber LJ *et al.* Methotrexate induces oxidative DNA damage and is selectively lethal to tumour cells with defects in the DNA mismatch repair gene MSH2. *EMBO Mol Med.* 2009 Sep; 1(6–7):323–37. doi: 10.1002/emmm.200900040. PMID:20049736.

Martincorena I, Campbell PJ. Somatic mutation in cancer and normal cells. *Science.* 2015 Sep 25;349(6255):1483–9. doi: 10.1126/science. aab4082. Epub 2015 Sep 24. Review. PMID:26404825.

Matamales M. Neuronal activity-regulated gene transcription: how are distant synaptic signals conveyed to the nucleus? F1000Res. 2012 Dec 19;1:69. doi: 10.12688/f1000research.1-69.v1. Review. PMID:24358817.

Matera AG, Wang Z. A day in the life of the spliceosome. *Nat Rev Mol Cell Biol.* 2014 Feb;15(2):108–21. doi: 10.1038/nrm3742. PMID:24452469.

Matsuoka S, Ballif BA, Smogorzewska A *et al.* ATM and ATR substrate analysis reveals extensive protein networks responsive to DNA damage. *Science.* 2007 May 25;316(5828):1160–6. PMID:17525332.

Mauri C, Bosma A. Immune regulatory function of B cells. *Annu Rev Immunol.* 2012;30:221–41. doi: 10.1146/annurev-immunol-020711-074934. PMID:22224776.

Mayer MP. Hsp70 chaperone dynamics and molecular mechanism. *Trends Biochem Sci.* 2013 Oct;38(10):507–14. doi: 10.1016/j.tibs.2013.08.001. PMID:24012426.

Mazurek S, Drexler HC, Troppmair J *et al.* Regulation of pyruvate kinase type M2 by A-Raf: a possible glycolytic stop or go mechanism. *Anticancer Res.* 2007 Nov-Dec;27(6B):3963–71.

McDonnell E, Peterson BS, Bomze HM *et al.* SIRT3 regulates progression and development of diseases of aging. *Trends Endocrinol Metab.* 2015 Sep;26(9):486–92. doi: 10.1016/j.tem.2015.06.001. Review. PMID: 26138757.

McGranahan N, Favero F, de Bruin EC, Birkbak NJ, Szallasi Z, Swanton C. Clonal status of actionable driver events and the timing of mutational processes in cancer evolution. *Sci Transl Med.* 2015 Apr 15;7(283):283ra54. doi: 10.1126/scitranslmed.aaa1408. PMID:25877892.

McKinnon PJ. ATM and the molecular pathogenesis of ataxia telangiectasia. *Annu Rev Pathol.* 2012;7:303–21. doi: 10.1146/annurev-pathol-011811-132509. Epub 2011 Oct 24. Review. PMID:22035194.

McLornan DP, List A, Mufti GJ. Applying synthetic lethality for the selective targeting of cancer. *N Engl J Med.* 2014 Oct 30;371(18):1725–35. doi: 10.1056/NEJMra1407390. PMID:25354106.

Meador CB, Lovly CM. Liquid biopsies reveal the dynamic nature of resistance mechanisms in solid tumors. *Nat Med.* 2015 Jul;21(7):663–5. doi: 10.1038/nm.3899. PMID:26151324.

Meir M, Galanty Y, Kashani L *et al.* The COP9 signalosome is vital for timely repair of DNA double-strand breaks. *Nucleic Acids Res.* 2015 May 19;43(9):4517–30. doi: 10.1093/nar/gkv270. PMID:25855810.

Melé M, Ferreira PG, Reverter F *et al.* Human genomics. The human transcriptome across tissues and individuals. *Science.* 2015 May 8;348 (6235):660–5. doi: 10.1126/science.aaa0355.

Mella P, Schumacher RF, Cranston T *et al.* Eleven novel JAK3 mutations in patients with severe combined immunodeficiency-including the first patients with mutations in the kinase domain. *Hum Mutat.* 2001 Oct;18(4):355–6. PMID:11668621.

Melo SA, Luecke LB, Kahlert C, *et al.* Glypican-1 identifies cancer exosomes and detects early pancreatic cancer. *Nature.* 2015 Jul 9;523(7559):177–82. doi:10.1038/nature14581. PMID:26106858.

Melton C, Reuter JA, Spacek DV *et al.* Recurrent somatic mutations in regulatory regions of human cancer genomes. *Nat Genet.* 2015 Jul; 47(7):710–6. doi: 10.1038/ng.3332. PMID:26053494.

Mencacci NE, Isaias IU, Reich MM *et al.* Parkinson's disease in GTP cyclohydrolase 1 mutation carriers. Brain. 2014 Sep;137(Pt 9):2480–92. doi: 10.1093/brain/awu179. PMID:24993959.

Mercer TR, Clark MB, Andersen SB *et al.* Genome-wide discovery of human splicing branchpoints. *Genome Res.* 2015 Feb;25(2):290–303. doi: 10.1101/gr.182899.114. PMID:25561518.

Mercer TR, Dinger ME, Mattick JS. Long non-coding RNAs: insights into functions. *Nat Rev Genet.* 2009 Mar;10(3):155–9. doi: 10.1038/ nrg2521. PMID:19188922.

Mercer TR, Mattick JS. Structure and function of long noncoding RNAs in epigenetic regulation. *Nat Struct Mol Biol.* 2013 Mar;20(3):300–7. doi: 10.1038/nsmb.2480. Review. PMID:23463315.

Mercimek-Mahmutoglu S, Patel J, Cordeiro D *et al*. Diagnostic yield of genetic testing in epileptic encephalopathy in childhood. *Epilepsia*. 2015 May;56(5):707–16. doi: 10.1111/epi.12954. PMID:25818041.

Miller JA, Ding SL, Sunkin SM, Smith KA *et al*. Transcriptional landscape of the prenatal human brain. *Nature*. 2014 Apr 10;508(7495): 199–206. doi: 10.1038/nature13185. PMID:24695229.

Mitchell C, Hobcraft J, McLanahan SS *et al*. Social disadvantage, genetic sensitivity, and children's telomere length. *Proc Natl Acad Sci USA*. 2014 Apr 22;111(16):5944–9. doi: 10.1073/pnas.1404293111. PMID: 24711381.

Mohan A, Goodwin M, Swanson MS. RNA-protein interactions in unstable microsatellite diseases. *Brain Res*. 2014 Oct 10;1584:3–14. doi: 10.1016/j.brainres.2014.03.039. Review. PMID:24709120.

Moore KL, Barr ML. Morphology of the nerve cell nucleus in mammals, with special reference to the sex chromatin. *J Comp Neurol*. 1953 Apr;98(2):213–31. PMID:13052743.

Morais VA, Haddad D, Craessaerts K *et al*. PINK1 loss-of-function mutations affect mitochondrial complex I activity via NdufA10 ubiquinone uncoupling. *Science*. 2014 Apr 11;344(6180):203–7. doi: 10.1126/science.1249161. PMID:24652937.

Morris HR, Waite AJ, Williams NM *et al*. Recent advances in the genetics of the ALS-FTLD complex. *Curr Neurol Neurosci Rep*. 2012 Jun;12(3): 243–50. doi: 10.1007/s11910-012-0268-5. Review. PMID:22477152.

Moussaud S, Jones DR, Moussaud-Lamodière EL *et al*. Alpha-synuclein and tau: teammates in neurodegeneration? Mol Neurodegener. 2014 Oct 29;9:43. doi: 10.1186/1750-1326-9-43. PMID:25352339.

Moustacchi Ethel, Fanconi's Anemia, in Orphanet Encyclopedia, October 2003.

Mueller KL. Cancer immunology and immunotherapy. Realizing the promise. Introduction. *Science*. 2015 Apr 3;348(6230):54–5. doi: 10.1126/science.348.6230.54. PMID:25838372.

Mullen SA, Carvill GL, Bellows S et al. Copy number variants are frequent in genetic generalized epilepsy with intellectual disability. *Neurology*. 2013 Oct 22;81(17):1507–14. doi: 10.1212/WNL.0b013e3182a95829. Review. PMID:24068782..

Müller-Rischart AK, Pilsl A, Beaudette P *et al*. The E3 ligase parkin maintains mitochondrial integrity by increasing linear ubiquitination of NEMO. *Mol Cell*. 2013 Mar 7;49(5):908–21. doi: 10.1016/j.molcel.2013.01.036. PMID:23453807.

Munnur RK, Cameron JD, Ko BS *et al.* Cardiac CT: atherosclerosis to acute coronary syndrome. *Cardiovasc Diagn Ther.* 2014 Dec;4(6):430–48. doi: 10.3978/j.issn.2223–3652.2014.11.03. PMID:25610801.

Najmabadi H, Hu H, Garshasbi M, Zemojtel T *et al.* Deep sequencing reveals 50 novel genes for recessive cognitive disorders. *Nature.* 2011 Sep 21;478(7367):57–63. doi: 10.1038/nature10423. PMID: 21937992.

Nakashima M, Miyajima M, Sugano H *et al.* The somatic GNAQ mutation c.548G>A (p.R183Q) is consistently found in Sturge-Weber syndrome. *J Hum Genet.* 2014 Dec;59(12):691–3. doi: 10.1038/jhg.2014.95. PMID:25374402.

Napoli C, Sessa M, Infante T *et al.* Unraveling framework of the ancestral Mediator complex in human diseases. *Biochimie.* 2012 Mar;94(3): 579–87. doi: 10.1016/j.biochi.2011.09.016. PMID:21983542.

Napoli M, Flores ER. The family that eats together stays together: new p53 family transcriptional targets in autophagy. *Genes Dev.* 2013 May 1;27(9):971–4. doi: 10.1101/gad.219147.113. PMID:23651851.

Narr KL, Leaver AM. Connectome and schizophrenia. *Curr Opin Psychiatry.* 2015 May;28(3):229–35. doi: 10.1097/YCO.0000000000000157. PMID:25768086.

Nathanson DA, Gini B, Mottahedeh J *et al.* Targeted therapy resistance mediated by dynamic regulation of extrachromosomal mutant EGFR DNA. *Science.* 2014 Jan 3;343(6166):72–6. doi: 10.1126/science. 1241328. PMID:24310612.

National Cancer Institute and Genetics Home Reference. Accessed November 2015. http://ghr.nlm.nih.gov/

National Human Genome Research Institute. The ENCODE Project: ENCyclopedia Of DNA Elements. Accessed November 2015. http://www.genome.gov/encode/

National Institutes of Health. About the Precision Medicine Initiative cohort program. Accessed November 2015. https://www.nih.gov/precision-medicine-initiative-cohort-program

Neul JL, Kaufmann WE, Glaze DG *et al.* Rett Search Consortium. Rett syndrome: revised diagnostic criteria and nomenclature. *Ann Neurol.* 2010 Dec;68(6):944–50. doi: 10.1002/ana.22124. PMID:21154482.

Neumann H, Daly MJ. Variant TREM2 as risk factor for Alzheimer's disease. *N Engl J Med.* 2013 Jan 10;368(2):182–4. doi: 10.1056/NEJMe1213157. PMID:23151315.

Ng AS, Rademakers R, Miller BL. Frontotemporal dementia: a bridge between dementia and neuromuscular disease. *Ann N Y Acad Sci.* 2015 Mar;1338:71–93. doi: 10.1111/nyas.12638. PMID:25557955.

Nguyen TA, Menendez D, Resnick MA *et al.* Mutant TP53 posttranslational modifications: challenges and opportunities. *Hum Mutat.* 2014 Jun;35(6):738–55. doi: 10.1002/humu.22506. PMID:24395704.

Nica AC, Dermitzakis ET. Expression quantitative trait loci: present and future. *Philos Trans R Soc Lond B Biol Sci.* 2013 May 6;368(1620):20120362. doi: 10.1098/rstb.2012.0362. Print 2013. Review. PMID:23650636.

Nilsson PM, Tufvesson H, Leosdottir M. Telomeres and cardiovascular disease risk: an update 2013. *Transl Res.* 2013 Dec;162(6):371–80. doi: 10.1016/j.trsl.2013.05.004. PMID:23748031.

Nixon RA. The role of autophagy in neurodegenerative disease. *Nat Med.* 2013 Aug;19(8):983–97. doi: 10.1038/nm.3232. Review. PMID: 23921753.

Noebels J. Pathway-driven discovery of epilepsy genes. *Nat Neurosci.* 2015 Mar;18(3):344–50. doi: 10.1038/nn.3933. Review. PMID:25710836

Noguchi M, Yi H, Rosenblatt HM *et al.* Interleukin-2 receptor gamma chain mutation results in X-linked severe combined immunodeficiency in humans. *Cell.* 1993 Apr 9;73(1):147–57. PMID:8462096.

Nomura K, Kanegane H, Karasuyama H *et al.* Genetic defect in human X-linked agammaglobulinemia impedes a maturational evolution of pro-B cells into a later stage of pre-B cells in the B-cell differentiation pathway. *Blood.* 2000 Jul 15;96(2):610–7. PMID:10887125.

Nunomura A, Moreira PI, Castellani RJ *et al.* Oxidative damage to RNA in aging and neurodegenerative disorders. *Neurotox Res.* 2012 Oct;22(3):231–48. doi: 10.1007/s12640-012-9331-x. Review. PMID:22669748.

Oeckinghaus A, Ghosh S. The NF-kappaB family of transcription factors and its regulation. *Cold Spring Harb Perspect Biol.* 2009 Oct;1(4):a000034. doi:10.1101/cshperspect.a000034. Review. PMID:20066092.

Ohno S, Makino S. The single-X nature of sex chromatin in man. *Lancet.* 1961 Jan 14;1(7168):78–9. PMID:13730522.

Olanow CW, Brundin P. Parkinson's disease and alpha synuclein: is Parkinson's disease a prion-like disorder? Mov Disord. 2013 Jan; 28(1):31–40. doi: 10.1002/mds.25373. Review. PMID:23390095.

Olivetti PR, Noebels JL. Interneuron, interrupted: molecular pathogenesis of ARX mutations and X-linked infantile spasms. *Curr Opin Neurobiol.* 2012 Oct;22(5):859–65. doi: 10.1016/j.conb.2012.04.006. PMID: 22565167.

Olzmann JA, Kopito RR, Christianson JC. The mammalian endoplasmic reticulum-associated degradation system. *Cold Spring Harb Perspect Biol.* 2013 Sep 1;5(9). pii: a013185. doi: 10.1101/cshperspect.a013185. PMID:23232094.

Pahnke J, Langer O, Krohn M. Alzheimer's and ABC transporters — new opportunities for diagnostics and treatment. *Neurobiol Dis.* 2014 Dec; 72 Pt A:54–60. doi: 10.1016/j.nbd.2014.04.001. Review. PMID:24746857.

Pal S, Gupta R, Kim H *et al.* Alternative transcription exceeds alternative splicing in generating the transcriptome diversity of cerebellar development. *Genome Res.* 2011 Aug;21(8):1260–72. doi: 10.1101/gr.120535. 111. PMID:21712398.

Palsuledesai CC, Distefano MD. Protein prenylation: enzymes, therapeutics, and biotechnology applications. *ACS Chem Biol.* 2015 Jan 16;10(1):51–62. doi: 10.1021/cb500791f. PMID:25402849.

Pan W, Gu W, Nagpal S *et al.* Brain tumor mutations detected in cerebral spinal fluid. *Clin Chem.* 2015 Mar;61(3):514–22. doi: 10.1373/clinchem.2014.235457. PMID 25605683.

Papadopoulos N, Nicolaides NC, Wei YF *et al.* Mutation of a mutL homolog in hereditary colon cancer. *Science* 1994 Mar 18;263(5153): 1625–9. PMID:8128251.

Pardoll DM. Immunology beats cancer: a blueprint for successful translation. *Nat Immunol.* 2012 Dec;13(12):1129–32. doi: 10.1038/ni.2392. PMID:23160205.

Parham P. Chapter 4: The development of B lymphocytes. *The Immune System*, 2nd Edition. 99–118, 2005, Garland Scientific Press.

Parikshak NN, Luo R, Zhang A *et al.* Integrative functional genomic analyses implicate specific molecular pathways and circuits in autism. *Cell.* 2013 Nov 21;155(5):1008–21. doi: 10.1016/j.cell.2013.10.031. PMID:24267887.

Park H, Suzuki T, Lennarz WJ. Identification of proteins that interact with mammalian peptide:N-glycanase and implicate this hydrolase in the proteasome-dependent pathway for protein degradation. *Proc Natl Acad Sci USA.* 2001 Sep 25;98(20):11163–8. PMID:11562482.

Parkes M, Cortes A, van Heel DA *et al.* Genetic insights into common pathways and complex relationships among immune-mediated diseases. *Nat Rev Genet.* 2013 Sep;14(9):661–73. doi: 10.1038/nrg3502. Review. PMID:23917628.

Pei YF, Zhang L, Liu Y *et al.* Meta-analysis of genome-wide association data identifies novel susceptibility loci for obesity. *Hum Mol Genet.* 2014 Feb 1;23(3):820–30. doi: 10.1093/hmg/ddt464. PMID:24064335

Peifer M, Hertwig F, Roels F *et al.* Telomerase activation by genomic rearrangements in high-risk neuroblastoma. *Nature.* 2015 Oct 14. doi: 10.1038/nature14980. PMID:26466568.

Pelechano V, Steinmetz LM. Gene regulation by antisense transcription. *Nat Rev Genet.* 2013 Dec;14(12):880–93. doi: 10.1038/nrg3594. Review. PMID:24217315.

Perry JA, Kiezun A, Tonzi P *et al.* Complementary genomic approaches highlight the PI3K/mTOR pathway as a common vulnerability in osteosarcoma. *Proc Natl Acad Sci USA.* 2014 Dec 23;111(51):E5564–73. doi: 10.1073/pnas.1419260111. PMID:25512523.

Petrovski S, Shashi C, Petrou S *et al.* Exome sequencing results in successful riboflavin treatment of a rapidly progressive neurological condition *Cold Spring Harb Mol Case Stud.* 2015 October 1: a000257.

Petrovski S, Wang Q, Heinzen EL *et al.* Genic intolerance to functional variation and the interpretation of personal genomes. *PLoS Genet.* 2013;9(8):e1003709. doi: 10.1371/journal.pgen.1003709. PMID: 23990802.

Placido AI, Pereira CM, Duarte AI *et al.* Modulation of endoplasmic reticulum stress: an opportunity to prevent neurodegeneration? CNS Neurol Disord Drug Targets. 2015;14(4):518–33. PMID:25921746.

Plácido AI, Pereira CM, Duarte AI *et al.* The role of endoplasmic reticulum in amyloid precursor protein processing and trafficking: implications for Alzheimer's disease. *Biochim Biophys Acta.* 2014. PMID:24832819.

PMID:18566288 Proliferating cells express mRNAs with shortened 3' untranslated regions and fewer microRNA target sites.

Pocklington AJ, Rees E, Walters JT, Han J *et al.* Novel Findings from CNVs Implicate Inhibitory and Excitatory Signaling Complexes in Schizophrenia. *Neuron.* 2015 Jun 3;86(5):1203–14. doi: 10.1016/j.neuron.2015.04.022. PMID:26050040.

Pohodich AE, Zoghbi HY. Rett syndrome: disruption of epigenetic control of postnatal neurological functions. *Hum Mol Genet.* 2015 Oct 15; 24(R1):R10–6. doi: 10.1093/hmg/ddv217. PMID:26060191.

Polani PE. Chromosome anomalies. *Annu Rev Med.* 1964;15:93–114. Review. available. PMID:14133859.

Porter JA, Young KE, Beachy PA. Cholesterol modification of hedgehog signaling proteins in animal development. *Science.* 1996 Oct 11; 274(5285):255–9. Erratum in: *Science* 1996 Dec 6;274(5293):1597. PMID:8824192.

Possemato R, Marks KM, Shaul YD *et al.* Functional genomics reveal that the serine synthesis pathway is essential in breast cancer. *Nature.*

2011 Aug 18;476(7360):346–50. doi: 10.1038/nature10350. PMID: 21760589.

Pourdehnad M, Truitt ML, Siddiqi IN *et al*. Myc and mTOR converge on a common node in protein synthesis control that confers synthetic lethality in Myc-driven cancers. *Proc Natl Acad Sci USA*. 2013 Jul 16;110(29): 11988–93. doi: 10.1073/pnas.1310230110. PMID:23803853.

Prager M, Büttner J, Büning C. PTGER4 modulating variants in Crohn's disease. *Int J Colorectal Dis*. 2014 Aug;29(8):909–15. doi: 10.1007/s00384-014-1881-3. PMID:24793213.

Price BD, D'Andrea AD. Chromatin remodeling at DNA double-strand breaks. *Cell*. 2013 Mar 14;152(6):1344–54. doi: 10.1016/j.cell.2013.02. 011. Review. PMID:23498941.

Prokocimer M, Barkan R, Gruenbaum Y. Hutchinson-Gilford progeria syndrome through the lens of transcription. *Aging Cell*. 2013 Aug;12(4):533–43. doi: 10.1111/acel.12070. PMID:23496208.

Prusiner SB. Biology and genetics of prions causing neurodegeneration. *Annu Rev Genet*. 2013;47:601–23. doi: 10.1146/annurev-genet-110711-155524. Review. PMID:24274755.

Puck JM. Laboratory technology for population-based screening for severe combined immunodeficiency in neonates: the winner is T-cell receptor excision circles. *J Allergy Clin Immunol*. 2012 Mar;129(3):607–16. doi: 10.1016/j.jaci.2012.01.032. PMID:22285280.

Ran FA, Cong L, Yan WX *et al*. *In vivo* genome editing using Staphylococcus aureus Cas9. *Nature*. 2015 Apr 9;520(7546):186–91. doi: 10.1038/nature14299. PMID:25830891.

Rastan S, Robertson EJ. X-chromosome deletions in embryo-derived (EK) cell lines associated with lack of X-chromosome inactivation. *J Embryol Exp Morphol*. 1985 Dec;90: 379–388. PMID:3834036.

Rastan S. Non-random X-chromosome inactivation in mouse X-autosome translocation embryos — location of the inactivation centre. *J Embryol Exp Morphol*. 1983 Dec;78:1–22. PMID:6198418.

Reddel RR. Telomere maintenance mechanisms in cancer: clinical implications. *Curr Pharm Des*. 2014;20(41):6361–74. Review. PMID:24975603.

Regad T. Targeting RTK Signaling Pathways in Cancer. *Cancers (Basel)*. 2015 Sep 3;7(3):1758–84. doi: 10.3390/cancers7030860. PMID:26404379.

Reitz C, Mayeux R. Genetics of Alzheimer's disease in Caribbean Hispanic and African American populations. *Biol Psychiatry*. 2014 Apr 1;75(7):534–41. doi: 10.1016/j.biopsych.2013.06.003. Review. PMID:23890735.

Renton AE, Chiò A, Traynor BJ. State of play in amyotrophic lateral sclerosis genetics. *Nat Neurosci.* 2014 Jan;17(1):17–23. doi: 10.1038/nn.3584. Epub 2013 Dec 26. Review. PMID:24369373.

Riant F, Bergametti F, Ayrignac X *et al.* Recent insights into cerebral cavernous malformations: the molecular genetics of CCM. *FEBS J.* 2010 Mar;277(5):1070–5. doi: 10.1111/j.1742-4658.2009.07535.x. Review. PMID:20096038.

Ricaño-Ponce I, Wijmenga C. Mapping of immune-mediated disease genes. *Annu Rev Genomics Hum Genet.* 2013;14:325–53. doi: 10.1146/annurev-genom-091212-153450. Review. PMID:23834318.

Richards S, Aziz N, Bale S, *et al.* Standards and guidelines for the interpretation of sequence variants: a joint consensus recommendation of the American College of Medical Genetics and Genomics and the Association for Molecular Pathology. *Genet Med.* 2015 May;17(5):405–24. doi: 10.1038/gim.2015.30. PMID:25741868.

Ricklin D, Hajishengallis G, Yang K *et al.* Complement: a key system for immune surveillance and homeostasis. *Nat Immunol.* 2010 Sep;11(9): 785–97. doi: 10.1038/ni.1923. PMID:20720586.

Rieser E, Cordier SM, Walczak H. Linear ubiquitination: a newly discovered regulator of cell signalling. *Trends Biochem Sci.* 2013 Feb;38(2):94–102. doi: 10.1016/j.tibs.2012.11.007. Review. PMID:23333406.

Rivers L, Gaspar HB Severe combined immunodeficiency: recent developments and guidance on clinical management. *Arch Dis Child.* 2015 Jul;100(7): 667–72. doi: 10.1136/archdischild-2014-306425. PMID:25564533.

Rizvi NA, Hellmann MD, Snyder A *et al.* Cancer immunology. Mutational landscape determines sensitivity to PD-1 blockade in non-small cell lung cancer. *Science.* 2015 Apr 3;348(6230):124–8. doi: 10.1126/science.aaa1348. PMID:25765070.

Robbins PF, Lu YC, El-Gamil M *et al.* Mining exomic sequencing data to identify mutated antigens recognized by adoptively transferred tumor-reactive T cells. *Nat Med.* 2013 Jun;19(6):747–52. doi: 10.1038/nm.3161. PMID:23644516.

Robinson EB, Samocha KE, Kosmicki JA *et al.* Autism spectrum disorder severity reflects the average contribution of de novo and familial influences. *Proc Natl Acad Sci USA.* 2014 Oct 21;111(42):15161–5. doi: 10.1073/pnas.1409204111. PMID:25288738.

Rodić N, Anders RA, Eshleman JR *et al.* PD-L1 expression in melanocytic lesions does not correlate with the BRAF V600E mutation. *Cancer*

Immunol Res. 2015 Feb;3(2):110–5. doi: 10.1158/2326-6066.CIR-14-0145. PMID:25370533.

Rohrer JD, Isaacs AM, Mizielinska S *et al.* C9orf72 expansions in fronto-temporal dementia and amyotrophic lateral sclerosis. *Lancet Neurol.* 2015 Mar;14(3):291–301. doi: 10.1016/S1474-4422(14)70233-9. Epub 2015 Jan 29. Review. Erratum in: *Lancet Neurol.* 2015 Apr;14(4): 350. PMID:25638642.

Rolland T, Taşan M, Charloteaux B *et al.* A proteome-scale map of the human interactome network. *Cell.* 2014 Nov 20;159(5):1212–26. doi: 10.1016/j.cell.2014.10.050. PMID:25416956.

Romaniello R, Arrigoni F, Bassi MT. Mutations in α- and β-tubulin encoding genes: implications in brain malformations. *Brain Dev.* 2015 Mar;37(3): 273–80. doi: 10.1016/j.braindev.2014.06.002. Review. PMID:25008804.

Romanoski CE, Glass CK, Stunnenberg HG *et al.* Epigenomics: Roadmap for regulation. *Nature.* 2015 Feb 19;518(7539):314–6. doi: 10.1038/518314a. PMID:25693562.

Ronan JL, Wu W, Crabtree GR. From neural development to cognition: unexpected roles for chromatin. *Nat Rev Genet.* 2013 May;14(5): 347–59. doi: 10.1038/nrg3413. Review. Erratum in: *Nat Rev Genet.* 2013 Jun;14(6):440. PMID:23568486.

Rosenson RS, Brewer HB Jr, Davidson WS *et al.* Cholesterol efflux and atheroprotection: advancing the concept of reverse cholesterol trans-port. *Circulation.* 2012 Apr 17;125(15):1905–19. doi: 10.1161/CIRCULATIONAHA.111.066589. PMID:22508840.

Rossor AM, Polke JM, Houlden H *et al.* Clinical implications of genetic advances in Charcot-Marie-Tooth disease. *Nat Rev Neurol.* 2013 Oct;9(10): 562–71. doi: 10.1038/nrneurol.2013.179. Review. PMID:24018473.

Rothblat GH, Phillips MC. High-density lipoprotein heterogeneity and fun-ction in reverse cholesterol transport. *Curr Opin Lipidol.* 2010 Jun;21(3): 229–38. PMID:20480549.

Roussel BD, Kruppa AJ, Miranda E *et al.* Endoplasmic reticulum dysfunc-tion in neurological disease. *Lancet Neurol.* 2013 Jan;12(1):105–18. doi: 10.1016/S1474-4422(12)70238-7. Review. PMID:23237905.

Rowley JD. Letter: A new consistent chromosomal abnormality in chronic myelogenous leukaemia identified by quinacrine fluorescence and Giemsa staining. *Nature.* 1973 Jun 1;243(5405):290–3. PMID:4126434.

Ruderman NB, Carling D, Prentki M *et al.* AMPK, insulin resistance, and the metabolic syndrome. *J Clin Invest.* 2013 Jul;123(7):2764–72. doi: 10.1172/JCI67227. Review. PMID:23863634.

Rulten SL, Caldecott KW. DNA strand break repair and neurodegeneration. *DNA Repair (Amst)*. 2013 Aug;12(8):558–67. doi: 10.1016/j.dnarep.2013.04.008. Review. PMID:23712058.

Russell RC, Tian Y, Yuan H *et al*. ULK1 induces autophagy by phosphorylating Beclin-1 and activating VPS34 lipid kinase. *Nat Cell Biol*. 2013 Jul;15(7):741–50. doi: 10.1038/ncb2757. PMID:23685627.

Rutter GA. Dorothy Hodgkin Lecture 2014. Understanding genes identified by genome-wide associationstudies for type 2 diabetes. *Diabet Med*. 2014 Dec;31(12):1480–7. doi: 10.1111/dme.12579. Review. PMID:25186316.

Samocha KE, Robinson EB, Sanders SJ *et al*. A framework for the interpretation of de novo mutation in human disease. *Nat Genet*. 2014 Sep;46(9):944–50. doi: 10.1038/ng.3050. PMID:25086666.

Sandberg R, Neilson JR, Sarma A *et al*. *Science*. 2008 Jun 20;320(5883): 1643–7. doi: 10.1126/science.1155390.

Sanders SJ, He X, Willsey AJ *et al*. Insights into Autism Spectrum Disorder Genomic Architecture and Biology from 71 Risk Loci. *Neuron*. 2015 Sep 23;87(6):1215–33. doi:10.1016/j.neuron.2015.09.016. PMID:26402605.

Sandin S, Lichtenstein P, Kuja-Halkola R *et al*. The familial risk of autism. *JAMA*. 2014 May 7;311(17):1770–7. doi: 10.1001/jama.2014.4144. PMID:24794370.

Sanz-Ortega J, Vocke C, Stratton P *et al*. Morphologic and molecular characteristics of uterine leiomyomas in hereditary leiomyomatosis and renal cancer (HLRCC) syndrome. *Am J Surg Pathol*. 2013 Jan;37(1): 74–80. doi: 10.1097/PAS.0b013e31825ec16f. PMID:23211287.

Saulin A, Savli M, Lanzenberger R. Serotonin and molecular neuroimaging in humans using PET. *Amino Acids*. 2012;42(6): 2039–2057. PMID: 21947614.

Savage PA. Tumor antigenicity revealed. *Trends Immunol*. 2014 Feb;35(2):47–8. doi: 10.1016/j.it.2014.01.001. PMID:24439426.

Saxena A, Sampson JR. Phenotypes associated with inherited and developmental somatic mutations in genes encoding mTOR pathway components. *Semin Cell Dev Biol*. 2014 Dec;36:140–6. doi: 10.1016/j.semcdb.2014.09.018. Review. PMID:25263008.

Scarpa ES, Fabrizio G, Di Girolamo M. A role of intracellular mono-ADP-ribosylation in cancer biology. *FEBS J*. 2013 Aug;280(15):3551–62. doi: 10.1111/febs.12290. Review. PMID:23590234.

Schadt EE Molecular networks as sensors and drivers of common human diseases. *Nature*. 2009 Sep 10;461(7261):218–23. doi: 10.1038/nature08454. PMID:19741703.

Schafer DP, Stevens B. Phagocytic glial cells: sculpting synaptic circuits in the developing nervous system. *Curr Opin Neurobiol.* 2013 Dec;23(6):1034–40. doi: 10.1016/j.conb.2013.09.012. Review. PMID: 24157239.

Schatz DG, Ji Y. Recombination centres and the orchestration of V(D)J recombination. *Nat Rev Immunol.* 2011 Apr;11(4):251–63. doi: 10.1038/nri2941. Review. PMID:21394103.

Schizophrenia Working Group of the Psychiatric Genomics Consortium. Biological insights from 108 schizophrenia-associated genetic loci. *Nature.* 2014 Jul 24;511(7510):421–7. doi: 10.1038/nature13595 PMID: 25056061.

Schmidt JC, Cech TR. Human telomerase: biogenesis, trafficking, recruitment, and activation. *Genes Dev.* 2015 Jun 1;29(11):1095–105. doi: 10.1101/gad.263863.115. Review. PMID:26063571.

Schneider JA, Arvanitakis Z, Bang W *et al.* Mixed brain pathologies account for most dementia cases in community-dwelling older persons. *Neurology.* 2007 Dec 11;69(24):2197–204. PMID:17568013.

Schneider JL, Cuervo AM. Autophagy and human disease: emerging themes. *Curr Opin Genet Dev.* 2014 Jun;26:16–23. doi: 10.1016/j. gde.2014.04.003. Review. PMID:24907664.

Schoch H, Abel T. Transcriptional co-repressors and memory storage. *Neuropharmacology.* 2014 May;80:53–60. doi: 10.1016/j.neuropharm.2014.01.003. Review. PMID:24440532.

Schramm EC, Clark SJ, Triebwasser MP *et al.* Genetic variants in the complement system predisposing to age-related macular degeneration: a review. *Mol Immunol.* 2014 Oct;61(2):118–25. doi: 10.1016/j. molimm.2014.06.032. Review. PMID:25034031.

Schubert CR, Xi HS, Wendland JR, O'Donnell P. Translating human genetics into novel treatment targets for schizophrenia. *Neuron.* 2014 Nov 5;84(3):537–41. doi: 10.1016/j.neuron.2014.10.037. PMID: 25442931.

Schuettengruber B, Martinez AM, Iovino N *et al.* Trithorax group proteins: switching genes on and keeping them active. *Nat Rev Mol Cell Biol.* 2011 Nov 23;12(12):799–814. doi: 10.1038/nrm3230. Review. PMID: 22108599.

Seabright M. Human chromosome banding. *Lancet.* 1972 Apr 29;1(7757):967. PMID:4112138.

Sekar A, Bialas AR, de Rivera H *et al.* Schizophrenia risk from complex variation of complement component 4. *Nature.* 2016 Feb 11; 530(7589):177–83. doi: 10.1038/nature16549. PMID:26814963.

Seltzer LE, Ma M, Ahmed S *et al.* Epilepsy and outcome in FOXG1-related disorders. *Epilepsia.* 2014 Aug;55(8):1292–300. doi: 10.1111/epi. 12648. PMID:24836831.

Sen SK, Boelte KC, Barb JJ *et al.* Integrative DNA, RNA, and protein evidence connects TREML4 to coronary artery calcification. *Am J Hum Genet.* 2014 Jul 3;95(1):66–76. doi: 10.1016/j.ajhg.2014.06.003. PMID: 24975946.

Senft D, Ronai ZA. UPR, autophagy, and mitochondria crosstalk underlies the ER stress response. *Trends Biochem Sci.* 2015 Mar;40(3):141–8. doi: 10.1016/j.tibs.2015.01.002. Review. PMID:25656104.

Sergin I, Razani B. Self-eating in the plaque: what macrophage autophagy reveals about atherosclerosis. *Trends Endocrinol Metab.* 2014 May;25(5):225–34. doi: 10.1016/j.tem.2014.03.010. Epub 2014 Apr 17. Review. PMID:24746519.

Serrano-Pozo A, Frosch MP, Masliah E *et al.* Neuropathological alterations in Alzheimer disease. *Cold Spring Harb Perspect Med.* 2011 Sep; 1(1):a006189. doi: 10.1101/cshperspect.a006189. Review. PMID: 22229116.

Shendure J, Akey JM. The origins, determinants, and consequences of human mutations. *Science.* 2015 Sep 25;349(6255):1478–83. doi: 10.1126/science.aaa9119. Review. PMID:26404824.

Shiang R, Ryan SG, Zhu YZ *et al.* Mutations in the alpha 1 subunit of the inhibitory glycine receptor cause the dominant neurologic disorder, hyperekplexia. *Nat Genet.* 1993 Dec;5(4):351–8. PMID:8298642.

Shirley MD, Tang H, Gallione CJ *et al.* Sturge-Weber syndrome and port-wine stains caused by somatic mutation in GNAQ. *N Engl J Med.* 2013 May 23;368(21):1971–9. doi: 10.1056/NEJMoa1213507. PMID: 23656586.

Shlyueva D, Stampfel G, Stark A. Transcriptional enhancers: from properties to genome-wide predictions. *Nat Rev Genet.* 2014 Apr;15(4): 272–86. doi: 10.1038/nrg3682. Epub 2014 Mar 11. Review. PMID: 24614317.

Shoubridge C, Fullston T, Gécz J. ARX spectrum disorders: making inroads into the molecular pathology. *Hum Mutat.* 2010 Aug;31(8):889–900. doi: 10.1002/humu.21288. Review. PMID:20506206.

Shoubridge C, Tan MH, Fullston T et al. Mutations in the nuclear localization sequence of the Aristaless related homeobox; sequestration of mutant ARX with IPO13 disrupts normal subcellular distribution of the transcription factor and retards cell division. *Pathogenetics*. 2010 Jan 5;3:1. doi: 10.1186/1755-8417-3-1. PMID:20148114.

Simpson JL. Genetics of female infertility due to anomalies of the ovary and mullerian ducts. *Methods Mol Biol*. 2014;1154:39–73. doi: 10.1007/978-1-4939-0659-8_3. PMID:24782005.

Singh RK, Cooper TA. Pre-mRNA splicing in disease and therapeutics. Pre-mRNA splicing in disease and therapeutics. *Trends Mol Med*. 2012 Aug;18(8):472–82. doi: 10.1016/j.molmed.2012.06.006. PMID: 22819011.

Siravegna G, Mussolin B, Buscarino M *et al*. Clonal evolution and resistance to EGFR blockade in the blood of colorectal cancer patients. *Nat Med*. 2015 Jul;21(7):827. doi:10.1038/nm0715-827b.PMID:26151329.

Skerka C, Chen Q, Fremeaux-Bacchi V *et al*. Complement factor H related proteins (CFHRs). *Mol Immunol*. 2013 Dec 15;56(3):170–80. doi: 10.1016/j.molimm.2013.06.001. PMID:23830046.

Slotkin W, Nishikura K. Adenosine-to-inosineRNA editing and human disease. *Genome Med*. 2013 Nov 29;5(11):105. doi: 10.1186/gm508. PMID:24289319.

Smale ST. Hierarchies of NF-κB target-gene regulation. *Nat Immunol*. 2011 Jul 19;12(8):689–94. doi: 10.1038/ni.2070. Review. PMID:21772277.

Small SA, Petsko GA. Retromer in Alzheimer disease, Parkinson disease and other neurological disorders. *Nat Rev Neurosci*. 2015 Mar;16(3): 126–32. doi: 10.1038/nrn3896. Review. PMID:25669742.

Smith MH, Ploegh HL, Weissman JS. Road to ruin: targeting proteins for degradation in the endoplasmic reticulum. *Science*. 2011 Nov 25;334(6059):1086–90. doi: 10.1126/science.1209235. Review. PMID:22116878.

Smith ZD, Meissner A. DNA methylation: roles in mammalian development. *Nat Rev Genet*. 2013 Mar;14(3):204–20. doi: 10.1038/nrg3354. Feb 12. Review. PMID:23400093.

Snyder A, Makarov V, Merghoub T *et al*. Genetic basis for clinical response to CTLA-4 blockade in melanoma. *N Engl J Med*. 2014 Dec 4;371(23):2189–99. doi: 10.1056/NEJMoa1406498. PMID: 25409260.

Soellner L, Monk D, Rezwan FI *et al*. Congenital imprinting disorders: Application of multilocus and high throughput methods to decipher new pathomechanisms and improve their management. *Mol Cell Probes*. 2015 Jun 10. pii: S0890-8508(15)00042-0. doi: 10.1016/j. mcp.2015.05.003. Review. PMID:26070988.

Son EY, Crabtree GR. The role of BAF (mSWI/SNF) complexes in mammalian neural development. *Am J Med Genet C Semin Med Genet.* 2014 Sep;166C(3):333–49. doi: 10.1002/ajmg.c.31416. PMID:25195934.

Sonenberg N, Hinnebusch AG. Regulation of translation initiation in eukaryotes: mechanisms and biologicaltargets. *Cell.* 2009 Feb 20; 136(4):731–45. doi: 10.1016/j.cell.2009.01.042. PMID:19239892.

Song SP, Hennig A, Schubert K *et al.* Ras palmitoylation is necessary for N-Ras activation and signal propagation in growth factor signalling. *Biochem J.* 2013 Sep 1;454(2):323–32. doi: 10.1042/BJ20121799. PMID: 23758196.

Sorrentino R. Genetics of autoimmunity: an update. *Immunol Lett.* 2014 Mar-Apr;158(1-2):116–9. doi: 10.1016/j.imlet.2013.12.005. PMID: 24370643.

Soussi T, Wiman KG. TP53: an oncogene in disguise. *Cell Death Differ.* 2015 Aug;22(8):1239–49. doi: 10.1038/cdd.2015.53. Epub 2015 May 29. Review. PMID:26024390.

Spaeth JM, Kim NH, Boyer TG. Mediator and human disease. *Semin Cell Dev Biol.* 2011 Sep;22(7):776–87. doi: 10.1016/j.semcdb.2011.07.024. Review. PMID:21840410.

Spiegelman BM. Transcriptional control of mitochondrial energy metabolism through the PGC1 coactivators. *Novartis Found Symp.* 2007; 287:60–3; discussion 63–9. Review. PMID:18074631.

Spillantini MG, Goedert M. Tau pathology and neurodegeneration. *Lancet Neurol.* 2013 Jun;12(6):609–22. doi: 10.1016/S1474-4422(13)70090-5. Review. PMID:23684085.

Spitz F, Furlong EE. Transcription factors: from enhancer binding to developmental control. *Nat Rev Genet.* 2012 Sep;13(9):613–26. doi: 10.1038/nrg3207. PMID:22868264.

Sporns O. Chapter 1. *Networks of the Brain.* 2010, 2. MIT Press (MA).

Stanford KI, Middelbeek RJ, Goodyear LJ. Exercise Effects on White Adipose Tissue: Beiging and Metabolic Adaptations. *Diabetes.* 2015 Jul;64(7):2361–8. doi: 10.2337/db15-0227. Review. Erratum in: Diabetes. 2015Sep;64(9):3334. PMID:26050668.

Steckel M, Molina-Arcas M, Weigelt B *et al.* Determination of synthetic lethal interactions in KRAS oncogene-dependent cancer cells reveals novel therapeutic targeting strategies. *Cell Res.* 2012 Aug;22(8): 1227–45. doi: 10.1038/cr.2012.82. PMID:22613949.

Steele MP, Schwartz DA. Molecular mechanisms in progressive idiopathic pulmonary fibrosis. *Annu Rev Med.* 2013;64:265–76. doi: 10.1146/annurev-med-042711-142004. Review. PMID:23020878.

Steinberg J, Webber C. The roles of FMRP-regulated genes in autism spectrum disorder: single- and multiple-hit genetic etiologies. *Am J Hum Genet.* 2013 Nov 7;93(5):825–39. doi: 10.1016/j.ajhg.2013.09.013. PMID:24207117.

Steinman RM, Banchereau J. Taking dendritic cells into medicine. *Nature.* 2007 Sep 27;449(7161):419–26. PMID:17898760.

Steward O, Levy WB. Preferential localization of polyribosomes under the base of dendritic spines in granule cells of the dentate gyrus. *J Neurosci.* 1982 Mar;2(3):284–91. PMID:7062109.

Stoffel EM, Mangu PB, Gruber SB *et al.* Hereditary colorectal cancer syndromes: American Society of Clinical Oncology Clinical Practice Guideline endorsement of the familial risk-colorectal cancer: European Society for Medical Oncology Clinical Practice Guidelines. American Society of Clinical Oncology; European Society for Medical Oncology Clinical Practice Guidelines. *J Clin Oncol.* 2015 Jan 10;33(2):209–17. doi: 10.1200/JCO.2014.58.1322. PMID:25452455.

Strathdee CA, Gavish H, Shannon WR, Buchwald M. Cloning of cDNAs for Fanconi's anaemia by functional complementation. *Nature* 1992 Apr 30;356(6372):763–7. Erratum in: *Nature.* 1992 Jul 30;358(6385): 434. PMID:1574115.

Strittmatter WJ, Saunders AM, Schmechel D *et al.* Apolipoprotein E: high-avidity binding to beta-amyloid and increased frequency of type 4 allele in late-onset familial Alzheimer disease. *Proc Natl Acad Sci USA.* 1993 Mar 1;90(5):1977–81. PMID:8446617.

Strong A, Rader DJ. Sortilin as a regulator of lipoprotein metabolism. *Curr Atheroscler Rep.* 2012 Jun;14(3):211–8. doi: 10.1007/s11883-012-0248-x. Review. PMID:22538429.

Sudmani P *et al.* 1000 Genomes. An integrates map of structural variation in 2,504 human genomes. *Nature.* Oct 2015, doi:10.1038/nature 15394.

Summermatter S, Shui G, Maag D *et al.* PGC-1α improves glucose homeostasis in skeletal muscle in an activity-dependent manner. *Diabetes.* 2013 Jan;62(1):85–95. doi: 10.2337/db12-0291. PMID:23086035.

Supattapone S. Expanding the prion disease repertoire. *Proc Natl Acad Sci USA.* 2015 Sep 1. pii: 201515143. PMID:26330608.

Suter B, Treadwell-Deering D, Zoghbi HY *et al.* Brief report: MECP2 mutations in people without Rett syndrome. *Hum Mol Genet.* 2015 Jun 9. pii: ddv217. PMID:26060191.

Suvà ML, Riggi N, Bernstein BE. Epigenetic reprogramming in cancer. *Science.* 2013 Mar 29;339(6127):1567–70. doi: 10.1126/science.1230184. Review. PMID:23539597.

Szafranski P, Dharmadhikari AV, Wambach JA *et al.* Two deletions overlapping a distant FOXF1 enhancer unravel the role of lncRNA LINC01081 in etiology of alveolar capillary dysplasia with misalignment of pulmonary veins. *Am J Med Genet A.* 2014 Aug;164A(8):2013–9. doi: 10.1002/ajmg.a.36606. PMID:24842713.

Szatkiewicz JP, O'Dushlaine C, Chen G, Chambert K *et al.* Copy number variation in schizophrenia in Sweden. *Mol Psychiatry.* 2014 Jul; 19(7):762–73. doi: 10.1038/mp.2014.40. PMID:24776740.

Takahashi K, Yamanaka S. Induction of pluripotent stem cells from mouse embryonic and adult fibroblast cultures by defined factors. *Cell.* 2006 Aug 25;126(4):663–76. PMID:16904174.

Tan HY, Chen AG, Kolachana B *et al.* Effective connectivity of AKT1-mediated dopaminergic working memory networks and pharmacogenetics of anti-dopaminergic treatment. *Brain.* 2012 May;135(Pt 5):1436–45. doi: 10.1093/brain/aws068. PMID:22525159.

Taylor BJ, Nik-Zainal S, Wu YL. DNA deaminases induce break-associated mutation showers with implication of APOBEC3B and 3A in breast cancer kataegis. *Elife.* 2013 Apr 16;2:e00534. doi: 10.7554/eLife.00534. PMID:23599896.

Tebas P[1], Stein D, Tang WW *et al.* Gene editing of CCR5 in autologous CD4 T cells of persons infected with HIV. *N Engl J Med.* 2014 Mar 6;370(10):901–10. doi: 10.1056/NEJMoa1300662. PMID:24597865.

The Cancer Genome Atlas (TCGA). Accessed November 2015. http://cancergenome.nih.gov/

Théry C. Cancer: Diagnosis by extracellular vesicles. *Nature.* 2015 Jul 9;523(7559):161–2. doi: 10.1038/nature14626. PMID:26106856.

Topalian SL, Drake CG, Pardoll DM. Immune checkpoint blockade: a common denominator approach to cancer therapy. *Cancer Cell.* 2015 Apr 13;27(4):450–61. doi: 10.1016/j.ccell.2015.03.001. Review. PMID:25858804.

Trowsdale J, Knight JC. Major histocompatibility complex genomics and human disease. *Annu Rev Genomics Hum Genet.* 2013;14:301–23. doi: 10.1146/annurev-genom-091212-153455. PMID:23875801.

Turturici G, Sconzo G, Geraci F. Hsp70 and its molecular role in nervous system diseases. *Biochem Res Int.* 2011;2011:618127. doi: 10.1155/2011/618127. PMID:21403864.

Tyburczy ME, Wang JA, Li S *et al.* Sun exposure causes somatic second-hit mutations and angiofibroma development in tuberous sclerosis complex. *Hum Mol Genet.* 2014 Apr 15;23(8):2023–9. doi: 10.1093/hmg/ddt597. PMID:24271014.

Vallot C, Huret C, Lesecque Y *et al.* XACT, a long noncoding transcript coating the active X chromosome in human pluripotent cells. *Nat Genet.* 2013 Mar;45(3):239–41. doi:10.1038/ng.2530.PMID:23334669.

Vallot C, Ouimette JF, Makhlouf M *et al.* Erosion of X Chromosome Inactivation in Human Pluripotent Cells Initiates with XACT Coating and Depends on a Specific Heterochromatin Landscape. *Cell Stem Cell.* 2015 May 7;16(5):533–46. doi: 10.1016/j.stem.2015.03.016. PMID: 25921272.

van Blitterswijk M, Gendron TF, Baker MC *et al.* Novel clinical associations with specific C9ORF72 transcripts in patients with repeat expansions in C9ORF72. *Acta Neuropathol.* 2015 Oct 5. PMID:26437865.

Van Cauwenberghe C, Van Broeckhoven C, Sleegers K. The genetic landscape of Alzheimer disease: clinical implications and perspectives. *Genet Med.* 2015 Aug 27. doi:10.1038/gim.2015.117 Review.PMID:26312828.

van Rooij N, van Buuren MM, Philips D *et al.* Tumor exome analysis reveals neoantigen-specific T-cell reactivity in an ipilimumab-responsive melanoma. *J Clin Oncol.* 2013 Nov 10;31(32):e439–42. doi: 10.1200/JCO.2012.47.7521. PMID:24043743.

Vance JE, Peake KB. Function of the Niemann-Pick type C proteins and their bypass by cyclodextrin. *Curr Opin Lipidol.* 2011 Jun;22(3):204–9. doi: 10.1097/MOL.0b013e3283453e69. Review. PMID:21412152.

Vannier JB, Sarek G, Boulton SJ. RTEL1: functions of a disease-associated helicase. *Trends Cell Biol.* 2014 Jul;24(7):416–25. doi: 10.1016/j.tcb.2014.01.004. PMID:24582487.

Vaquerizas JM, Kummerfeld SK, Teichmann SA *et al.* A census of human transcription factors: function, expression and evolution. *Nat Rev Genet.* 2009 Apr;10(4):252–63.doi:10.1038/nrg2538.PMID:19274049.

Varjosalo M, Taipale J. Hedgehog: functions and mechanisms. *Genes Dev.* 2008 Sep 15;22(18):2454–72. doi: 10.1101/gad.1693608. Review. PMID:18794343.

Velasquez-Manoff M. Genetics: Relative risk. *Nature.* 2015 Nov 18;527(7578):S116–7. doi: 10.1038/527S116a. PMID:26580161.

Verbsky JW, Chatila TA. Immune dysregulation, polyendocrinopathy, enteropathy, X-linked (IPEX) and IPEX-related disorders: an evolving web of heritable autoimmune diseases. *Curr Opin Pediatr.* 2013 Dec;25(6):708–14. doi: 10.1097/MOP.0000000000000029. PMID:24240290.

Vittal V, Stewart MD, Brzovic PS *et al.* Regulating the Regulators: Recent Revelations in the Control of E3 Ubiquitin Ligases. *J Biol Chem.* 2015 Aug 28;290(35):21244–51. doi: 10.1074/jbc.R115.675165. Review. PMID:26187467.

Vogan JM, Collins K. Dynamics of Human Telomerase Holoenzyme Assembly and Subunit Exchange across the Cell Cycle. *J Biol Chem.* 2015 Aug 28;290(35):21320–35. doi: 10.1074/jbc.M115.659359. PMID:26170453.

Vogelstein B, Papadopoulos N, Velculescu VE *et al.* Cancer genome landscapes. *Science.* 2013 Mar 29;339(6127):1546–58. doi: 10.1126/science.1235122. Review. PMID:23539594.

Voineagu I, Wang X, Johnston P, Lowe JK *et al.* Transcriptomic analysis of autistic brain reveals convergent molecular pathology. *Nature.* 2011 May 25;474(7351):380–4. doi: 10.1038/nature10110. PMID: 21614001.

Wade-Martins R. Genetics: the MAPT locus-a genetic paradigm in disease susceptibility. *Nat Rev Neurol.* 2012 Sep;8(9):477–8. doi: 10.1038/nrneurol.2012.169. PMID:22940644.

Wagner H. Innate immunity's path to the Nobel Prize 2011 and beyond. *Eur J Immunol.* 2012 May;42(5):1089–92. doi: 10.1002/eji.201242404. PMID:22539282.

Waite AJ, Bäumer D, East S *et al.* Reduced C9orf72 protein levels in frontal cortex of amyotrophic lateral sclerosis and frontotemporal degeneration brain with the C9ORF72 hexanucleotide repeat expansion. *Neurobiol Aging.* 2014 Jul;35(7):1779.e5-1779.e13. doi: 10.1016/j.neurobiolaging.2014.01.016. PMID:24559645.

Walden H, Deans AJ. The Fanconi anemia DNA repair pathway: structural and functional insights into a complex disorder. *Annu Rev Biophys.* 2014;43:257–78. doi: 10.1146/annurev-biophys-051013-022737. PMID:24773018.

Walter P, Ron D. The unfolded protein response: from stress pathway to homeostatic regulation. *Science.* 2011 Nov 25;334(6059):1081–6. doi: 10.1126/science.1209038. Review. PMID:22116877.

Wang ET, Sandberg R, Luo S, Khrebtukova I *et al.* Alternative isoform regulation in human tissue transcriptomes. *Nature.* 2008 Nov 27; 456(7221):470–6. doi: 10.1038/nature07509. PMID:18978772.

Wang J, Tao Y, Song F *et al.* Common Regulatory Variants of CYFIP1 Contribute to Susceptibility for Autism Spectrum Disorder (ASD) and Classical Autism. *Ann Hum Genet.* 2015 Jun 19. doi: 10.1111/ahg.12121. PMID:26094621.

Wang L, McLeod HL, Weinshilboum RM. Genomics and drug response. *N Engl J Med.* 2011 Mar 24;364(12):1144–53. doi: 10.1056/NEJMra 1010600. Review. PMID:21428770.

Wang Y, Springer S, Zhang M *et al.* Detection of tumor-derived DNA in cerebrospinal fluid of patients with primary tumors of the brain and spinal cord. *Proc Natl Acad Sci USA.* 2015 Aug 4;112(31):9704–9. doi: 10.1073/pnas.151169411 PMID:26195750.

Warburg, O. Ueber Milchsaurebildung beira Vi'achstum. *Biochem. Z.,* 1925,160:307–311.

Ward PS, Thompson CB. Signaling in control of cell growth and metabolism. *Cold Spring Harb Perspect Biol.* 2012 Jul 1;4(7):a006783. doi: 10.1101/cshperspect.a006783. PMID:22687276.

Watson CT, Marques-Bonet T, Sharp AJ *et al.* The genetics of microdeletion and microduplication syndromes: an update. *Annu Rev Genomics Hum Genet.* 2014;15:215–44. doi: 10.1146/annurev-genom-091212-153408. PMID:24773319.

Weaver AN, Cooper TS, Rodriguez M *et al.* DNA double strand break repair defect and sensitivity to poly ADP-ribose polymerase (PARP) inhibition in human papillomavirus 16-positive head and neck squamous cell carcinoma. *Oncotarget.* 2015 Sep 29;6(29):26995–7007. doi: 10.18632/oncotarget.4863. PMID:26336991.

Weckselblatt B, Rudd MK. Human structural variation: mechanisms of chromosome rearrangements. *Trends Genet.* 2015 Jul 22. pii: S0168–9525 (15)00110–9. doi: 10.1016/j.tig.2015.05.010. Review. PMID:26209074.

Weinberg F, Hamanaka R, Wheaton WW *et al.* Mitochondrial metabolism and ROS generation are essential for Kras-mediated tumorigenicity. *Proc Natl Acad Sci USA.* 2010 May 11;107(19):8788–93. doi: 10.1073/pnas.1003428107. PMID:20421486.

Werner ER, Blau N, Thöny B. Tetrahydrobiopterin: biochemistry and pathophysiology. *Biochem J.* 2011 Sep 15;438(3):397–414. doi: 10.1042/BJ20110293. PMID:21867484.

Whiley L, Sen A, Heaton J *et al.* Evidence of altered phosphatidylcholine metabolism in Alzheimer's disease. *Neurobiol Aging.* 2014 Feb;35(2):271–8. doi: 10.1016/j.neurobiolaging.2013.08.001. PMID:2404197.

Whittington RA, Bretteville A, Dickler MF, Planel E. Anesthesia and tau pathology. *Prog Neuropsychopharmacol Biol Psychiatry.* 2013 Dec 2;47:147–55. doi: 10.1016/j.pnpbp.2013.03.004. PMID:23535147.

Wiegert JS, Bading H. Activity-dependent calcium signaling and ERK-MAP kinases in neurons: a link to structural plasticity of the nucleus and gene transcription regulation. *Cell Calcium.* 2011 May;49(5):296–305. doi: 10.1016/j.ceca.2010.11.009. PMID:21163523.

Will CL, Lührmann R. Spliceosome structure and function. *Cold Spring Harb Perspect Biol.* 2011 Jul 1;3(7). pii: a003707. doi: 10.1101/cshperspect.a003707. Review. PMID:21441581.

Williams CA, Battaglia A. Molecular biology of epilepsy genes. *Exp Neurol.* 2013 Jun;244:51–8. doi: 10.1016/j.expneurol.2011.12.001. PMID:22178301.

Williams GH, Stoeber K. The cell cycle and cancer. *J Pathol.* 2012 Jan;226(2):352–64. doi: 10.1002/path.3022. Review. PMID:21990031.

Wingender E, Schoeps T, Haubrock M *et al. Nucleic Acids Res.* 2015 Jan;43(Database issue):D97-102. doi: 10.1093/nar/gku1064. PMID: 25361979.

Wodak SJ, Vlasblom J, Turinsky AL *et al.* Protein-protein interaction networks: the puzzling riches. *Curr Opin Struct Biol.* 2013 Dec;23(6): 941–53. doi: 10.1016/j.sbi.2013.08.002. Review. PMID:24007795.

Wolda SL, Glomset JA. Evidence for modification of lamin B by a product of mevalonic acid. *J Biol Chem.* 1988 May 5;263(13):5997–6000. PMID:3283116.

Wong YC, Holzbaur EL. Optineurin is an autophagy receptor for damaged mitochondria in parkin-mediated mitophagy that is disrupted by an ALS-linked mutation. *Proc Natl Acad Sci USA.* 2014 Oct 21;111(42): E4439–48. doi: 10.1073/pnas.1405752111. PMID:25294927.

Wright PE, Dyson HJ. Intrinsically disordered proteins in cellular signalling and regulation. *Nat Rev Mol Cell Biol.* 2015 Jan;16(1):18–29. doi: 10.1038/nrm3920. Review. PMID:25531225.

Wu H, Luo J, Yu H *et al.* Cellular resolution maps of X chromosome inactivation: implications for neural development, function, and disease. *Neuron.* 2014 Jan 8;81(1):103–19. doi: 10.1016/j.neuron.2013.10.051. PMID:24411735.

Wutz A, Rasmussen TP, Jaenisch R. Chromosomal silencing and localization are mediated by different domains of Xist RNA. *Nat Genet.* 2002 Feb;30(2):167–74. PMID:11780141.

Xiao Q, Yan P, Ma X *et al.* Neuronal-targeted TFEB accelerates lysosomal degradation of APP, reducing aβ generation and amyloid plaque pathogenesis. *J Neurosci.* 2015 Sep 2;35(35):12137–51. doi: 10.1523/JNEUROSCI.0705-15.2015. PMID:26338325.

Xiong HY, Alipanahi B, Lee LJ *et al.* RNA splicing. The human splicing code reveals new insights into the genetic determinants of disease. *Science.* 2015 Jan 9;347(6218):1254806. doi: 10.1126/science.1254806. PMID:25525159.

Xu B, Konze KD, Jin J *et al.* Targeting EZH2 and PRC2 dependence as novel anticancer therapy. *Exp Hematol.* 2015 Aug;43(8):698–712. doi:10.1016/j.exphem.2015.05.001. PMID:26027790.

Yang L, Chen LL. Microexons go big. *Cell.* 2014 Dec 18;159(7):1488–9. doi: 10.1016/j.cell.2014.12.004. PMID:25525868.

Youle RJ, van der Bliek AM. Mitochondrial fission, fusion, and stress. *Science.* 2012 Aug 31;337(6098):1062–5. doi: 10.1126/science.1219855. Review. PMID:22936770.

Yu JT, Tan L, Hardy J. Apolipoprotein E in Alzheimer's disease: an update. *Annu Rev Neurosci.* 2014;37:79–100. doi: 10.1146/annurev-neuro-071013-014300. Review. PMID:24821312.

Yuen RK, Thiruvahindrapuram B, Merico D *et al.* Whole-genome sequencing of quartet families with autism spectrum disorder. *Nat Med.* 2015 Feb;21(2):185–91. doi: 10.1038/nm.3792. PMID:25621899.

Yun M, Wu J, Workman JL *et al.* Readers of histone modifications. *Cell Res.* 2011 Apr;21(4):564–78. doi:10.1038/cr.2011.42. PMID:21423274.

Zampese E, Fasolato C, Pozzan T, Pizzo P. Presenilin-2 modulation of ER-mitochondria interactions: FAD mutations, mechanisms and pathological consequences. *Commun Integr Biol.* 2011 May;4(3):357–60. doi: 10.4161/cib.4.3.15160. PMID:21980580.

Zaret KS, Carroll JS. Pioneer transcription factors: establishing competence for gene expression. *Genes Dev.* 2011 Nov 1;25(21):2227–41. doi: 10.1101/gad.176826.111. PMID:22056668.

Zauri M, Berridge G, Thézénas ML *et al.* CDA directs metabolism of epigenetic nucleosides revealing a therapeutic window in cancer. *Nature.* 2015 Aug 6;524(7563):114–8. doi: 10.1038/nature14948. PMID:26200337.

Zeevi D, Korem T, Zmora N *et al.* Personalized nutrition by prediction of glycemic responses. *Cell.* 2015 Nov 19;163(5):1079–94. doi: 10.1016/j.cell.2015.11.001. PMID:26590418.

Zhang CZ, Leibowitz ML, Pellman D. Chromothripsis and beyond: rapid genome evolution from complex chromosomal rearrangements. *Genes Dev.* 2013 Dec 1;27(23):2513–30. doi: 10.1101/gad.229559.113. PMID:24298051.

Zhang CZ, Spektor A, Cornils H *et al.* Chromothripsis from DNA damage in micronuclei. *Nature.* 2015 Jun 11;522(7555):179–84. doi: 10.1038/nature14493. PMID:26017310.

Zhang F, Lupski JR. Non-coding genetic variants in human disease. *Hum Mol Genet.* 2015 Oct 15;24(R1):R102-10. doi: 10.1093/hmg/ddv259. Review. PMID:26152199.

Zhang J, Manley JL. Misregulation of pre-mRNA alternative splicing in cancer. *Cancer Discov*. 2013 Nov;3(11):1228–37. doi: 10.1158/2159-8290.CD-13-0253. PMID:24145039.

Zhang K, Donnelly CJ, Haeusler AR *et al*. The C9orf72 repeat expansion disrupts nucleocytoplasmic transport. *Nature*. 2015 Sep 3;525(7567):56–61. doi: 10.1038/nature14973. PMID:26308891.

Ziller MJ, Edri R, Yaffe Y *et al*. Dissecting neural differentiation regulatory networks through epigenetic footprinting. *Nature*. 2015 Feb 19; 518(7539): 355–9. doi: 10.1038/nature13990. PMID:25533951.

Zimmermann R, Eyrisch S, Ahmad M *et al*. Protein translocation across the ER membrane. *Biochim Biophys Acta*. 2011 Mar;1808(3):912–24. doi: 10.1016/j.bbamem.2010.06.015. PMID:20599535.

Zuo L, Motherwell MS. The impact of reactive oxygen species and genetic mitochondrial mutations in Parkinson's disease. Gene. 2013 Dec 10;532(1):18–23. doi: 10.1016/j.gene.2013.07.085. PMID:23954870.

Zuo L, Zhou T, Pannell BK *et al*. Biological and physiological role of reactive oxygen species — the good, the bad and the ugly. *Acta Physiol* (Oxf). 2015 Jul;214(3):329–48. doi: 10.1111/apha.12515. PMID:25912260.

Zylka MJ, Simon JM, Philpot BD. Gene length matters in neurons. *Neuron*. 2015 Apr 22; 86(2):353–5. doi: 10.1016/j.neuron.2015.03.059. PMID:25905808.

Index

www.ingramcontent.com/pod-product-compliance
Lightning Source LLC
Chambersburg PA
CBHW072256210326
41458CB00074B/1786